REACTION MECHANISMS FOR RARE ISOTOPE BEAMS

To learn more about the AIP Conference Proceedings, including the
Conference Proceedings Series, please visit the webpage
http://proceedings.aip.org/proceedings

REACTION MECHANISMS FOR RARE ISOTOPE BEAMS

2nd Argonne/MSU/JINA/INT RIA Workshop

East Lansing, Michigan   9 – 12 March 2005

EDITOR
B. Alex Brown
*National Superconducting Cyclotron
Laboratory at Michigan State University*

SPONSORING ORGANIZATIONS
National Superconducting Cyclotron Laboratory
Argonne National Laboratory
Joint Institute for Nuclear Astrophysics
Institute for Nuclear Theory

Melville, New York, 2005
AIP CONFERENCE PROCEEDINGS ■ VOLUME 791

Editor:

B. Alex Brown
Department of Physics and Astronomy
and the National Superconducting Cyclotron Laboratory
Michigan State University
E. Lansing, MI 48824

E-mail: brown@nscl.msu.edu

Authorization to photocopy items for internal or personal use, beyond the free copying permitted under the 1978 U.S. Copyright Law (see statement below), is granted by the American Institute of Physics for users registered with the Copyright Clearance Center (CCC) Transactional Reporting Service, provided that the base fee of $22.50 per copy is paid directly to CCC, 222 Rosewood Drive, Danvers, MA 01923, USA. For those organizations that have been granted a photocopy license by CCC, a separate system of payment has been arranged. The fee code for users of the Transactional Reporting Services is: ISBN/0-7354-0280-9/05/$22.50.

© 2005 American Institute of Physics

Permission is granted to quote from the AIP Conference Proceedings with the customary acknowledgment of the source. Republication of an article or portions thereof (e.g., extensive excerpts, figures, tables, etc.) in original form or in translation, as well as other types of reuse (e.g., in course packs) require formal permission from AIP and may be subject to fees. As a courtesy, the author of the original proceedings article should be informed of any request for republication/reuse. Permission may be obtained online using Rightslink. Locate the article online at http://proceedings.aip.org, then simply click on the Rightslink icon/ "Permission for Reuse" link found in the article abstract. You may also address requests to: AIP Office of Rights and Permissions, Suite 1NO1, 2 Huntington Quadrangle, Melville, NY 11747-4502, USA; Fax: 516-576-2450; Tel.: 516-576-2268; E-mail: rights@aip.org.

L.C. Catalog Card No. 2005932387
ISBN 0-7354-0280-9
ISSN 0094-243X

Printed in the United States of America

CONTENTS

Preface vii

Review of Semiclassical Calculations for Breakup 1
    D. Baye

Time-dependent Analysis of the Nuclear and Coulomb Dissociation of ^{11}Be 12
    P. Capel, G. Goldstein, and D. Baye

A Transport Model for Nuclear Reactions Induced by Radioactive Beams 22
    B.-A. Li, L.-W. Chen, C. B. Das, S. Das Gupta, C. Gale, C. M. Ko, G.-C. Yong, and W. Zuo

No-core Shell Model and Reactions 32
    P. Navrátil, W. E. Ormand, E. Caurier, and C. Bertulani

Combined Method to Extract the Spectroscopic Factors from Transfer Reactions 40
    A. M. Mukhamedzhanov and F. M. Nunes

Ground State Neutron Spectroscopic Factors for Z=3-24 Isotopes from Transfer Reactions 49
    M. B. Tsang and H. C. Lee

Contribution to the Wednesday Afternoon Discussion on Spectroscopic Factors 57
    C. Barbieri

Direct Reactions with Exotic Nuclei 61
    G. Baur and S. Typel

On the Splitting of Nucleon Effective Masses at High Isospin Density: Reaction Observables .. 70
    M. Di Toro, M. Colonna, and J. Rizzo

Isoscaling and Symmetry Energy in Dynamical Fragment Formation 83
    A. Ono

Determining Cross Sections for Reactions on Unstable Nuclei: A Consideration of Indirect Approaches 93
    J. Escher and F. S. Dietrich

Recent Progress on Understanding "Pasta" Phases in Dense Stars 101
    G. Watanabe and H. Sonoda

Production of Complex Particles in Low-energy Spallation and in Fragmentation Reactions by In-medium Random Clusterization 112
    D. Lacroix and D. Durand

Isospin Effects on Fragmentation Mechanisms 119
    M. Colonna, V. Baran, M. Di Toro, and R. Lionti

Adiabatic Approximation for Nucleus-Nucleus Scattering 128
    R. C. Johnson

Near-barrier Elastic Scattering of Weakly Bound Nuclei and the Threshold Anomaly 140
    M. S. Hussein, L. C. Chamon, and P. R. S. Gomes

Is the Optical Model Valid for the Scattering of Exotic Nuclei? 146
    J. Gómez-Camacho, M. Alvarez, A. Moro, I. Martel, A. Sánchez-Benítez, D. Escrig, and M. J. G. Borge

Coulomb Breakup for Spectroscopy 154
    I. J. Thompson

Three-body Decay of Many-body Resonances 164
    A. S. Jensen, D. V. Fedorov, H. O. U. Fynbo, and E. Garrido

Continuum-discretized Coupled-channels Method for Four-body Breakup Reactions 174
    M. Kamimura, T. Matsumoto, E. Hiyama, K. Ogata, Y. Iseri, and M. Yahiro

Special Relativity and Reactions with Unstable Nuclei 185
    C. A. Bertulani

Effective Interactions in Neutron-rich Matter 193
    F. Sammarruca, P. Krastev, and W. Barredo

Program of the Workshop 203
Author Index 205

PREFACE

This volume contains the Proceedings of the *Reaction Mechanisms for Rare Isotope Beams*, held on March 9-12, 2005, at National Superconducting Cyclotron Laboratory, Michigan State University, East Lansing, Michigan. This was the second workshop in the Argonne/MSU/JINA/INT RIA Workshop series.

The workshop focused on theoretical descriptions of the nuclear reactions that can be studied at present and future radioactive beam facilities, from low to relativistic energies. These facilities allow for the study of reactions with exotic nuclei that have never been accessible before, providing challenges for reaction theory as well as important input for nuclear astrophysics. Overview of the standard reaction formalisms was an important component of the meeting, keeping in mind the old issues that arose from reactions with stable nuclei which help us understand the challenges of applying existing approaches to radioactive beams.

The workshop covered several types of reactions that are of interest for different reasons. For example, radiative capture, breakup and transfer reactions can provide single-particle spectroscopic information. While fusion reactions at low energy and more violent central collisions at high energy can provide information about nuclear properties that are crucial to understanding the conditions inside neutron stars. The descriptions that have commonly been used to analyze reactions between stable nuclei are often unrealistic for reactions of weakly bound, unstable nuclei. Some progress has been made in reformulating reaction theories to incorporate the principal features of the reaction mechanisms and apply them to a variety of new phenomena. This work is typically done in terms of simplified few-body models, so an open problem for the future is how to implement the full many-body dynamics. A particular challenge is to perform reaction calculations that exploit the full complexity of the ab-initio structure models, an effort that has grown rapidly in recent years. Contributions to this workshop focus on the progress being made in these directions. The discussion sessions were particularly useful in defining the issues and problems that need to be addressed. A few of the contributions to this volume are based on the material presented at the discussion sessions.

The Workshop was organized by Alex Brown, Pawel Danielewicz, Henning Esbensen, Ken Nollett and Filomena Nunes. We thank Shari Conroy, Chasity Fudella and Jean McIntyre for their smooth coordination of the workshop arrangements. We would like to acknowledge support from the NSCL and JINA for their help in the organization of the Workshop and publication of the Proceedings. We thank the discussion leaders for several very interesting discussion sessions. Finally, we are grateful to the participants of the Workshop for their active response, extensive work during the meeting and excellent contributions to this volume.

Alex Brown

NSCL/MSU

Review of semi-classical calculations for breakup

Daniel Baye

Physique Quantique, C.P. 165/82, and Physique Nucléaire Théorique et Physique Mathématique, C.P. 229, Université Libre de Bruxelles, B 1050 Brussels, Belgium

Abstract. In semi-classical approximations, the relative motion between target and projectile is represented by a classical trajectory but the projectile internal motion is treated quantum mechanically. A time-dependent Schrödinger equation describes the breakup of exotic nuclei induced by the Coulomb and nuclear forces. Different accurate techniques of resolution of this time-dependent equation are reviewed for one space dimension. The respective merits of their extensions to three dimensions are compared. Applications to the breakup of the ^{11}Be, ^{15}C, and ^{19}C halo nuclei are presented and discussed. The first-order perturbation theory is compared with the time-dependent method and its relevance for the Coulomb breakup determination of the astrophysical S factor is analyzed.

Keywords: breakup, halo nuclei, time-dependent Schrödinger equation, semi-classical, Lagrange mesh, astrophysical S factor
PACS: 24.10.-i, 25.60.Gc, 25.70.De, 03.65.Sq

1. INTRODUCTION

The semi-classical approximation is a well known tool for the study of collisions [1]. It was developed for the description of Coulomb excitation about half a century ago. At high enough velocities, the relative motion between the target and projectile can be treated in a classical way. In the target frame of reference, the projectile center of mass follows a classical trajectory which can be well approximated by a straight line or a Rutherford trajectory. The projectile internal structure is treated quantum mechanically. When following the trajectory, the projectile experiences Coulomb and nuclear fields from the target that vary with time. This time variation may induce excitation and breakup.

The semi-classical approximation leads to the resolution of a time-dependent Schrödinger equation. It has been widely used both in atomic and nuclear physics. More recently, it has been extended to transitions toward the continuum i.e., ionization and transfer in atomic physics [2–4] and dissociation in nuclear physics [5–8]. It is a fully dynamical theory where all couplings, not only between the bound states and the continuum but also inside the continuum, are properly taken into account. Its application to the Coulomb and nuclear breakups of halo nuclei is of particular importance because of its relevance for the determination of the physical properties of these nuclei. Coulomb breakup is also at the basis of an indirect technique of determination of the astrophysical S factor for radiative-capture reactions [9].

This dynamical method was however developed in a context where a natural basis of square-integrable states is available. The need for descriptions of ionization and transfer in atomic physics and of breakup and stripping in nuclear physics has led to the development of methods independent of a basis choice. They were rendered possible by the steady increase of computer power. The time-dependent Schrödinger equation is now solved numerically in three dimensions [7, 8, 10–18]. This allows studying the breakup of halo nuclei described in a simple two-body (or potential) model. A purely numerical resolution of the time-dependent Schrödinger equation presents the advantage that no simplifying assumptions need to be done about the description of the continuum. Nevertheless, the validity of the physical results heavily relies on the convergence and the accuracy of the solution.

An approximate solution of the time-dependent Schrödinger equation is given by perturbation theory [19–21]. Of particular interest is the first-order perturbation approximation because of its simplicity and its use in the Coulomb-breakup extraction of the astrophysical S factor [9]. Its comparison with the dynamical results allows evaluating the importance of higher-order effects [8, 10, 11, 14, 19, 21–24]. The role of the different multipoles is also often studied.

After summarizing the study of breakup by time-dependent methods (Sec. 2), different approaches to the numerical resolution of the time-dependent Schrödinger equation are summarized in one dimension (Sec. 3) and then in three dimensions (Sec. 4). Some results concerning the breakup of ^{11}Be, ^{15}C, and ^{19}C are discussed in Sec. 5. The validity of the first-order perturbation approximation is then evaluated and the role of couplings in the continuum is analyzed in Sec. 6. Section 7 contains concluding remarks.

2. BREAKUP BY TIME-DEPENDENT METHODS

We consider the breakup by the interaction with a target (with mass m_T and charge $Z_T e$) of a two-body projectile made up of a structureless core (with mass m_c and charge $Z_c e$) and a structureless fragment (with mass m_f and charge $Z_f e$). We assume that the conditions of a semi-classical approximation are verified. In the projectile rest frame, the time-dependent Schrödinger equation can be written as

$$i\hbar \frac{\partial}{\partial t}\Psi(\vec{r},t) = H(t)\Psi(\vec{r},t) = [H_0(\vec{r}) + V(\vec{r},t)]\Psi(\vec{r},t). \quad (1)$$

The Hamiltonian of the projectile reads

$$H_0(\vec{r}) = -\frac{\hbar^2}{2\mu}\Delta + V_{cf}(r), \quad (2)$$

where μ is the reduced mass of the core and fragment, $\vec{r} \equiv (\Omega, r)$ is the relative coordinate between them and V_{cf} is the internal interaction between the core and fragment of the projectile. We assume that the spin I_c of the core is zero and that the fragment spin I is fixed. In other words, the coupling with the core spin and core excitations are neglected.

The time-dependent potential is given by

$$V(\vec{r},t) = V_{cT}[r_{cT}(t)] + V_{fT}[r_{fT}(t)] - \frac{(Z_c + Z_f)Z_T e^2}{R(t)}, \quad (3)$$

where $\vec{R}(t)$ is the time-dependent target position with respect to the projectile center of mass.

At high energies, trajectories are very close to straight lines which are usually employed. Some calculations with Rutherford trajectories have however been performed [14, 15]. Trajectories are defined by the initial velocity v and the impact parameter b.

For partial wave lj, the projectile eigenstates at a given relative-motion energy E are defined for bound and scattering states by

$$H_0 \phi_{ljm}(E,\vec{r}) = E \phi_{ljm}(E,\vec{r}). \quad (4)$$

The projectile angular momentum j with projection m results from the coupling of the orbital momentum l for the relative motion between core and fragment with the fragment spin I. Negative energy states are normed and describe either the physical bound states of the projectile or states forbidden by the Pauli principle. Positive energy states are normalized as $\cos\delta_{lj} F_l + \sin\delta_{lj} G_l$, where F_l and G_l are the usual Coulomb functions and δ_{lj} is a phase shift. They correspond to core-fragment scattering which may involve resonances. They are necessary to analyze the final state of the system.

As initial condition at $t = -\infty$, the system is in its ground state $l_0 j_0 m_0$ with energy $E_0 < 0$, $\Psi^{(m_0)}(\vec{r},-\infty) = \phi_{l_0 j_0 m_0}(E_0, \vec{r})$. The time-dependent wave function $\Psi^{(m_0)}(\vec{r},t)$ is calculated by solving numerically Eq. (1) for the different m_0 values. It also depends on the impact parameter b which is understood. The breakup probability distribution is given by

$$\frac{dP}{dE}(E,b) = \frac{2\mu}{\pi\hbar^2 k} \frac{1}{2j_0+1} \sum_{m_0} \sum_{ljm} \left| \langle \phi_{ljm}(E,\vec{r}) | \Psi^{(m_0)}(\vec{r},+\infty) \rangle \right|^2, \quad (5)$$

where k is the wavenumber [13]. It includes full distortion of the scattering eigenstates of the projectile. The breakup cross section reads

$$\frac{d\sigma_{bu}}{dE}(E) = 2\pi \int_0^\infty \frac{dP}{dE}(E,b) \, b \, db. \quad (6)$$

When the nuclear interactions are neglected in Eq. (3), the lower bound in Eq. (6) is replaced by a cutoff b_{min} simulating these nuclear effects. Some authors [11, 13, 15, 16] extract the breakup component from $\Psi^{(m_0)}(\vec{r},+\infty)$ before calculating the matrix elements appearing in Eq. (5). With exact scattering states, this additional projection is not necessary [14].

Momentum distributions can also be calculated [7, 8]. Differential cross sections can be evaluated [14, 15] but interference effects are missing in a semi-classical treatment.

3. TIME-DEPENDENT METHODS FOR ONE DIMENSION

3.1. Treatment of space coordinate

The principal algorithms used for the resolution of the time-dependent Schrödinger equation in Coulomb breakup are exemplified in the simpler one-dimensional case. The Schrödinger equation reads

$$i\hbar \frac{\partial}{\partial t}\psi(x,t) = \left(-\frac{\partial^2}{\partial x^2} + V(x,t)\right)\psi(x,t). \tag{7}$$

Different methods, *expansion on a basis*, *discretization*, and *Lagrange mesh*, are described below.

In the method of *expansion on a basis*, the wave function is expanded over N orthonormal functions φ_j as

$$\psi(x,t) = \sum_{j=1}^{N} c_j(t)\varphi_j(x), \tag{8}$$

where the unknown coefficients depend on time. The Schrödinger equation (7) leads to the system of coupled equations

$$i\hbar \frac{dc_i}{dt} = \sum_{j=1}^{N} [T_{ij} + V_{ij}(t)]c_j, \tag{9}$$

($i = 1, \ldots, N$). In this system, the $T_{ij} = \langle \varphi_i | T | \varphi_j \rangle$ are the matrix elements of the kinetic energy and the $V_{ij} = \langle \varphi_i | V | \varphi_j \rangle$ are those of the potential energy. In general, both matrices are full.

An interesting particular case is obtained when the basis states φ_j are eigenstates of a Hamiltonian $H_0 = T + V_0$ corresponding to the eigenenergies E_{0j}. Then by posing $c_j(t) = \exp(-iE_{0j}t/\hbar)b_j(t)$, one obtains the simpler system

$$i\hbar \frac{db_i}{dt} = \sum_{j=1}^{N} e^{i\omega_{ij}t}[V_{ij}(t) - V_{0ij}]b_j \tag{10}$$

with $\omega_{ij} = (E_{0i} - E_{0j})/\hbar$. The resolution of system (10) has been widely used in Coulomb excitation [1].

The *discretization method* consists in discretizing the spatial interval where the wave function is not negligible with equidistant mesh points jh ($j = 1$ to N) where h is some step size. The values of the wave function at mesh points are $\psi_j = \psi(jh)$. The $(2n+1)$-point approximation of the second derivative of the wave function reads

$$\left(\frac{d^2\psi}{dx^2}\right)_j \approx h^{-2} \sum_{k=-n}^{n} c_k^{(2)} \psi_{j+k}, \tag{11}$$

where the coefficients $c_k^{(2)}$ are given e.g. in Ref. [17]. It has an accuracy of order $O(h^{2n+2})$. The Schrödinger equation (7) can be discretized as

$$i\hbar \frac{d\psi_i}{dt} = \sum_{j=i-n}^{i+n} [T_{ij} + V_{ij}(t)]\psi_j \tag{12}$$

($i = 1, \ldots, N$). The kinetic-energy coefficients are given according to (11) as

$$T_{ij} = \begin{cases} -h^{-2} c_{i-j}^{(2)} & |i-j| \leq n, \\ 0 & |i-j| > n, \end{cases} \tag{13}$$

while the potential-energy coefficients simply read

$$V_{ij} = V(jh,t)\delta_{ij}. \tag{14}$$

In this case, the kinetic-energy matrix is a band matrix while the potential-energy matrix is diagonal. The step size h must be small enough to ensure sufficient accuracy.

The *Lagrange-mesh method* combines the accuracy of an expansion on a basis with the simplicity of a discretization technique. The Lagrange functions f_j are N infinitely differentiable orthonormal functions. With N mesh points x_j, they satisfy the Lagrange conditions [25, 26]

$$f_j(x_i) = \lambda_j^{-1/2} \delta_{ij}. \tag{15}$$

In other words, they vanish at all mesh points but one. Property (15) is inspired by Lagrange interpolation but the Lagrange functions can be more general than polynomials. The coefficients λ_j are the weights of an associated Gauss quadrature formula,

$$\int_a^b g(x)\,dx \approx \sum_{k=1}^N \lambda_k g(x_k). \tag{16}$$

The orthonormality of the basis is automatically satisfied at the Gauss approximation.

The wave function is approximated by the expansion

$$\psi(x,t) = \sum_{j=1}^N c_j(t) f_j(x) \tag{17}$$

where the coefficients depend on time. Property (15) indicates that these coefficients are proportional to the values of the wave function at mesh points, $c_j(t) = \lambda_j^{1/2} \psi(x_j,t)$. The system of equations (9) remains valid but takes a simpler form. At the Gauss approximation, the matrix elements of the kinetic energy are given by

$$T_{ij} = \langle f_i|T|f_j\rangle \approx -\sum_k \lambda_k f_i(x_k) f_j''(x_k) = -\lambda_i^{1/2} f_j''(x_i) \tag{18}$$

and those of the potential energy by the discretization-like expressions

$$V_{ij} = \langle f_i|V|f_j\rangle \approx \sum_k \lambda_k f_i(x_k) V(x_k,t) f_j(x_k) = V(x_i,t)\delta_{ij}. \tag{19}$$

Here the kinetic-energy matrix is full but the potential-energy matrix is diagonal. Moreover the remaining potential matrix elements are extremely simple as they only require an evaluation of the potential at mesh points. Remarkably, the accuracy of the Lagrange-mesh technique is close to the accuracy of a variational calculation performed with functions f_j in spite of the Gauss approximation [26].

3.2. Time evolution

The evolution of the wave function after a time interval Δt is obtained with the evolution operator U according to

$$\psi(t+\Delta t) = U(t+\Delta t,t)\psi(t). \tag{20}$$

For a small enough time step, approximations of this operator can be defived from the Magnus expansion [27, 28]

$$U = \exp\left[-i\int_t^{t+\Delta t} dt_1 H(t_1) - \tfrac{1}{2}\int_t^{t+\Delta t} dt_1 \int_t^{t_1} dt_2 [H(t_1),H(t_2)] + O(\Delta t^5)\right]. \tag{21}$$

The integrals must be approximated in a consistent way [29]. At second order, one obtains for example

$$U = \exp\left[-i\int_t^{t+\Delta t} dt_1 H(t_1) + O(\Delta t^3)\right] = \exp\left[-i\Delta t H(t+\tfrac{1}{2}\Delta t)\right] + O(\Delta t^3). \tag{22}$$

The exponential operator can also be approximated at the same accuracy with a [1, 1] Padé approximant,

$$\exp(-i\Delta t A) \approx (1+\tfrac{1}{2}i\Delta t A)^{-1}(1-\tfrac{1}{2}i\Delta t A) + O(\Delta t^3). \tag{23}$$

Equations (22) and (23) are known as the Crank-Nicholson approximation.

In the preceeding method, the Hamiltonian $H(t)$ is used as a whole. In order to take advantage of some special properties of the kinetic or potential matrix representations, it is useful to employ splitting methods. The Hamiltonian is written as $H(t) = H_0 + V(t)$, where H_0 is independent of time and contains the kinetic energy and a possible time-independent part of the potential. Splitting methods make use of separate exponentials of H_0 and $V(t)$.

A simple first-order splitting method is obtained from Eq. (21) as

$$U = e^{-i\Delta t H_0} e^{-i\Delta t V(t)} + O(\Delta t^2). \tag{24}$$

It may seem advantageous to work at second order with the symmetric expression

$$U = e^{-i\frac{1}{2}\Delta t V(t+\Delta t)} e^{-i\Delta t H_0} e^{-i\frac{1}{2}\Delta t V(t)} + O(\Delta t^3). \tag{25}$$

However, after iteration, approximations (24) and (25) become essentially equivalent as shown by grouping the exponentials of the potential. Eq. (24) thus provides in fact the second-order accuracy $O(\Delta t^3)$.

Higher-order approximations can be derived from the Magnus expansion (21) [29]. More accurate approximations are obtained with a technique proposed in Ref. [30]. For example, a 4th-order approximation reads [31]

$$U = e^{-i\frac{1}{6}\Delta t V(t+\Delta t)} e^{-i\frac{1}{2}\Delta t H_0} e^{-i\frac{2}{3}\Delta t \widetilde{V}(t+\frac{1}{2}\Delta t)} e^{-i\frac{1}{2}\Delta t H_0} e^{-i\frac{1}{6}\Delta t V(t)} + O(\Delta t^5) \tag{26}$$

where $\widetilde{V}(t) = V(t) - \frac{1}{48}\Delta t^2 [\nabla V(t)]^2$. A sixth-order factorization $[O(\Delta t^7)]$ proposed in Ref. [32] shows that 4th-order factorizations might be optimal with respect to computational time except when very high accuracies are needed.

4. NUMERICAL TREATMENT OF 3D BREAKUP

After pioneering 1D [5] and 2D [6] calculations, several groups have developed 3D methods. Among the techniques presented in the previous section, the expansion on a basis requires a discretization of the continuum. In time-dependent calculations, it was only used for semi-quantitative analyses [33, 34] and for studying inelastic scattering [35].

A purely numerical resolution of the time-dependent Schrödinger equation presents the advantage that no simplifying assumptions need to be done about the description of the continuum. Codes are usually developed in spherical coordinates [7, 8, 12, 13, 17]. A 3D breakup study with a cartesian mesh based on algorithms developed for the time-dependent Hartree-Fock method is performed in Ref. [16]. Though conceptually simple, it has the drawback that angular-momentum-dependent interactions cannot be employed.

The time-dependent Schrödinger equation in spherical coordinates is solved with two techniques described below differing in the treatment of the angular part: an expansion over spherical harmonics or an angular Lagrange-mesh method. The radial coordinate is treated with a discretization technique. For simplicity, spins are dropped in the following.

In the standard method, the 3D wave function is expanded in spherical harmonics as

$$\Psi(\vec{r},t) = r^{-1} \sum_{lm} \psi_{lm}(r,t) Y_l^m(\Omega). \tag{27}$$

Putting the radial components $\psi_{lm}(r,t)$ in a column matrix $\hat{\Psi}(r,t)$, one obtains the matrix differential equation

$$i\hbar \frac{\partial}{\partial t} \hat{\Psi}(r,t) = [\hat{H}_0(r) + \hat{V}(r,t)] \hat{\Psi}(r,t). \tag{28}$$

The matrix elements of the projectile Hamiltonian have a simple diagonal form,

$$H_{0\,lm,l'm'}(r) = \left[-\frac{\hbar^2}{2\mu} \frac{d^2}{dr^2} + \frac{\hbar^2}{2\mu} \frac{l(l+1)}{r^2} + V_0(r) \right] \delta_{ll'} \delta_{mm'}. \tag{29}$$

The potential matrix on the contrary is non diagonal with elements

$$V_{lm,l'm'}(r,t) = \langle lm | V(\vec{r},t) | l'm' \rangle. \tag{30}$$

Its calculation requires an expansion of the potential in multipoles which may be tedious and time consuming. This method has been combined with the splitting method [7, 11] or used without it [8, 12].

An expansion in angular Lagrange functions is also possible. Let us consider N angular mesh points $\Omega_j = (\theta_{j'}, \varphi_{j''})$ as defined by Melezhik [36] and N Gauss quadrature weights λ_j associated with the mesh. Then one can define N angular Lagrange functions f_j with the Lagrange property $f_j(\Omega_i) = \lambda_j^{-1/2} \delta_{ij}$. These functions read

$$f_j(\Omega) = \lambda_j^{1/2} \sum_{\nu=1}^{N} Y_\nu^*(\Omega_j) Y_\nu(\Omega), \qquad (31)$$

where the Y_ν are ususal spherical harmonics complemented by modified spherical harmonics [17]. At the Gauss quadrature approximation, the Lagrange functions are orthonormal and the potential matrix elements are diagonal [Eq. (19)].

The 3D wave function is expanded as

$$\Psi(\vec{r},t) = r^{-1} \sum_{j=1}^{N} \lambda_j^{1/2} f_j(\Omega) \psi_j(r,t). \qquad (32)$$

The components $\psi_j(r,t)$ represent the value of the 3D wave function along the direction Ω_j, $\psi_j(r,t) = r\Psi(\vec{r},t)|_{\Omega=\Omega_j}$. Putting the radial components $\lambda_j^{1/2} \psi_j(r,t)$ in a column matrix $\hat{\Psi}(r,t)$, one obtains a matrix differential equation similar to Eq. (28) but with different matrix properties. The matrix elements of the projectile Hamiltonian have a simple diagonal form in the basis Y_ν,

$$H_{0\nu\nu'}(r) = \left[-\frac{\hbar^2}{2\mu} \frac{d^2}{dr^2} + \frac{\hbar^2}{2\mu} \frac{l(l+1)}{r^2} + V_0(r) \right] \delta_{\nu\nu'}. \qquad (33)$$

According to Eq. (19), the potential matrix is diagonal in the Lagrange basis with elements

$$V_{jj'}(r,t) \approx V(r,\Omega_j,t) \delta_{jj'}. \qquad (34)$$

These matrix elements are easily calculated. No multipole expansion is needed. The Lagrange-mesh approach requires using two different bases according to the type of matrix element. This can be combined with a splitting method such as Eq. (25), as employed in Ref. [17].

In both methods, a radial discretization must also be performed. The evaluation of the exponential operators is then simple with Eq. (23) but two changes of basis are required in the Lagrange-mesh method at each time step. Uniform meshes $r_j = jh$ with constant step size h and a 5-point difference formula (11) are used in Refs. [7, 8, 12]. A quasiuniform mesh $r_j = r_{\max} g(jh)$ where $g(x)$ is a mapping of $[0, 1]$ on itself is employed in Refs. [13, 15, 17]. A discussion and examples can be found in Ref. [17].

All calculations in the following make use of straight-line trajectories.

5. BREAKUP OF HALO NUCLEI

The best studied one-nucleon halo nucleus is ^{11}Be. It has two weakly bound states, $1/2^+$ at -0.503 MeV and $1/2^-$ at -0.183 MeV, and a resonance $5/2^+$ at 1.27 MeV with width 0.10 ± 0.02 MeV. The bound states are fitted as $1s1/2$ and $0p1/2$ with ^{10}Be + n potentials. The $0s1/2$ and $0p3/2$ bound states are non physical but play almost no role [37].

Breakup calculations involve either target-core and target-neutron optical potentials or a cutoff on impact parameters. Introducing nuclear potentials is very simple in the Lagrange-mesh approach since it only requires values of these potentials at mesh points. The role of the optical potentials is illustrated for ^{11}Be breakup on ^{208}Pb in Fig. 1 at different relative energies E [17]. The breakup probability obtained with only Coulomb potentials monotonically decreases as a function of b. It diverges when b tends to 0. A cutoff must be introduced near $b = 12$ fm. In the calculations with projectile-target optical potentials, the breakup probability is negligible for small b because of strong absorption. From $b = 20\text{-}25$ fm, all results exhibit the same behavior. Fig. 1 shows that impact-parameter cutoffs cannot simulate the results with optical potentials at all relative energies E.

The ^{11}Be $\rightarrow$ ^{10}Be + n breakup on ^{208}Pb is nevertheless dominated by Coulomb effects. Different authors reach very similar results and conclusions [15, 17], as shown in Fig. 2. The projectile-target nuclear forces play an increasing role

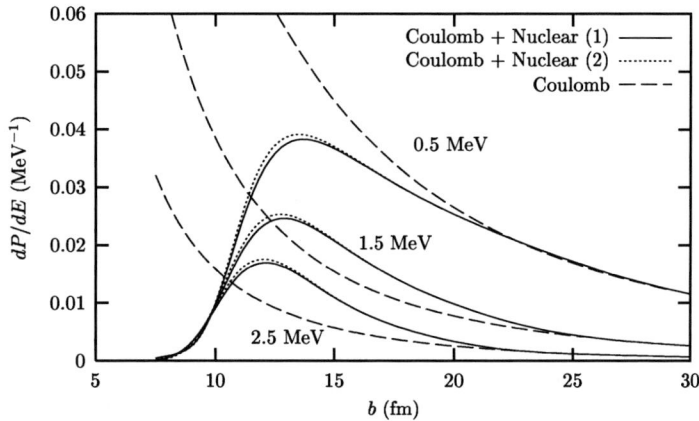

FIGURE 1. ^{11}Be $\rightarrow$ ^{10}Be + n breakup on ^{208}Pb at 72 MeV/nucleon ($v = 0.37c$) calculated with only Coulomb interactions (dashed lines) and with two optical potentials (full and dotted lines) [17].

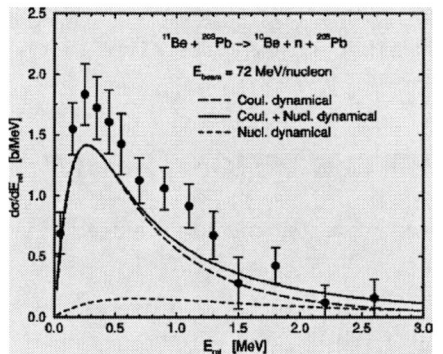 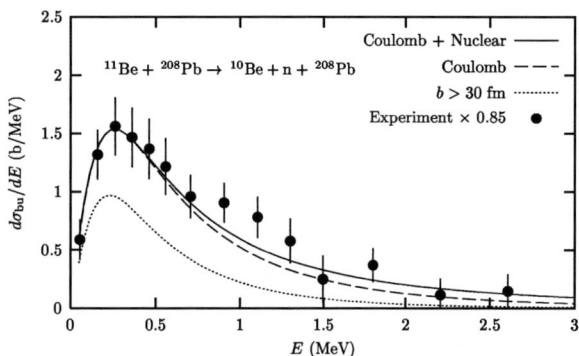

FIGURE 2. Comparison of Coulomb (dashed lines) and Coulomb + nuclear (full lines) time-dependent calculations of the ^{11}Be breakup on ^{208}Pb at 72 MeV/nucleon: Ref. [15] (left) and Ref. [17] (right). Notice that experimental data from Ref. [38] are multiplied by 0.85 in the right panel.

for $E > 1$ MeV. A good agreement with the data of Ref. [38] is obtained when they are scaled by a factor 0.85, as in the right panel, according to a reanalysis of the original experiment [39]. The description of ^{11}Be seems to be valid with a spectroscopic factor close to unity.

The breakup of ^{11}Be on ^{12}C is discussed in Refs. [18, 40]. A significant contribution from the ^{11}Be $5/2^+$ resonance is obtained (see Ref. [24] and P. Capel's contribution) in agreement with experiment [41].

The ^{15}C halo nucleus has two bound states, $1/2^+$ at -1.218 MeV and $5/2^+$ at -0.478 MeV. These bound states are fitted as $1s1/2$ and $0d5/2$ with an angular-momentum dependent ^{14}C + n potential. The $0s1/2$, $0p1/2$ and $0p3/2$ bound states are non physical. The breakup of ^{15}C on ^{208}Pb is illustrated by Fig. 3 [17]. The left panel displays the time evolution of $s1/2$ (a) and $p3/2$ (b) partial waves as a function of time in the projectile frame of reference. Significant modifications occur in the wave function near the time of closest approach ($t = 0$). At later times, components of the wave packet are dragged by the moving off target. The right panel shows a good agreement with experiment [39] for the integrated cross sections over all impact parameters and over $b > 30$ fm. Here also the spectroscopic factor seems to be close to unity although the $b > 30$ fm results appear to be somewhat overestimated.

The ^{19}C breakup on lead has been studied in Ref. [15]. The neutron separation energy of this nucleus is not accurately known, $S_n = 0.53 \pm 0.13$ MeV. The ground-state spin and parity should be $1/2^+$. The authors employ breakup experimental data [42] to try to fix the value of this energy but this approach relies on the accuracy of the experimental normalization and on assuming a spectroscopic factor equal to 1.

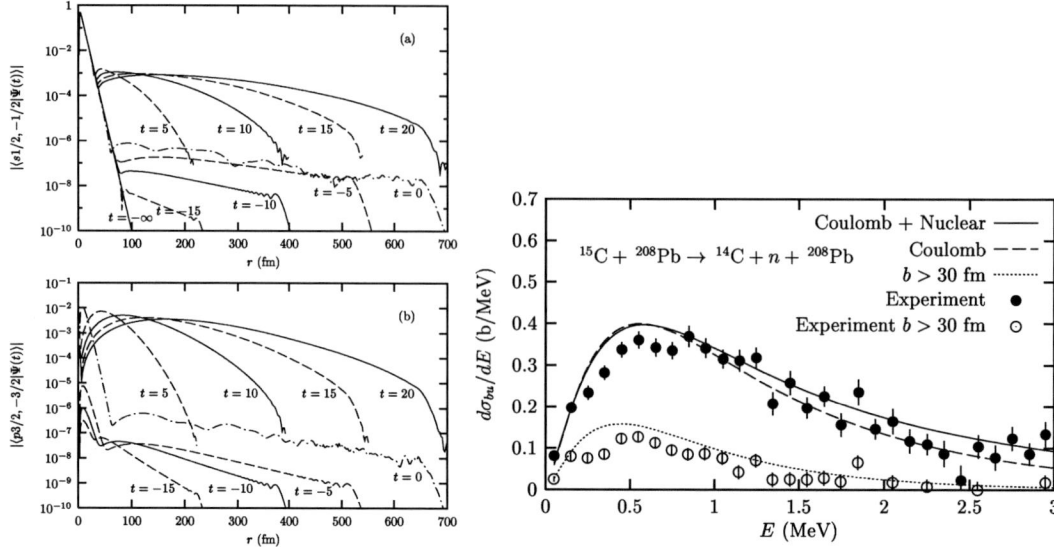

FIGURE 3. Left: time evolution of the moduli of the $s1/2 - 1/2$ (a) and $p3/2 - 3/2$ (b) partial waves during ^{15}C breakup on ^{208}Pb starting with an $1s1/2 - 1/2$ initial state for $v = 0.36c$ and $b = 30$ fm [17]. Right: total (full line) and partial ($b > 30$ fm, dotted line) breakup cross sections at 68 MeV/nucleon [17] compared with data from Ref. [39].

6. COMPARISON WITH FIRST-ORDER PERTURBATION THEORY

The validity of first-order perturbation is discussed by comparing at fixed impact parameters the numerical probability dP/dE with its first-order approximation

$$\frac{dP^{(1)}}{dE}(E,b) = \frac{2\mu}{\pi\hbar^4 k} \frac{1}{2j_0+1} \sum_{m_0} \sum_{ljm} \left| \int_{-\infty}^{\infty} e^{i\omega t} \langle \phi_{ljm}(E,\vec{r}) | V(\vec{r},t) | \phi_{l_0 j_0 m_0}(E_0,\vec{r}) \rangle dt \right|^2, \quad (35)$$

where $\omega = (E - E_0)/\hbar$. At large impact parameters, the Coulomb potential is expanded in multipoles and the expansion can be restricted to E1 for neutron emission or to E1 + E2 for proton emission. The integral over time is then usually evaluated with the simplifying approximation $r_< = r$, $r_> = R(t)$, sometimes known as the 'far-field approximation'. This leads to the appearance of effective charges

$$Z_{\text{eff}}^{(E\lambda)} = Z_c \left(-\frac{m_N}{M}\right)^{\lambda} + Z_N \left(\frac{m_c}{M}\right)^{\lambda} \quad (36)$$

which are convenient to estimate the role of E2.

First-order perturbation is an essential ingredient for indirect determinations of the astrophysical S factor by Coulomb breakup. Its validity in this context is discussed for ^{17}F in Refs. [14, 22]. and for ^{8}B in Refs. [23, 24].

Let us first consider the Coulomb breakup of ^{11}Be on ^{208}Pb [24]. The ratio of the numerical and first-order breakup probabilities is displayed for various large impact parameters in the left panel of Fig. 4. The initial state corresponds to $s1/2$. Both calculations are performed under the conditions of Ref. [17]. The first-order probability is restricted to E1. At all velocities shown, assuming that the first-order approximation is valid would introduce an error (decreasing with increasing projectile energies) and a distortion of the relative energy dependence.

A decomposition of the ratio of probabilities into partial waves is displayed for $b = 100$ fm in the right panel of Fig. 4. With an initial s state, the first-order E1 transition only populates the p waves. One observes that $l = 1$ is indeed the dominant contribution in the numerical solution. However, the s and d waves which can only be reached by higher-order processes, since E2 is negligible here, are not very small. Surprisingly, the first-order purely p-wave breakup is a better approximation of the sum of s, p, d, ... breakups than of the corresponding p components alone.

For understanding the origin of this effect, the evolution of the breakup probability and of its components is followed as a function of time in Fig. 5. This probability is negligible until near closest approach ($t \approx 0$) and then steeply

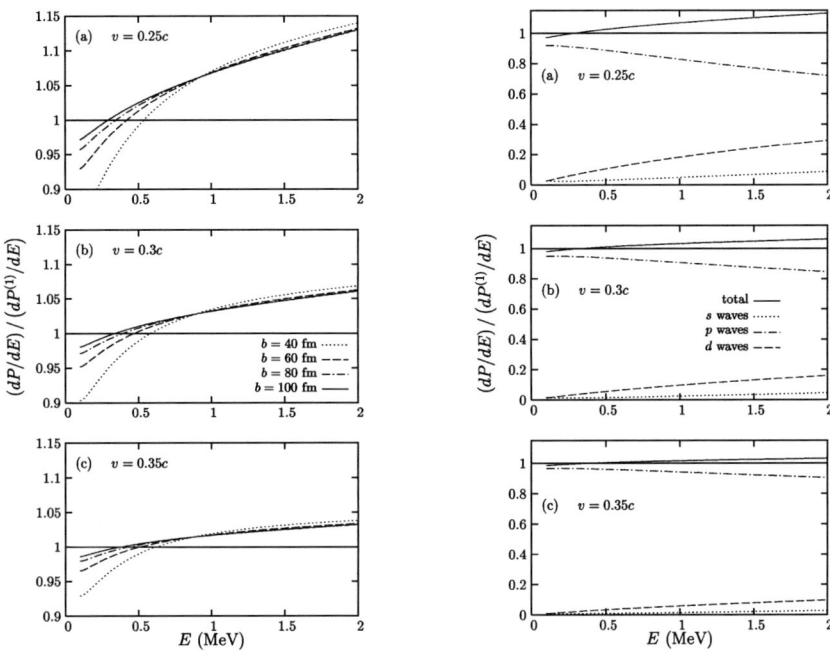

FIGURE 4. Ratio of the dynamical and E1 first-order probabilities for the breakup of ^{11}Be on ^{208}Pb as a function of the energy E of the relative motion between the fragments (in MeV) at three projectile velocities $v = 0.25c$ (a), $0.3c$ (b), and $0.35c$ (c) for impact parameters $b = 40$, 60, 80, and 100 fm (left) and decomposition into partial waves at $b = 100$ fm (right) [24].

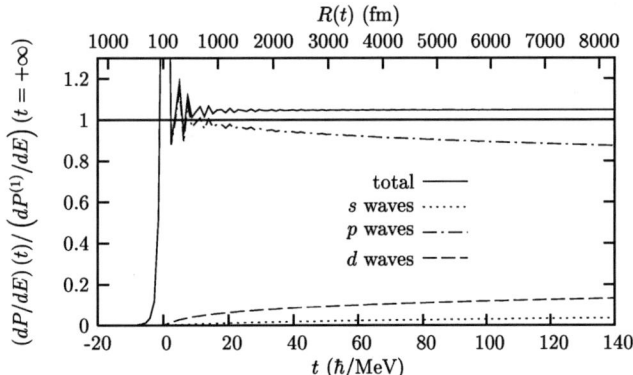

FIGURE 5. Time evolution of partial-wave populations at $v = 0.3c$, $E = 1.5$ MeV, and $b = 100$ fm in ^{11}Be breakup on ^{208}Pb [24].

rises. Breakup first leads to p waves with small d components, indicating dominant first-order effects. After damped oscillations, the total breakup probability remains remarkably constant beyond $t = 30$ $\hbar$/MeV. However, beyond that time, p waves are still depleted toward s and d waves. The s and d components increase only after the p waves have been populated. Since the total breakup component is extremely stable, these transitions take place in the continuum. This effect reflects the long range nature of the Coulomb interaction.

The Coulomb breakup of ^{8}B into ^{7}Be and proton on a ^{208}Pb target is useful for astrophysics. The ratio of the numerical and first-order breakup probabilities [24] is displayed in Fig. 5 for a large impact parameter ($b = 100$ fm). The first-order probability (36) here contains E1 and E2. The E1 contribution to the first-order probability is higher than 80% (thick curve). The total breakup probability obtained with the time-dependent approach is always smaller than the first-order E1+E2 approximation. It is larger than the purely E1 approximation. Assuming the first-order E1+E2 approximation to be valid when extracting the ^{7}Be(p,γ)^{8}B astrophysical S factor from ^{8}B breakup data thus leads to an underestimation of the S factor after correction for E2. It also introduces a distortion of the energy dependence, but

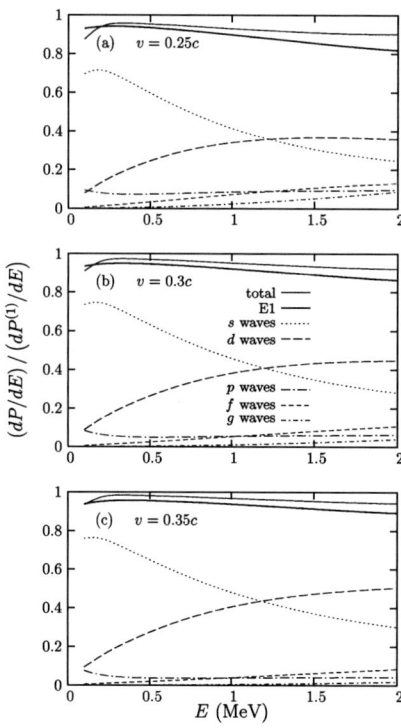

FIGURE 6. Ratio of the dynamical and E1+E2 first-order probabilities for the breakup of ^{8}B on ^{208}Pb at impact parameter $b = 100$ fm and projectile velocities $v = 0.25c$ (a), $0.3c$ (b), and $0.35c$ (c) (full curves) [24]. Different curves represent the s (dots), p (long dashes and dots), d (long dashes), f (short dashes), and g (short dashes and dots) contributions. The thick curves display the proportion of E1 transitions in the first-order E1+E2 approximation.

much weaker than in the ^{11}Be case. A similar conclusion is reached in Ref. [23].

7. CONCLUSION

The semi-classical approximation provides a dynamical treatment at all orders of the breakup of halo nuclei. Couplings in the continuum are fully taken into account. Accurate numerical solutions of the time-dependent Schrödinger equation are available by different methods. Nevertheless various improvements remain possible: higher-order treatment of time evolution, Lagrange mesh for the radial coordinate, ...

This method has been applied to the breakup of different one-neutron halo nuclei: ^{11}Be, ^{15}C, ^{19}C, ... Both Coulomb and nuclear breakups can be treated. Comparisons with first-order perturbation theory aimed at assessing the validity of the extraction of astrophysical S factors from Coulomb breakup predict systematic errors for this method.

Many questions remain open. A technical problem concerns the comparison with experiment which can not easily be performed because the experimental resolution and acceptance are usually not available in accessible references. Uncertainties in the normalization affect the extraction of spectroscopic factors or even indirect determinations of the binding energy as in the ^{19}C case [15].

A significant effort remains to be done in order to compare the merits of the time-dependent method with those of purely quantal approaches: the continuum discretized coupled-channels method [43, 44], the adiabatic approximation [45], the distorted-wave Born approximation [46, 47] and the eikonal approximation [40, 48]. A comparison of the semi-classical method with the eikonal approximation can be found in Ref. [40].

In my opinion, the main challenge in this type of dynamical calculation is now to better describe the one-neutron halo nuclei. The simple potential model should be replaced by models involving core excitation or even by microscopic models. A realistic description of the breakup of two-neutron halo nuclei in a time-dependent model is still a distant goal.

ACKNOWLEDGMENTS

I wish to thank Pierre Capel, Gérald Goldstein, and Vladimir Melezhik for many interesting discussions and for their calculations. This text presents research results of the Belgian program P5/07 on interuniversity attraction poles initiated by the Belgian-state Federal Services for Scientific, Technical and Cultural Affairs (FSTC).

REFERENCES

1. K. Alder, and A. Winther, *Electromagnetic excitation* (North-Holland, Amsterdam, 1975).
2. K. Sakimoto, *J. Phys.* B **33**, 5165 (2000).
3. X.-M. Tong, D. Kato, T. Watanabe, and S. Ohtani, *J. Phys.* B **33**, 5585 (2000).
4. V.S. Melezhik, J.S. Cohen, and C.-Y. Hu, *Phys. Rev.* A **69**, 032709 (2004).
5. G.F. Bertsch, and C.A. Bertulani, *Nucl. Phys.* **A556**, 136 (1993).
6. C.A. Bertulani, and G.F. Bertsch, *Phys. Rev.* C **49**, 2839 (1994).
7. T. Kido, K. Yabana, and Y. Suzuki, *Phys. Rev.* C **50**, R1276 (1994).
8. H. Esbensen, G.F. Bertsch, and C.A. Bertulani, *Nucl. Phys.* **A581**, 107 (1995).
9. G. Baur, C.A. Bertulani, and H. Rebel, *Nucl. Phys.* **A458**, 188 (1986).
10. H. Esbensen, and G.F. Bertsch, *Nucl. Phys.* **A600**, 37 (1996).
11. T. Kido, K. Yabana, and Y. Suzuki, *Phys. Rev.* C **53**, 2296 (1996).
12. S. Typel, and H.H. Wolter, *Z. Naturforsch.* A **54**, 63 (1999).
13. V.S. Melezhik, and D. Baye, *Phys. Rev.* C **59**, 3232 (1999).
14. V.S. Melezhik, and D. Baye, *Phys. Rev.* C **64**, 054612 (2001).
15. S. Typel, and R. Shyam, *Phys. Rev.* C **64**, 024605 (2001).
16. M. Fallot, J.A. Scarpaci, D. Lacroix, P. Chomaz, and J. Margueron, *Nucl. Phys.* **A700**, 70 (2002).
17. P. Capel, D. Baye, and V.S. Melezhik, *Phys. Rev.* C **68**, 014612 (2003).
18. P. Capel, G. Goldstein, and D. Baye, *Phys. Rev.* C **70**, 064605 (2004).
19. S. Typel, and G. Baur, *Phys. Rev. C* **50**, 2104 (1994).
20. S. Typel, H.H. Wolter, and G. Baur, *Nucl. Phys.* **A613**, 147 (1997).
21. S. Typel, and G. Baur, *Phys. Rev.* C **64**, 024601 (2001).
22. H. Esbensen, and G.F. Bertsch, *Nucl. Phys.* **A706**, 383 (2002).
23. H. Esbensen, G.F. Bertsch, and K.A. Snover, *Phys. Rev. Lett.* **94**, 042502 (2005).
24. P. Capel, and D. Baye, *Phys. Rev.* C (2005).
25. D. Baye, and P.-H. Heenen, *J. Phys.* A **19**, 2041 (1986).
26. D. Baye, M. Hesse, and M. Vincke, *Phys. Rev.* E **65**, 026701 (2002).
27. W. Magnus, Commun. *Pure Appl. Math.* **7**, 649 (1954).
28. R.M. Wilcox, *J. Math. Phys.* **8**, 962 (1967).
29. D. Baye, G. Goldstein, and P. Capel, *Phys. Lett.* A **317**, 337 (2003).
30. M. Suzuki, *Proc. Japan Acad.* **69** Ser. B, 161 (1993).
31. S.A. Chin, and C.R. Chen, *J. Chem. Phys.* **117**, 1409 (2002).
32. G. Goldstein, and D. Baye, *Phys. Rev.* E **70**, 056703, (2004).
33. C.A. Bertulani, and L.F. Canto, *Nucl. Phys.* **A539**, 163 (1992).
34. H. Esbensen, and G.F. Bertsch, *Phys. Rev.* C **59**, 3240 (1999).
35. C.A. Bertulani, L.F. Canto, and M.S. Hussein, *Phys. Lett.* **B353**, 413 (1995).
36. V.S. Melezhik, *Phys. Lett.* A **230**, 203 (1997).
37. P. Capel, D. Baye, and V.S. Melezhik, *Phys. Lett.* B **552**, 145 (2003).
38. T. Nakamura, S. Shimoura, T. Kobayashi, T. Teranishi, K. Abe, et al., *Phys. Lett.* B **331**, 296 (1994).
39. T. Nakamura, N. Fukuda, N. Aoi, H. Iwasaki, T. Kobayashi, et al., *Nucl. Phys.* **A722**, 301c (2003).
40. H. Esbensen, and G.F. Bertsch, *Phys. Rev.* C **64**, 014608 (2001).
41. N. Fukuda, T. Nakamura, N. Aoi, N. Imai, M. Ishihara, et al., *Phys. Rev.* C **70**, 054606 (2004).
42. T. Nakamura, N. Fukuda, T. Kobayashi, N. Aoi, H. Iwasaki, et al., *Phys. Rev. Lett.* **83**, 1112 (1999).
43. N. Austern, Y. Iseri, M. Kamimura, M. Kawai, G. Rawitscher, and M. Yahiro, *Phys. Rep.* **154**, 125 (1987).
44. J.A. Tostevin, F.M. Nunes, and I.J. Thompson, *Phys. Rev.* C **63**, 024617 (2001).
45. J.A. Tostevin, S. Rugmai, and R.C. Johnson, *Phys. Rev.* C **57**, 3225 (1998).
46. R. Chatterjee, P. Banerjee, and R. Shyam, *Nucl. Phys.* **A675**, 477 (2000).
47. M. Zadro, *Phys. Rev.* C **70**, 044605 (2004).
48. R.J. Glauber, *High energy collision theory* in Lectures in Theoretical Physics, Vol. 1, Editors W.E. Brittin and L.G. Dunham (Interscience, New York, 1959) p. 315.

Time-dependent analysis of the nuclear and Coulomb dissociation of ^{11}Be

Pierre Capel*, Gérald Goldstein† and Daniel Baye†

*TRIUMF, 4004 Wesbrook Mall, Vancouver, B.C., Canada V6T 2A3
†Physique Quantique, C.P. 165/82 and Physique Nucléaire Théorique et Physique Mathématique, C.P. 229, Université Libre de Bruxelles, B-1050 Brussels, Belgium

Abstract. The breakup of ^{11}Be on carbon and lead targets around 70 MeV/nucleon is investigated within a semiclassical framework. The role of the $\frac{5}{2}^+$ resonance is analyzed in both cases. It induces a narrow peak in the nuclear-induced breakup cross section, while its effect on Coulomb breakup is small. The nuclear interactions between the projectile and the target is responsible for the transition toward this resonant state. The influence of the parametrization of the ^{10}Be-n potential that simulates ^{11}Be is also addressed. The breakup calculation is found to be dependent on the potential choice. This leads us to question the reliability of this technique to extract spectroscopic factors.

Keywords: halo nuclei, semiclassical approximation, time-dependent Schrödinger equation, dissociation, ^{11}Be
PACS: 25.60.Gc, 25.70.De, 25.70.Ef, 27.20.+n

1. INTRODUCTION

The ^{11}Be nucleus is one of the best known one-neutron halo nuclei. Its halo structure has thus been the subject of many theoretical and experimental analyzes [1]. In particular, breakup reactions are used as tools to extract its structure properties [2, 3]. Various theoretical models have been developed to interpret the experimental data [4]: perturbation expansion, adiabatic approximation [5], Eikonal model [6], coupled channel with a discretized continuum (CDCC) [7, 8], and numerical resolution of a three-dimensional time-dependent Schrödinger equation [9].

Recently, the breakup of ^{11}Be on both lead and carbon targets has been measured at RIKEN around 70 MeV/nucleon [3]. In the present talk, we investigate these reactions with a time-dependent technique. This reaction model is based on a semiclassical approximation [9, 10] in which the relative motion of the projectile and the target is approximated by a classical trajectory. Therefore, the projectile is seen as evolving in a time-dependent potential that simulates its interaction with the target. This approximation leads to the resolution of a time-dependent Schrödinger equation. Different techniques have been developed to solve this equation [11–16]. We use the technique described in Ref. [16].

Up to now, ^{11}Be is described in all reaction models as a two-body system: a halo neutron loosely bound to a structureless ^{10}Be core. The interaction between the neutron and the core is modeled by a simple local potential. This ^{10}Be-n potential is usually adjusted to reproduce the bound states of ^{11}Be [11–15]. It is of course important to analyze the accuracy of that description. In particular, one needs to know what is reproducible using such a simple model, and what is not. In a recent paper, we studied the breakup of ^{11}Be on a ^{12}C target [17]. For that study, we developed a new ^{10}Be-n potential that reproduces not only the bound states of ^{11}Be, but also its first resonant state above the one-neutron threshold. That resonance is found to induce a narrow peak in the breakup cross section. A similar peak is observed in the experimental data. This suggests that the resonance can be fairly well reproduced in the two-body description, and that its presence in reaction models is required to reproduce the experimental data. In this talk, we present the results of this analysis. We also present recent calculations of the Coulomb breakup of ^{11}Be. In particular, we discuss the role played by the resonance in that reaction and compare it to its role played in the dissociation on ^{12}C.

Besides the capability of this simple two-body description to reproduce physical levels of ^{11}Be, the sensitivity of the calculations to the parametrization of the ^{10}Be-n potential must also be assessed. In particular, the breakup cross section should not be too sensitive to the potential choice if one wishes to reliably extract spectroscopic information from measurements. In this talk, we present the first results of such an analysis. The results of calculations of the Coulomb breakup of ^{11}Be performed with different ^{10}Be-n potentials are discussed.

The talk is structured as follows. After a brief description of the time-dependent model and the parametrizations of the potential that describes ^{11}Be, we present, in Sec. 3, the results we have obtained in the breakup on ^{12}C [17]. The analysis of the Coulomb breakup of ^{11}Be is discussed in Sec. 4. The final section contains our concluding remarks.

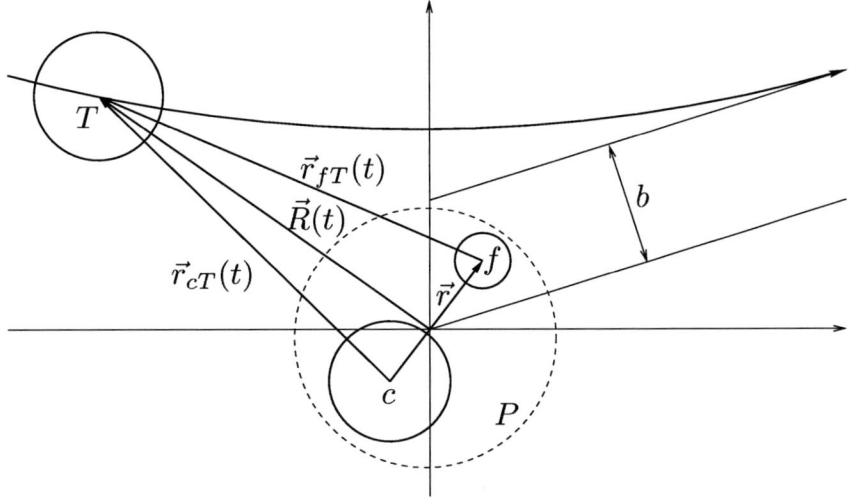

FIGURE 1. Semiclassical scheme of the reaction. In the projectile (*P*) rest frame, the target (*T*) follows a classical trajectory described by the time-dependent relative coordinate $\vec{R}$. The projectile is assumed to have a two-body structure: a fragment *f* loosely bound to a core *c*. Their relative coordinate is $\vec{r}$.

2. THEORETICAL FRAMEWORK

2.1. Time-dependent model

We consider the breakup of a projectile *P* by a target *T*. The projectile *P* is assumed to have a two-body structure: a pointlike and structureless fragment *f* (of mass m_f and charge $Z_f e$) loosely bound to a structureless core *c* (of mass m_c and charge $Z_c e$). The target is seen as a structureless particle of mass m_T and charge $Z_T e$. In the semiclassical approximation [9, 10], the *P-T* relative motion is treated classically: in the projectile rest frame, the target is assumed to follow a classical trajectory (see Fig. 1). Therefore, the interaction between the projectile and the target is simulated by a time-dependent potential. The internal motion of the projectile, however, is treated quantum mechanically. The wave function Ψ describing this motion is solution of the following time-dependent Schrödinger equation:

$$\begin{aligned} i\hbar \frac{\partial}{\partial t} \Psi(\vec{r},t) &= [H_0(\vec{r}) + V(\vec{r},t)] \Psi(\vec{r},t) \\ &= \left\{ -\frac{\hbar^2}{2\mu}\Delta + V_{cf}(r) + V_{cT}[r_{cT}(\vec{r},t)] + V_{fT}[r_{fT}(\vec{r},t)] - \frac{(Z_c+Z_f)Z_T e^2}{R(t)} \right\} \Psi(\vec{r},t), \end{aligned} \quad (1)$$

where $\vec{r}$ is the relative coordinate of the fragment to the core, $\vec{R}$ is the time-dependent coordinate that describes the trajectory of the target in the projectile rest frame, and $\vec{r}_{cT}$ and $\vec{r}_{fT}$ are respectively the core-target and fragment-target relative coordinates as illustrated in Fig. 1.

In Eq. (1), the Hamiltonian H_0 describes the internal structure of the two-body projectile. It is the sum of the kinetic term and the local potential V_{cf}, which simulates the interaction between the core and the fragment (see Sec. 2.2). The potentials V_{cT} and V_{fT} model the core-target and fragment-target interactions, respectively. They comprise a Coulomb term and a short-range optical potential, which simulates the nuclear interaction. The latter is usually chosen in the literature.

Eq. (1) is solved with the initial condition that at time $t \to -\infty$ the projectile is in its ground state. The wave function Ψ at time t is then obtained iteratively using the evolution algorithm described in Ref. [16]. The calculation is performed for different trajectories parametrized by the impact parameter *b*. For each trajectory, we deduce the breakup probability by projecting the output wave function $\Psi(\vec{r}, t \to +\infty)$ onto the positive eigenstates of H_0 that describe the continuum of the projectile. The breakup cross section is then obtained by summing this probability over all impact parameters.

TABLE 1. Parameters of the ^{10}Be-n potentials [see Eqs. (3)-(5)]. Note that R_0 used in (3)-(5) is parametrized as $r_0 A_c^{1/3}$.

Potential	V_{leven} (MeV)	V_{lodd} (MeV)	V_{LS} (MeV fm^2)	a (fm)	r_0 (fm)
V1	62.52	39.74	21.0	0.6	1.2
V2	66.325	38.37	12.44	0.5	1.2
V3	58.905	40.025	27.68	0.7	1.2
V4	71.28	49.015	29.95	0.6	1.1
V5	55.25	32.515	12.86	0.6	1.3
V6	59.05	59.05	0	0.62	1.236

2.2. ^{11}Be description

As done in previous works [11–17], we describe ^{11}Be as a neutron loosely bound to a ^{10}Be core. The ^{10}Be core is assumed to be in its 0^+ ground state, and the spectroscopic factor associated to that configuration is set equal to unity. The potential which simulates the ^{10}Be-n interaction is composed of a central part plus a spin-orbit coupling term

$$V_{cf}(r) = V_0(r) + \vec{L} \cdot \vec{I} V_{LI}(r), \qquad (2)$$

where $\vec{L}$ is the orbital momentum of the ^{10}Be-n relative motion, and $\vec{I}$ is the spin of the neutron. The central part of V_{cf} has a Woods-Saxon form factor

$$V_0(r) = -V_l f(r, R_0, a), \qquad (3)$$

where

$$f(r, R_0, a) = \left[1 + \exp\left(\frac{r - R_0}{a}\right)\right]^{-1}. \qquad (4)$$

The spin-orbit coupling term has the usual Thomas form factor

$$V_{LI}(r) = V_{LS} \frac{1}{r} \frac{d}{dr} f(r, R_0, a). \qquad (5)$$

The radius of the form factor is parametrized as usual: $R_0 = r_0 A_c^{1/3}$.

The depths of the potential are adjusted to reproduce the energies of the low-lying states of ^{11}Be. The well known shell inversion observed between the bound states is reproduced by using a parity-dependent depth of the central part of the potential V_l. The $\frac{1}{2}^+$ ground state is modeled by a $1s1/2$ state, the $\frac{1}{2}^-$ excited state by a $0p1/2$ state, and the first $\frac{5}{2}^+$ resonance is reproduced in the $d5/2$ wave.

In order to study the sensitivity of our calculations to the potential choice, we developed five sets of parameters that reproduce the physical states mentioned above. They are summarized in Table 1. The first potential (V1) has been devised for our recent calculation of the breakup of ^{11}Be on ^{12}C [17] (see also Sec. 3). The next four (V2 to V5) have been obtained by varying either the diffuseness or the radius of the Woods-Saxon form factor. The values were chosen to encompass those used by most other groups [11, 12, 14, 15]. Besides the three physical levels, these potentials all exhibit two unphysical bound states: $0s1/2$ and $0p3/2$. These states correspond to the shells occupied by the neutrons in the core and are forbidden by the Pauli principle. Their energies have not been adjusted, and thus vary from one potential to the other. Each potential also displays a $d3/2$ resonance. This resonance does not correspond to any known physical state. Therefore it has not been fitted. Its location and width vary with the potential choice. Since it is very broad and located at high energy, we doubt this resonance might play any significant role in our calculations. In Table 1, we also list a sixth potential (V6) developed by Fukuda et al. [3]. It reproduces only the ground state energy and does not contain a spin-orbit coupling term.

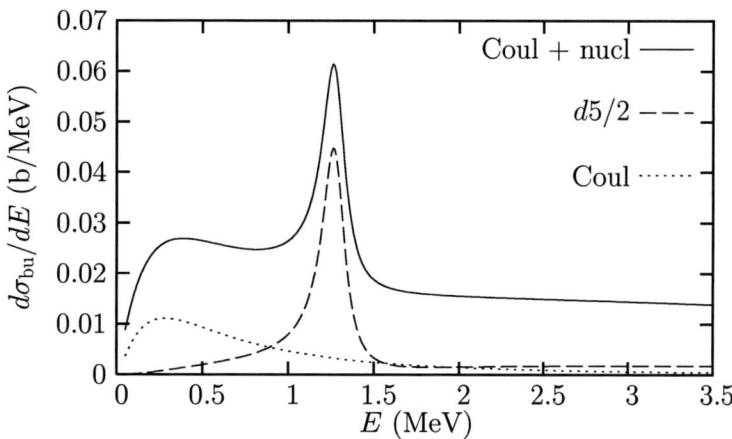

FIGURE 2. Breakup cross section of ^{11}Be on ^{12}C at 67 MeV/nucleon as a function of the ^{10}Be-n relative energy E after breakup. The full line corresponds to the calculation considering both the nuclear and Coulomb interactions between the projectile and the target. The dashed line displays its $d5/2$ component, responsible for the peak at the resonance energy. The result obtained with a pure Coulomb P-T interaction is displayed as the dotted line.

3. BREAKUP OF ^{11}BE ON ^{12}C

3.1. Breakup cross section

Recently, the breakup of ^{11}Be on ^{12}C at 67 MeV/nucleon has been measured at RIKEN [3]. We analyze this reaction within the semiclassical framework described in the previous section. In Fig. 2, the breakup cross section is displayed as a function of the relative energy E between the ^{10}Be core and the neutron after breakup. The full line corresponds to the results obtained considering both nuclear and Coulomb interactions between the projectile and the target. The conditions of the calculations are those described in Ref. [17]. In particular, ^{11}Be is described by the ^{10}Be-n potential V1 of Table 1 developed in that previous work. For the ^{10}Be-^{12}C interaction, we first use the optical potential developed by Al-Khalili, Tostevin and Brook, which has been adjusted to reproduce ^{10}Be-^{12}C scattering data [18]. The n-^{12}C interaction is simulated by the Becchetti and Greenlees parametrization [19]. For comparison, we also display the cross section computed with a pure Coulomb interaction between the projectile and the target (dotted line). This emphasizes the strong dominance of the nuclear interactions in this dissociation reaction.

First, the breakup cross section is significantly enhanced when optical potentials are considered. This is true on the entire energy range, but is particularly striking at high energy. The presence of nuclear P-T interactions leads to a gentle decrease of the cross section with energy, while a pure Coulomb interaction induces a rapid drop of the cross section beyond 0.5 MeV.

Second, a narrow peak is observed in the breakup cross section obtained with optical potentials. This peak is due to the $d5/2$ resonance present in our description of ^{11}Be. It is indeed located at the same energy and exhibits the same width as that resonance. Moreover, it appears solely in the contribution of the $d5/2$ partial wave to the cross section (dashed line). The absence of peak in the purely Coulomb result indicates that P-T nuclear interactions are necessary to populate that resonant state.

3.2. Comparison with experiment and analysis of the influence of the optical potentials

In Fig. 3, we compare the results of our calculation with the breakup cross section measured at RIKEN [3]. The full line (labeled ATB+BG) corresponds to the full line of Fig. 2 convoluted with the experimental energy resolution. The main effect of this convolution is to significantly broaden the resonance peak and slightly shift it toward lower energies.

We observe a very good agreement between theory and experiment. Note that all the parameters have been fixed prior to the calculation; there is no adjustment of our results to the experimental data. At low energy, theory and

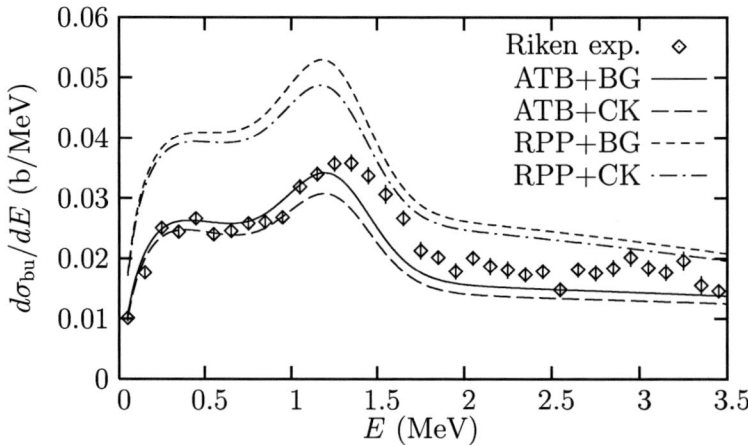

FIGURE 3. Theoretical and experimental breakup cross sections of ^{11}Be on ^{12}C as a function of energy. The four curves correspond to the calculations performed with various combinations of optical potentials simulating the nuclear P-T interactions. The theoretical results have been convoluted with energy resolution. Experimental data are from Ref. [3].

experiment exhibit the same behavior. In particular, they both display a peak in the vicinity of the $\frac{5}{2}^+$ resonance. Moreover, these peaks have approximately the same shape (height and width). This confirms that the low-lying resonance in the ^{11}Be spectrum has a significant influence on the nuclear induced breakup.

The discrepancy between theory and experiment observed at larger energies is most likely due to the fact that our ^{11}Be model does not reproduce any known physical state above the $\frac{5}{2}^+$ one. The experimental ^{11}Be spectrum indeed includes two other low-lying resonances, which should have an influence on the breakup as well. The first is located at 2.2 MeV and is probably responsible for the underestimation of the theory with respect to the experiment. The second is located at 2.9 MeV and may explain the small peak observed in the RIKEN data.

With the aim of analyzing the influence of the parametrization of the nuclear P-T interactions on those results, we perform the same calculation using different sets of optical potentials. Besides the potential of Al-Khalili, Tostevin and Brook [18] (labeled ATB), we choose another potential to simulate the ^{10}Be-^{12}C nuclear interaction. Following Chatterjee [20], we use a parametrization listed in the Perey and Perey compilation [21] which is a simplified expression of a potential developed by Robson [22] to reproduce the scattering of ^{10}B on ^{12}C at 18 MeV (labeled RPP). As an alternative to the Becchetti and Greenlees potential (BG) [19] for simulating the n-^{12}C interaction, we consider the potential developed by Comfort and Karp (CK) to reproduce scattering data of protons impinging on ^{12}C [23].

The breakup cross sections obtained with the four possible combinations of those potentials are displayed in Fig. 3 after convolution with the energy resolution. All curves exhibit the same pattern. In particular, they all display similar peaks near the $\frac{5}{2}^+$ resonance energy. This result shows that the optical potential choice has but little influence on the shape of that peak. It therefore confirms that the peak reflects the presence of the low-lying resonance in our ^{11}Be model.

The main difference between the four calculations is due to the ^{10}Be-^{12}C potential. The amplitude of the breakup cross section is indeed multiplied by almost 2 when the ATB potential is substituted by the RPP parametrization. This increase is due to the much smaller imaginary part of RPP. On the other hand, it seems that both n-^{12}C interactions are equivalent to describing breakup reactions. The difference between the cross sections obtained with the BG and CK potentials is indeed rather small.

A detailed analysis of the breakup probability as a function of the impact parameter b confirms these results (see Ref. [17]). In particular, it shows that the internal structure of the projectile, like the presence of the $\frac{5}{2}^+$ resonance, is probed only when the nuclear P-T interactions are taken into account.

These results show that, as expected, the nuclear P-T interactions play a dominant role in the dissociation of halo nuclei on light targets. In particular, these interactions emphasize the presence of low-lying resonances in the projectile spectrum. These resonances must therefore be taken into account in order to reproduce the experimental data. This suggests that nuclear induced breakup can be used as a probe of the continuum spectrum of the projectile.

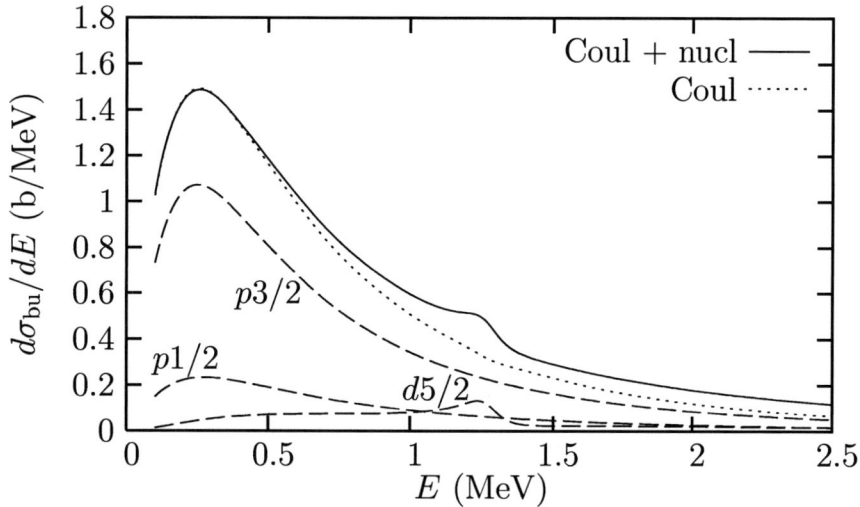

FIGURE 4. Breakup cross section of ^{11}Be on ^{208}Pb at 69 MeV/nucleon as a function of energy. The full line corresponds to the calculation performed with both Coulomb and nuclear *P-T* interactions Some major contributions of the partial waves to the cross section are displayed as dashed lines. The dotted line corresponds to the result obtained with a pure Coulomb *P-T* interaction.

4. BREAKUP OF ^{11}BE ON ^{208}PB

4.1. Breakup cross section

We now turn to the Coulomb breakup of ^{11}Be on ^{208}Pb. This reaction has been recently remeasured at RIKEN at 69 MeV/nucleon [3]. Moreover, it is interesting to see how the $\frac{5}{2}^+$ resonance in the ^{11}Be spectrum will affect the dissociation reaction when it is Coulomb dominated. The breakup cross section is displayed in Fig. 4 as a function of the energy E. The calculations were performed with the same conditions as in Ref. [16], but for the ^{10}Be-n potential, which is the same as in Sec. 3.1 (i.e. V1 of Table 1).

The full line corresponds to the calculation performed including both Coulomb and nuclear *P-T* interactions. In this case, the optical potential simulating the nuclear interaction between ^{10}Be and ^{208}Pb is adapted from an α-^{208}Pb potential [24] as explained in Ref. [16]. The n-^{208}Pb potential is chosen to be the Becchetti and Greenlees parametrization [19]. The cross section obtained with a purely Coulomb potential between ^{11}Be and ^{208}Pb is displayed as a dotted line. In that case, the nuclear interactions are simulated by an impact parameter parameter cutoff.

As expected, the breakup of ^{11}Be on ^{208}Pb is strongly dominated by the Coulomb interaction. The discrepancy between the cross sections computed with and without the nuclear *P-T* interactions is indeed small. However, there remain some interesting differences. As in the breakup on ^{12}C, but to a much smaller extent, the use of optical potentials leads to an increase of the breakup cross section at high energy. This has already been observed by Typel and Shyam [14]. As explained in Ref. [16], the effect of the nuclear interactions, though small, cannot be fully reproduced by a mere impact parameter cutoff. We also observe that the nuclear *P-T* interactions induce a small bump in the breakup cross section. As in the previous case, this bump is due to the presence of the $\frac{5}{2}^+$ state in our description of ^{11}Be: it is located at the resonance energy and is due only to the contribution of the $d5/2$ partial wave (lowest dashed line in Fig. 4). It is much smaller than in the previous case (cf. Fig. 2). Although the Coulomb field is very strong in this case, it appears only when optical potentials are used. This effect confirms that only the nuclear interactions can significantly populate the $\frac{5}{2}^+$ resonant state. Therefore, this low-lying resonance does not affect much the Coulomb dissociation of ^{11}Be.

Note that, since the breakup of ^{11}Be on ^{208}Pb is Coulomb dominated, the calculation of its cross section is much less sensitive to the optical potential choice than in the nuclear induced breakup. A variation of 20% in the amplitude of the optical potentials leads to only 2% variation in the breakup cross section.

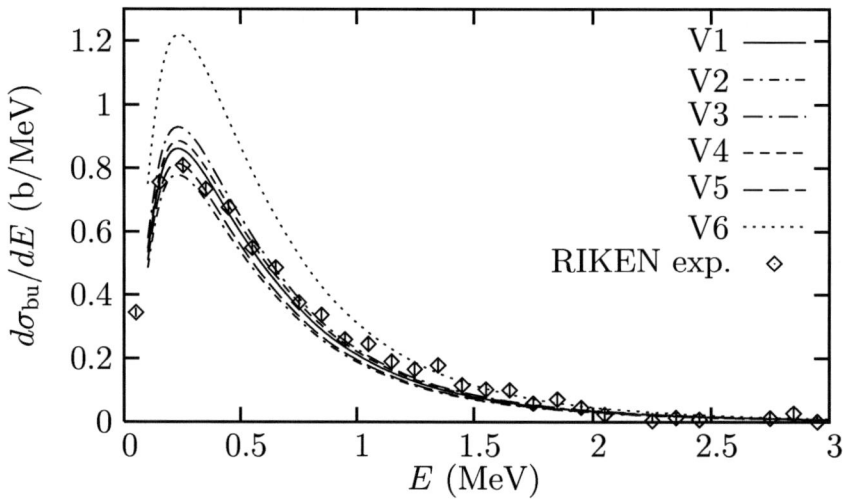

FIGURE 5. Breakup cross sections of ^{11}Be on ^{208}Pb at 69 MeV/nucleon obtained for $b > 30$ fm. The six curves correspond to calculations performed with the different potentials of Table 1. Experimental data are from Ref. [3].

4.2. Comparison with experiment and analysis of the sensitivity to the ^{10}Be-n potentials

When we compare the results of our calculations with the experimental breakup cross sections measured at RIKEN [3], we find a rather good agreement between theory and experiment. This is illustrated in Fig. 5, where the Coulomb breakup cross section corresponding to $b > 30$ fm is plotted as a function of the energy. At these impact parameters, the nuclear interactions between projectile and target are completely negligible. This enables us to get rid of the problem of their simulation. The cross section obtained from the calculation presented in the previous section (i.e. using potential V1 of Table 1) is displayed by the full line. This result is indeed very close to the experiment. Note that no parameter has been adjusted to fit the data. In particular, the theoretical cross section has not been scaled by any factor. This suggests that ^{11}Be is well described by a neutron loosely bound to a ^{10}Be core in its 0^+ ground state. The spectroscopic factor of that configuration should therefore be close to unity.

With the aim of testing the sensitivity of our results to the ^{10}Be-n potential, we perform the same calculation with different potentials. These potentials are obtained by varying either the radius or the diffuseness of the Woods-Saxon form factor (see Sec. 2.2). They are the potentials V2 to V5 given in Table 1. The corresponding breakup cross sections are displayed in Fig. 5. All curves exhibit the same shape. The only difference lies in their amplitude, which varies by about 15%. Surprisingly, these variations are not due to the asymptotic normalization constant of the initial ground state. For example, V4 leads to a larger breakup cross section than V5 (by approximately 9%), although its ANC (0.82) is smaller than that of V5 (0.87). This puzzling feature is currently under investigation. Up to now, it seems that this difference is due to the scattering properties of the potentials (e.g. scattering length), which differ from one potential to the other. Nevertheless, all these results confirm that the spectroscopic factor of the ^{10}Be(0^+)-n configuration should be close to 1.

For the analysis of their measurements, Fukuda et al. use another ^{10}Be-n potential (V6 in Table 1) [3]. It is a Woods-Saxon potential whose depth is adjusted only to the ground state energy of ^{11}Be. It does not include any spin-orbit coupling term. From that analysis, they deduce a spectroscopic factor of 0.7, much lower than what we get from our calculations. In order to understand the discrepancy between our value and theirs, we perform a calculation using potential V6 within our model. The corresponding cross section is displayed as a dotted line in Fig. 5. It lies approximately 30% above the V1 curve, which explains the lower spectroscopic factor.

The reason for this difference is still to be analyzed. However, the discrepancy obtained within the same reaction model using different V_{cf} supposed to describe the same nucleus is very large. Therefore, we wonder whether this technique is reliable for extracting spectroscopic factors. This result indicates that a strong effort should be made to improve the description of a halo nucleus used in reaction models. At least, the core-fragment potential should be constrained by other experimental data or predictions from precise structure models.

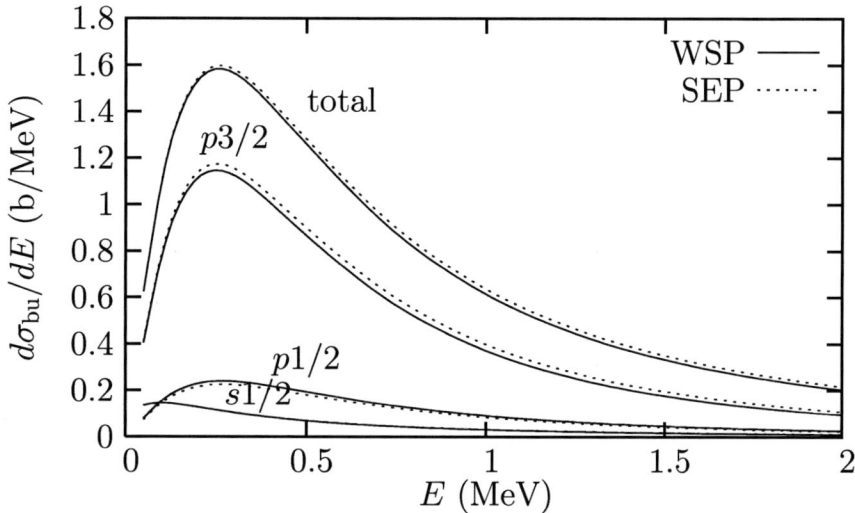

FIGURE 6. Influence of Pauli-forbidden states upon the breakup cross section of ^{11}Be on ^{208}Pb. The full lines correspond to the calculation performed with a Woods-Saxon potential (WSP), while the dotted lines display the results obtained with the supersymmetric equivalent potential (SEP). The latter exhibits the same scattering properties and the same bound spectrum as the former but for the unphysical bound states $0s1/2$ and $0p3/2$, which have been removed.

4.3. Influence of Pauli-forbidden states

In the preceding section, we saw that the V_{cf} potential used to describe ^{11}Be has a significant influence on the breakup cross section. A thorough analysis of this influence is therefore necessary in order to find a convenient way to constrain the potential choice. A first step in that direction has been done in Ref. [25]. In that previous work, we have studied the influence of the Pauli-forbidden states on the Coulomb breakup of ^{11}Be. As explained earlier, ^{11}Be is usually described by deep potentials that exhibit spurious bound states besides the adjusted physical levels. Those unphysical states simulate orbitals occupied by the neutrons of the core. They are thus forbidden to the halo neutron by the Pauli principle. Their presence is usually ignored [11–15]. However, in reaction models, nothing prevents the transfer of the halo neutron toward one of those spurious states. It is therefore interesting to test the influence of those states upon our calculations.

It is possible to modify a potential in order to remove one of its bound states [26]. The modification consists in a pair of supersymmetric transformations that keeps all the other spectrum properties of the potential unchanged. It means that the supersymmetric partner of the potential exhibits the same scattering properties (i.e. phase shifts) and the same bound spectrum (i.e. energy levels) as the initial potential, except for the bound state that has been removed.

In Ref. [25], we perform two evolution calculations: one with a usual Woods-Saxon potential (WSP), and one with its supersymmetric equivalent potential (SEP) in which both the unphysical $0s1/2$ and $0p3/2$ states have been suppressed. The results are shown in Fig. 6. The full lines correspond to the breakup cross section computed with the WSP, and the dotted lines show the result obtained with the SEP. Contributions of the major partial waves are pictured as well. The dependence of the cross section on the potential is very weak. The difference is only 1% in the peak region. Note that the effect differs according to the partial wave. We observe a slight increase (2.5%) of the $p3/2$ contribution when the SEP is used. The opposite is obtained for the $p1/2$ component: the SEP cross section is smaller by 5%. The $s1/2$ contribution remains practically unchanged.

This result shows that the Pauli-forbidden states in the projectile spectrum do not play any significant role in the breakup reaction. They may be ignored. The use of deep potentials seems therefore fully justified in such calculations. Moreover, the tiny difference observed between WSP and SEP suggests that the discrepancy between the various ^{10}Be-n potentials are not directly due to the differences in energies of their unphysical-states.

5. CONCLUSION

In this talk, we have presented the results of time-dependent calculations of the nuclear and Coulomb breakup of ^{11}Be [16, 17]. The calculations are performed around 70 MeV/nucleon in order to compare them to recent experimental data [3].

The description of ^{11}Be is improved in comparison to previous works [11–16]. We developed a new ^{10}Be-n potential that reproduces not only the bound states of ^{11}Be but also its low-lying $\frac{5}{2}^+$ resonance. This resonance has a significant influence on the breakup on ^{12}C [17]: it induces a narrow peak in the breakup cross section. The very good agreement obtained with experiment [3] confirms the validity of the model, and the ability of the time-dependent technique to simulate nuclear induced breakup. The spectrum of ^{11}Be includes other low-lying resonances, which should influence the breakup as well. However, the analysis of their actual effect on the cross section requires a better description of the projectile since these resonances cannot be reproduced with such a simple two-body model. In the dissociation on ^{208}Pb, however, the $\frac{5}{2}^+$ resonance is found to play a rather minor role. It only induces a small bump in the breakup cross section at the resonance energy. In this case also, we obtain a rather good agreement with experimental data [3].

In our reaction model, the nuclear interactions between the projectile and the target are described by optical potentials [16]. In the breakup on ^{12}C, they are dominant. In particular, they are found to be responsible for the strong population of the resonant state, which causes the peak in the cross section. The nuclear induced breakup is therefore rather sensitive to the optical potential choice. Using different optical potentials leads to significant variations in the breakup cross section. However, these variations only affect the amplitude of the cross section and not its general pattern. In particular the location and the shape of the resonance peak remain the same for all choices of optical-potential. The breakup of ^{11}Be on ^{208}Pb is Coulomb dominated. The nuclear P-T interactions are therefore much less significant, and the cross section is much less dependent on the optical potentials.

The sensitivity of the model onto the ^{11}Be description has also been presented. We performed time-dependent calculations of the Coulomb breakup of ^{11}Be using various ^{10}Be-n potentials. These potentials lead to cross sections which exhibit the same shape, but differ by up to 30% in amplitude. This variation is not directly related to the asymptotic normalization constant of the initial ground state. Neither is it to the presence of Pauli-forbidden states in the ^{11}Be spectrum. The role played by these unphysical states in the breakup is indeed negligible [25]. Another effect is thus at play here. Its analysis requires further investigations. Anyway, because of this significant variation in the amplitude of the cross section, using Coulomb breakup as a tool for extracting spectroscopic factors seems questionable.

From this analysis, it seems that we have now reached the limit of the simple two-body description of halo nuclei used in reaction theory. In order to improve our results in the nuclear induced breakup of ^{11}Be we need a more precise model that reproduces the other low-lying resonances. Moreover, the sensitivity of the breakup cross section to the current ^{11}Be model might be too large to extract accurate structure information. Therefore, a reaction model including a more precise description of halo nuclei should be developed in order to improve the theoretical predictions.

ACKNOWLEDGMENTS

This text presents research results of the Belgian program P5/07 on interuniversity attraction poles initiated by the Belgian-state Federal Services for Scientific, Technical and Cultural Affairs. P.C. acknowledges the support of the Natural Sciences and Engineering Research Council of Canada (NSERC). G.G. acknowledges the support of the FRIA (Belgium).

REFERENCES

1. I. Tanihata, *J. Phys. G* **22**, 157 (1996).
2. T. Nakamura, S. Shimoura, T. Kobayashi, T. Teranishi, K. Abe, N. Aoi, Y. Doki, M. Fujimaki, N. Inabe, N. Iwasa, K. Katori, T. Kubo, H. Okuno, T. Suzuki, I. Tanihata, Y. Watanabe, A. Yoshida, and M. Ishihara, *Phys. Lett. B* **331**, 296 (1994).
3. N. Fukuda, T. Nakamura, N. Aoi, N. Imai, M. Ishihara, T. Kobayashi, H. Iwasaki, T. Kubo, A. Mengoni, M. Notani, H. Otsu, H. Sakurai, S. Shimoura, T. Teranishi, Y. X. Watanabe, and K. Yoneda, *Phys. Rev. C* **70** 054606 (2004).
4. J. Al-Khalili, and F. M. Nunes, *J. Phys. G* **29**, R89 (2003).
5. R. Johnson, contribution to this workshop (2005).
6. J. Tostevin, contribution to this workshop (2005).

7. I. Thompson, contribution to this workshop (2005).
8. M. Kamimura, contribution to this workshop (2005).
9. D. Baye, contribution to this workshop (2005).
10. K. Alder, and A. Winther, *Electromagnetic Excitation*, North-Holland, Amsterdam, 1975.
11. T. Kido, K. Yabana, and Y. Suzuki, *Phys. Rev. C* **50**, R1276 (1994).
12. H. Esbensen, G. F. Bertsch, and C. A. Bertulani, *Nucl. Phys.* **A581**, 107 (1995).
13. V. S. Melezhik, and D. Baye, *Phys. Rev. C* **59**, 3232 (1999).
14. S. Typel, and R. Shyam, *Phys. Rev. C* **64**, 024605 (2001).
15. M. Fallot, J. A. Scarpaci, D. Lacroix, P. Chomaz, and J. Margueron, *Nucl. Phys.* **A700**, 70 (2002).
16. P. Capel, D. Baye, and V. S. Melezhik, *Phys. Rev. C* **68**, 014612 (2003).
17. P. Capel, G. Goldstein, and D. Baye, *Phys. Rev. C* **70**, 064605 (2004).
18. J. S. Al-Khalili, J. A. Tostevin, and J. M. Brooke, *Phys. Rev. C* **55**, R1018 (1997).
19. F. D. Becchetti, Jr., and G. W. Greenlees, *Phys. Rev.* **182**, 1190 (1969).
20. R. Chatterjee, *Phys. Rev. C* **68**, 044604 (2003).
21. C. M. Perey, and F. G. Perey, *At. Data Nucl. Data Tables* **17**, 1 (1976).
22. D. Robson, in *Proceedings of the Symposium on Heavy-ion Scattering*, edited by R. H. Siemmsen, G. C. Morrison, and J. P. Schiffer, ANL-7837, Argonne National Laboratory, 1971, p. 239.
23. J. R. Comfort, and B. C. Karp, *Phys. Rev. C* **21**, 2162 (1980), *ibid.* **22**, 1809 (E) (1980).
24. B. Bonin, N. Alamanos, B. Berthier, G. Brugea, H. Faraggi, J. C. Lugola, W. Mittigc, L. Papineaub, A. I. Yavind, J. Arvieux, L. Farvacque, M. Buenerd, and W. Bauhoff, *Nucl. Phys.* **A445**, 381 (1985).
25. P. Capel, D. Baye, and V. S. Melezhik, *Phys. Lett. B* **552**, 145 (2003).
26. D. Baye, *J. Phys. A* **20**, 5529 (1987).

A Transport Model for Nuclear Reactions Induced by Radioactive Beams

Bao-An Li[*], Lie-Wen Chen[†], Champak B. Das[**], Subal Das Gupta[**], Charles Gale[**], Che Ming Ko[‡], Gao-Chan Yong[§] and Wei Zuo[§]

[*]*Department of Chemistry and Physics, Arkansas State University, State University, AR 72467-0419, USA*
[†]*Institute of Theoretical Physics, Shanghai Jiao Tong University, Shanghai 200240, P.R. China*
[**]*Physics Department, McGill University, Montreal, Canada H3A 2T8*
[‡]*Cyclotron Institute and Physics Department, Texas A&M University, College Statio, TX 77843, USA*
[§]*Institute of Modern Physics, Chinese Academy of Science, Lanzhou 730000, P.R. China*

Abstract. Major ingredients of an isospin and momentum dependent transport model for nuclear reactions induced by radioactive beams are outlined. Within the IBUU04 version of this model we study several experimental probes of the equation of state of neutron-rich matter, especially the density dependence of the nuclear symmetry energy. Comparing with the recent experimental data from NSCL/MSU on isospin diffusion, we found a nuclear symmetry energy of $E_{sym}(\rho) \approx 31.6(\rho/\rho_0)^{1.05}$ at subnormal densities. Predictions on several observables sensitive to the density dependence of the symmetry energy at supranormal densities accessible at GSI and the planned Rare Isotope Accelerator (RIA) are also made.

Keywords: Radioactive beams, Equation of State, Neutron-Rich Matter, Nuclear Symmetry Energy, Transport Models, Heavy-Ion Reactions, Neutron Stars
PACS: 25.70.-z,21.30.Fe,21.65.+f, 24.10.Lx

INTRODUCTION

The Equation of State (EOS) of isospin asymmetric nuclear matter can be written within the well-known parabolic approximation, which has been verified by all many-body theories, as

$$E(\rho, \delta) = E(\rho, \delta = 0) + E_{sym}(\rho)\delta^2 + \mathcal{O}(\delta^4), \tag{1}$$

where $\delta \equiv (\rho_n - \rho_p)/(\rho_p + \rho_n)$ is the isospin asymmetry and $E_{sym}(\rho)$ is the density-dependent nuclear symmetry energy. The latter is very important for many interesting astrophysical problems[1], the structure of radioactive nuclei[2, 3] and heavy-ion reactions[4, 5, 6, 7]. Unfortunately, the density dependence of symmetry energy $E_{sym}(\rho)$, especially at supranormal densities, is still poorly known. Predictions based on various many-body theories diverge widely at both low and high densities. In fact, even the sign of the symmetry energy above $3\rho_0$ remains uncertain[8]. Fortunately, heavy-ion reactions, especially those induced by radioactive beams, provide a unique opportunity to pin down the density dependence of nuclear symmetry energy in terrestrial laboratories. Significant progress in determining the symmetry energy at subnormal densities has been made recently both experimentally and theoretically[9, 10]. High energy radioactive beams to be available at GSI and RIA will allow us to determine the symmetry energy at supranormal densities.

To extract information about the EOS of neutron-rich matter from nuclear reactions induced by radioactive beams, one needs reliable theoretical tools. Transport models are especially useful for this purpose. Especially for central, energetic reactions at RIA and GSI, transport models are the most useful tool for understanding the role of isospin degree of freedom in the reaction dynamics and extract information about the EOS of neutron-rich matter. Significant progresses have been made recently in improving semi-classical transport models for nuclear reactions. While developing practically implementable quantum transport theories is a long term goal, applications of the semi-classical transport models have enabled us to learn a great deal of interesting physics from heavy-ion reactions. In the following we outline the major ingredients of an isospin and momentum dependent transport model applicable for heavy-ion reactions induced by both stable and radioactive beams[11]. This model has been found very useful in understanding a number

---
[1] Present address: Variable Energy Cyclotron Center, 1/AF, Bidhannagar, Kolkata 700064, India

of new phenomena associated with the isospin degree of freedom in heavy-ion reactions. Based on applications of this model, we highlight here the most recent progress in determining the symmetry energy at subnormal densities and present our predictions on several most sensitive probes of the symmetry energy at supranormal densities.

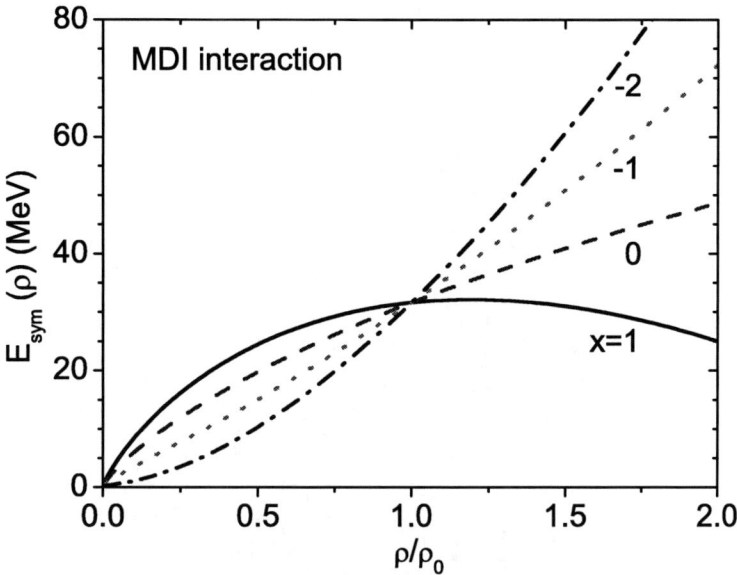

FIGURE 1. Density dependence of the symmetry energy for four x parameters.

IBUU04 VERSION OF THE MODEL

Crucial to the extraction of critical information about the $E_{\text{sym}}(\rho)$ is to compare experimental data with transport model calculations. We outline here the major ingredients of the version IBUU04 of an isospin- and momentum-dependent transport model for nuclear reactions induced by radioactive beams[11]. The single nucleon potential is one of the most important inputs to all transport models. Both the isovector (symmetry potential) and isoscalar parts of this potential should be momentum dependent due to the non-locality of strong interactions and the Pauli exchange effects in many-fermion systems. In the IBUU04, we use a single nucleon potential derived from the Hartree-Fock approximation using a modified Gogny effective interaction (MDI)[12], i.e.,

$$\begin{aligned} U(\rho,\delta,\vec{p},\tau,x) &= A_u(x)\frac{\rho_{\tau'}}{\rho_0} + A_l(x)\frac{\rho_\tau}{\rho_0} \\ &+ B(\frac{\rho}{\rho_0})^\sigma(1-x\delta^2) - 8\tau x \frac{B}{\sigma+1}\frac{\rho^{\sigma-1}}{\rho_0^\sigma}\delta\rho_{\tau'} \\ &+ \frac{2C_{\tau,\tau}}{\rho_0}\int d^3p' \frac{f_\tau(\vec{r},\vec{p}')}{1+(\vec{p}-\vec{p}')^2/\Lambda^2} \\ &+ \frac{2C_{\tau,\tau'}}{\rho_0}\int d^3p' \frac{f_{\tau'}(\vec{r},\vec{p}')}{1+(\vec{p}-\vec{p}')^2/\Lambda^2}. \end{aligned} \quad (2)$$

In the above $\tau = 1/2$ $(-1/2)$ for neutrons (protons) and $\tau \neq \tau'$; $\sigma = 4/3$; $f_\tau(\vec{r},\vec{p})$ is the phase space distribution function at coordinate $\vec{r}$ and momentum $\vec{p}$. The parameters $A_u(x), A_l(x), B, C_{\tau,\tau}, C_{\tau,\tau'}$ and Λ were obtained by fitting the momentum-dependence of the $U(\rho,\delta,\vec{p},\tau,x)$ to that predicted by the Gogny Hartree-Fock and/or the Brueckner-Hartree-Fock calculations, the saturation properties of symmetric nuclear matter and the symmetry energy of 30 MeV at normal nuclear matter density $\rho_0 = 0.16$ fm^{-3}[12]. The incompressibility K_0 of symmetric nuclear matter at ρ_0 is set to be 211 MeV. The parameters $A_u(x)$ and $A_l(x)$ depend on the x parameter according to

$$A_u(x) = -95.98 - x\frac{2B}{\sigma+1}, \quad A_l(x) = -120.57 + x\frac{2B}{\sigma+1}. \quad (3)$$

The parameter x can be adjusted to mimic predictions on the $E_{sym}(\rho)$ by microscopic and/or phenomenological many-body theories. The last two terms contain the momentum-dependence of the single-particle potential. The momentum dependence of the symmetry potential stems from the different interaction strength parameters $C_{\tau,\tau'}$ and $C_{\tau,\tau}$ for a nucleon of isospin τ interacting, respectively, with unlike and like nucleons in the background fields. More specifically, we use $C_{unlike} = -103.4$ MeV and $C_{like} = -11.7$ MeV. As an example, shown in Fig. 1 is the density dependence of the symmetry energy for $x = -2, -1, 0$ and 1.

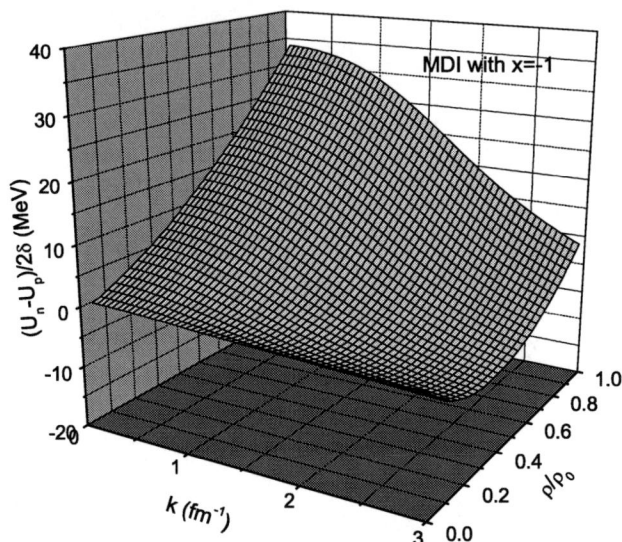

FIGURE 2. Symmetry potential as a function of momentum and density for MDI interaction with $x = -1$.

Systematic analyses of a large number of nucleon-nucleus and (p,n) charge exchange scattering experiments at beam energies below about 100 MeV indicate undoubtedly that the symmetry potential at ρ_0, i.e., the Lane potential, decreases approximately linearly with increasing beam energy E_{kin} according to $U_{Lane} = a - bE_{kin}$ where $a \simeq 22 - 34$ MeV and $b \simeq 0.1 - 0.2$[13, 14]. This provides a stringent constraint on the symmetry potential. The potential in eq.2 meets this requirement very well as seen in Fig. 2 where the symmetry potential $(U_n - U_p)/2\delta$ as a function of momentum and density for the parameter $x = -1$ is displayed.

One characteristic feature of the momentum dependence of the symmetry potential is the different effective masses for neutrons and protons in isospin asymmetric nuclear matter, i.e.,

$$\frac{m_\tau^*}{m_\tau} = \left\{ 1 + \frac{m_\tau}{\hbar^2 k} \frac{dU_\tau}{dk} \right\}^{-1}_{k=k_\tau^F}, \quad (4)$$

where k_τ^F is the nucleon Fermi wave number. With the potential in eq. 2, since the momentum-dependent part of the nuclear potential is independent of the parameter x, the nucleon effective masses are independent of the x parameter too. Shown in Fig. 3 are the nucleon effective masses as a function of density (upper window) and isospin asymmetry (lower window). It is seen that the neutron effective mass is higher than the proton effective mass and the splitting between them increases with both the density and isospin asymmetry of the medium[11]. We notice here that the momentum dependence of the symmetry potential and the associated splitting of nucleon effective masses in isospin asymmetric matter is still highly contoversial[15, 16]. The experimental determination of both the density and momentum dependence of the symmetry potential is required. Please see also ref.[17] on this point.

Since both the incoming current in the initial state and the level density of the final state in nucleon-nucleon (NN) scatterings depend on the effective masses of colliding nucleons in medium, the in-medium nucleon-nucleon cross sections are expected to be reduced by a factor

$$\sigma_{NN}^{medium}/\sigma_{NN} = (\mu_{NN}^*/\mu_{NN})^2 \quad (5)$$

where μ_{NN} and μ_{NN}^* are the reduced mass of the colliding nucleon pairs in free-space and in medium, respectively[18, 19, 20]effective masses is consistent with predictions based on more microscopic many-body theories[21]. Thus, be-

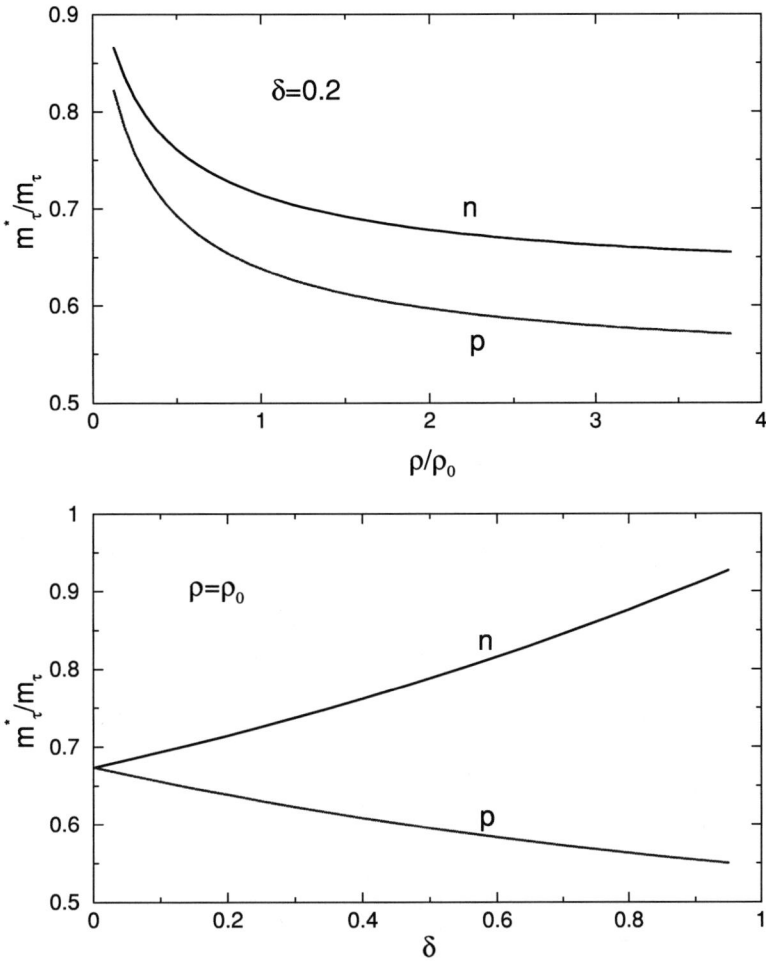

FIGURE 3. Nucleon effective masses in asymmetric matter as a function of density (upper window) and isospin asymmetry (lower window).

cause of the reduced in-medium nucleon effective masses and their dependence on the density and isospin asymmetry of the medium, the in-medium NN cross sections are not only reduced compared to their free-space values, the nn and pp cross sections are split and the difference between them grows in more asymmetric matter as shown in Fig. 4. The in-medium NN cross sections are also independent of the parameter x. The isospin-dependence of the in-medium NN cross sections is expected to play an important role in nuclear reactions induced by neutron-rich nuclei[22]

Other details, such as the initialization of colliding nuclei in phase space, Pauli blocking, etc can be found in our earlier publication[4, 5, 11].

APPLICATION OF THE MODEL

The model outlined above have many applications in heavy-ion reactions induced by both stable and radioactive beams. In this section, we illustrate several examples of studying the role of isospin degree of freedom in the reaction dynamics and extracting the density dependence of nuclear symmetry energy. Besides probes of symmetry energy at subnormal densities, several probes sensitive to the high density behavior of the symmetry energy have been also proposed largely based on transport model simulations.

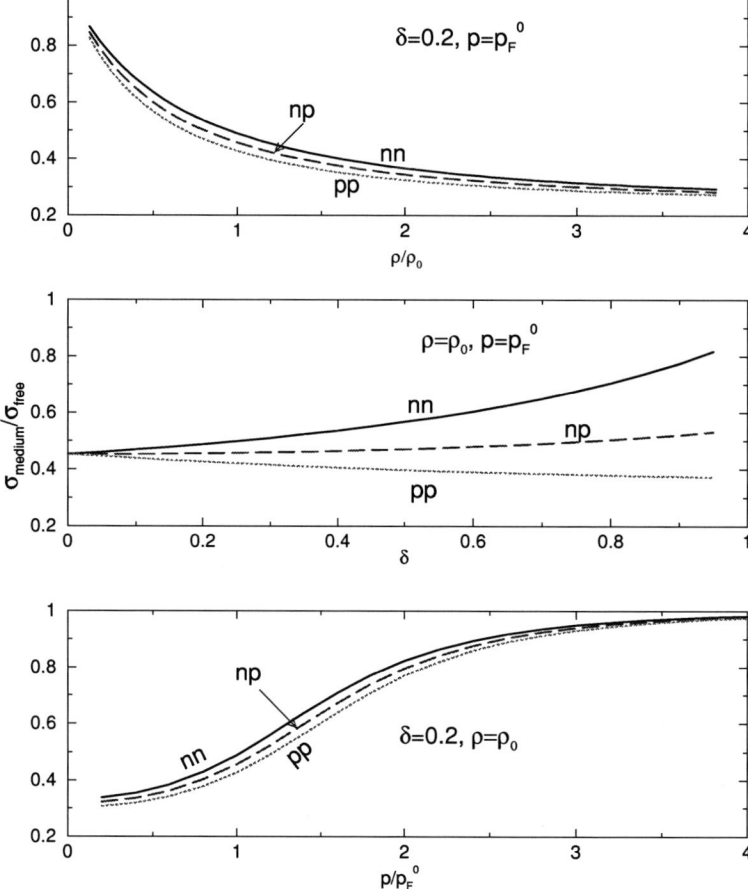

FIGURE 4. The reduction factor of the in-medium nucleon-nucleon cross sections with respect to the free space ones as a function of density (top), isospin asymmetry (middle) and momentum (bottom).

Probing the symmetry energy at subnormal densities with isospin diffusion

Tsang et al.[9] recently studied the degree of isospin diffusion in the reaction ^{124}Sn + ^{112}Sn by measuring[23]

$$R_i = \frac{2X_{^{124}Sn+^{112}Sn} - X_{^{124}Sn+^{124}Sn} - X_{^{112}Sn+^{112}Sn}}{X_{^{124}Sn+^{124}Sn} - X_{^{112}Sn+^{112}Sn}} \qquad (6)$$

where X is the average isospin asymmetry $\langle \delta \rangle$ of the ^{124}Sn-like residue. The data are indicated in Fig. 5 together with the IBUU04 predictions about the time evolutions of R_i and the average central densities calculated with $x = -1$ using both the MDI and the soft Bertsch-Das Gupta-Kruse (SBKD) interactions. It is seen that the isospin diffusion process occurs mainly from about 30 fm/c to 80 fm/c corresponding to an average central density from about $1.2\rho_0$ to $0.3\rho_0$. The experimental data from MSU are seen to be reproduced nicely by the MDI interaction with $x = -1$, while the SBKD interaction with $x = -1$ leads to a significantly lower R_i value[10].

Effects of the symmetry energy on isospin diffusion were also studied by varying the parameter x[10]. Only with the parameter $x = -1$ the data can be well reproduced. The corresponding symmetry energy can be parameterized as $E_{sym}(\rho) \approx 31.6(\rho/\rho_0)^{1.05}$. In the present study on isospin diffusion, only the free-space NN cross sections are used and thus effects completely due to the different density dependence of symmetry energy are investigated. As the next step we are currently investigating effects of the in-medium NN cross sections on the isospin diffusion.

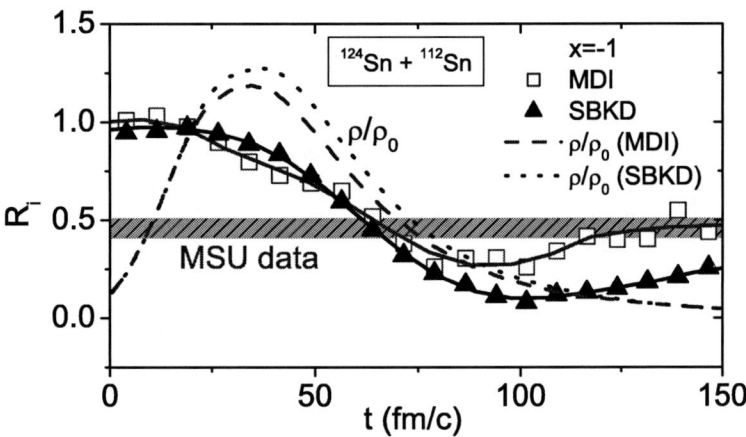

FIGURE 5. The degree of isospin diffusion as a function of time with the MDI and SBKD interactions. The corresponding evolutions of central density are also shown.

Isospin asymmetry of dense matter formed in high energy heavy-ion reactions

What are the maximum baryon density and isospin asymmetry that can be achieved in central heavy-ion collisions at the highest beam energy expected at RIA? This is an interesting question relevant to the study of the EOS of asymmetric nuclear matter. To answer this question we show in Fig. 6 the central baryon density (upper window) and the average $(n/p)_{\rho \geq \rho_0}$ ratio (lower window) of all regions with baryon densities higher than ρ_0 in the reaction of $^{132}Sn + ^{124}Sn$ at a beam energy of 400 MeV/nucleon and an impact parameter of 1 fm. It is seen that the maximum baryon density is about 2 times normal nuclear matter density. Moreover, the compression is rather insensitive to the symmetry energy because the latter is relatively small compared to the EOS of symmetric matter around this density. The high density phase lasts for about 15 fm/c from 5 to 20 fm/c for this reaction. It is interesting to see that the isospin asymmetry of the high density region is quite sensitive to the symmetry energy. The soft (e.g., $x = 1$) symmetry energy leads to a significantly higher value of $(n/p)_{\rho \geq \rho_0}$ than the stiff one (e.g., $x = -2$). This is consistent with the well-known isospin fractionation phenomenon. Because of the $E_{sym}(\rho)\delta^2$ term in the EOS of asymmetric nuclear matter, it is energetically more favorable to have a higher isospin asymmetry δ in the high density region with a softer symmetry energy functional $E_{sym}(\rho)$. In the supranormal density region, as shown in Fig. 1, the symmetry energy changes from being soft to stiff when the parameter x varies from 1 to -2. Thus the value of $(n/p)_{\rho \geq \rho_0}$ becomes lower as the parameter x changes from 1 to -2. It is worth mentioning that the initial value of the quantity $(n/p)_{\rho \geq \rho_0}$ is about 1.4 which is less than the average n/p ratio of 1.56 of the reaction system. This is because of the neutron-skins of the colliding nuclei, especially that of the projectile ^{132}Sn. In the neutron-rich nuclei, the n/p ratio on the low-density surface is much higher than that in their interior. It is clearly seen that the dense region can become either neutron-richer or neutron-poorer with respect to the initial state depending on the symmetry energy functional $E_{sym}(\rho)$ used.

Pions yields and π^-/π^+ ratio as a probe of the symmetry energy at supranormal densities

At the highest beam energy at RIA, pion production is significant. Pions may thus carry interesting information about the EOS of dense neutron-rich matter[24, 25]. Shown in Fig.7 are the π^- and π^+ yields as a function of the x parameter. It is interesting to see that the π^- multiplicity depends more sensitively on the symmetry energy. The π^- multiplicity increases by about 20% while the π^+ multiplicity remains about the same when the x parameter is changed from -2 to 1. The multiplicity of π^- is about 2 to 3 times that of π^+. This is because the π^- mesons are mostly produced from neutron-neutron collisions. Moreover, with the softer symmetry energy the high density region is more neutron-rich due to isospin fractionation[25]. The π^- mesons are thus more sensitive to the isospin asymmetry of the reaction system and the symmetry energy. However, one should notice that it is well known that pion yields are also sensitive to the symmetric part of the nuclear EOS. It is thus hard to get reliable information about the symmetry

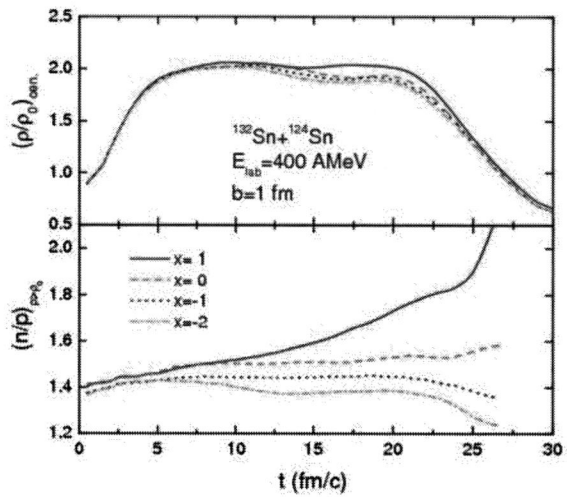

FIGURE 6. Central baryon density (upper window) and isospin asymmetry (lower window) of high density region for the reaction of $^{132}Sn + ^{124}Sn$ at a beam energy of 400 MeV/nucleon and an impact parameter of 1 fm.

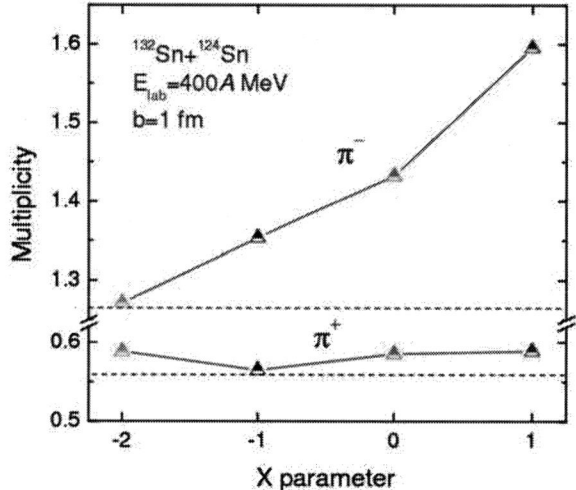

FIGURE 7. The π^- and π^+ yields as functions of the x parameter.

energy from π^- yields alone. Fortunately, the π^-/π^+ ratio is a better probe since statistically this ratio is only sensitive to the difference in the chemical potentials for neutrons and protons[26]. This expectation is well demonstrated in Fig. 8. It is seen that the pion ratio is quite sensitive to the symmetry energy, especially at low transverse momenta. Thus, it is promising that the high density behavior of nuclear symmetry energy $E_{sym}(\rho)$ can be probed using the π^-/π^+ ratio.

Isospin fractionation and n-p differential flow at RIA and GSI

The degree of isospin equilibration or translucency can be measured by the rapidity distribution of nucleon isospin asymmetry $\delta_{free} \equiv (N_n - N_p)/(N_n + N_p)$ where N_n (N_p) is the multiplicity of free neutrons (protons)[27]. Although it might be difficult to measure directly δ_{free} because it requires the detection of neutrons, similar information can be extracted from ratios of light clusters, such as, $^3H/^3He$, as demonstrated recently within a coalescence model[28, 29].

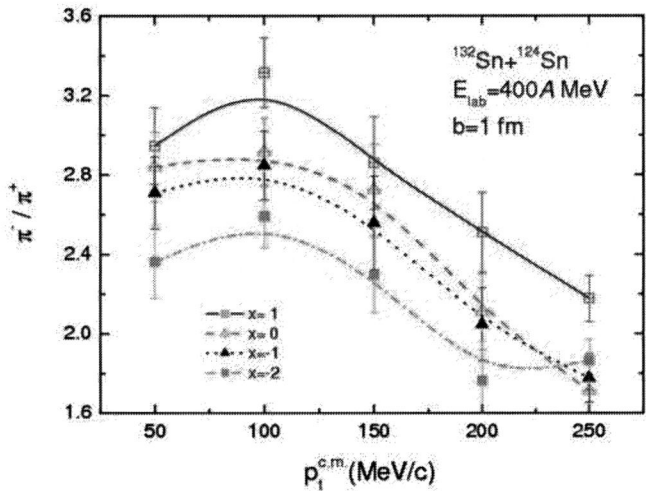

FIGURE 8. The π^-/π^+ ratio as a function of transverse momentum.

Shown in Fig. 9 are the rapidity distributions of δ_{free} with (upper window) and without (lower window) the Coulomb potential. It is interesting to see that the δ_{free} at midrapidity is particularly sensitive to the symmetry energy. As the parameter x increases from -2 to 1 the δ_{free} at midrapidity decreases by about a factor of 2. Moreover, the forward-backward asymmetric rapidity distributions of δ_{free} with all four x parameters indicates the apparent nuclear translucency during the reaction[30].

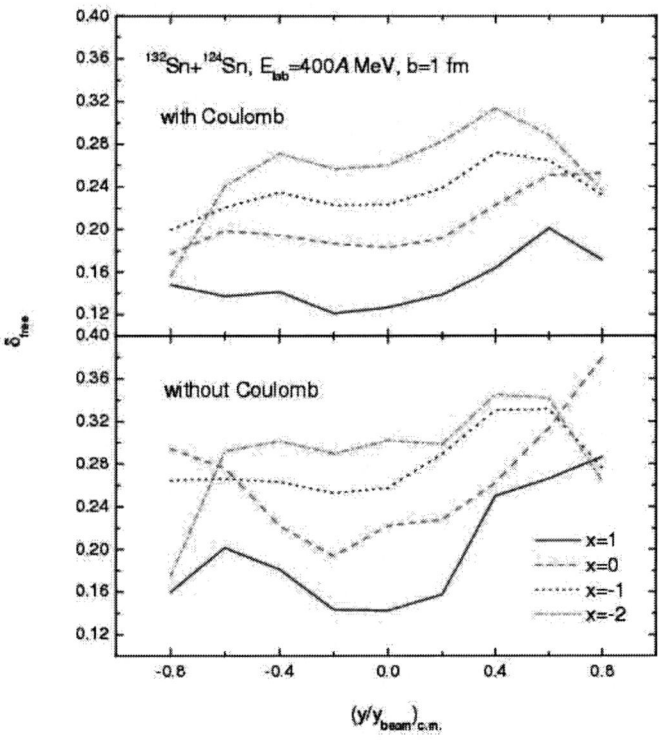

FIGURE 9. Isospin asymmetry of free nucleons with and without the Coulomb force.

Another observable that is sensitive to the high density behavior of symmetry energy is the neutron-proton differential flow[31]

$$F_{n-p}^x(y) \equiv \sum_{i=1}^{N(y)} (p_i^x w_i)/N(y), \tag{7}$$

where $w_i = 1(-1)$ for neutrons (protons) and $N(y)$ is the total number of free nucleons at rapidity y. The differential flow combines constructively effects of the symmetry potential on the isospin fractionation and the collective flow. It has the advantage of maximizing the effects of the symmetry potential while minimizing those of the isoscalar potential. Shown in Fig. 10 is the n-p differential flow for the reaction of $^{132}Sn + ^{124}Sn$ at a beam energy of 400 MeV/nucleon and an impact parameter of 5 fm. Effects of the symmetry energy are clearly revealed by changing the parameter x.

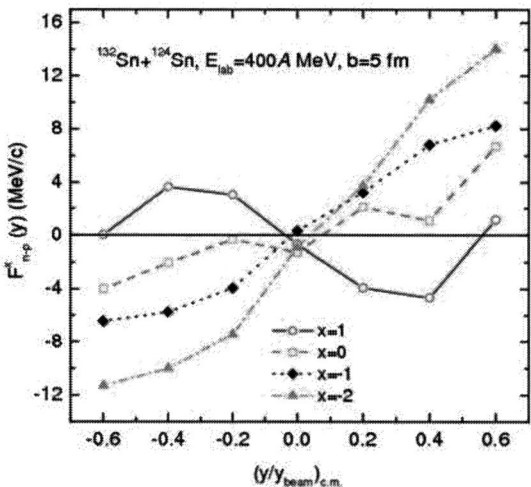

FIGURE 10. Neutron-proton differential flow at RIA and GSI energies

CONCLUSIONS

Transport models are powerful tools for investigating especially central heavy-ion reactions induced by neutron-rich nuclei. Applications of these models will help us understand the isospin dependence of in-medium nuclear effective interactions. Comparing with the experimental data, we can extract the isospin dependence of thermal, mechanical and transport properties of asymmetric nuclear matter playing important roles in nuclei, neutron stras and supernovae. Currently, the most important issue is the density dependence of the nuclear symmetry energy. The latter is very important for both nuclear physics and astrophysics. Significant progress has been made recently by the heavy-ion community in determining the density dependence of the nuclear symmetry energy. Based on transport model calculations, a number of sensitive probes of the symmetry energy have been found. The momentum dependence in both the isoscalar and isovector parts of the nuclear potential was found to play an important role in extracting accurately the density dependence of the symmetry energy. Comparing with recent experimental data on isospin diffusion from NSCL/MSU, we have extracted a symmetry energy of $E_{sym}(\rho) \approx 31.6(\rho/\rho_0)^{1.05}$ at subnormal densities. It would be interesting to compare this conclusion with those extracted from studying other observables. More experimental data including neutrons with neutron-rich beams in a broad energy range are needed. Looking forward to experiments at RIA and GSI with high energy radioactive beams, we hope to pin down the symmetry energy at supranormal densities in the near future. Theoretically, the development of a practically implementable quantum transport theory for nuclear reactions induced by rdioactive beams remains a big challenge.

This work was supported in part by the US National Science Foundation of the under grant No. PHY 0098805, PHYS-0243571 and PHYS0354572, Welch Foundation grant No. A-1358, and the NASA-Arkansas Space Grants Consortium award ASU15154. C.B. Das, S. Das Gupta and C. Gale were supported in part by the Natural Sciences and Engineering Research Council of Canada, and the Fonds Nature et Technologies of Quebec. L.W. Chen was supported by the National Natural Science Foundation of China grant No. 10105008. The work of G.C. Yong and W. Zuo was supported in part by the Chinese Academy of Science Knowledge Innovation Project (KECK2-SW-N02), Major State Basic Research Development Program (G2000077400), the National Natural Science Foundation of China (10235030) and the Important Pare-Research Project (2002CAB00200) of the Chinese Ministry of Science and Technology.

REFERENCES

1. J.M. Lattimer and M. Prakash, Phys. Rep., **333**, 121 (2000); Astr. Phys. Jour. **550**, 426 (2001); Science Vol. **304**, 536 (2004); A. W. Steiner, M. Prakash, J.M. Lattimer and P.J. Ellis, nucl-th/0410066, Phys. Rep. (2005) in press.
2. B.A. Brown, Phys. Rev. Let. **85**, 5296 (2000).
3. J.R. Stone at al., Phys. Rev. **C68**, 034324 (2004).
4. B.A. Li, C.M. Ko and W. Bauer, topical review, Int. J. Mod. Phys. **E7**, 147 (1998).
5. *Isospin Physics in Heavy-Ion Collisions at Intermediate Energies*, Eds. B. A. Li and W. Uuo Schrödear (Nova Science Publishers, Inc, New York, 2001).
6. P. Danielewicz, R. Lacey and W.G. Lynch, Science 298, 1592 (2002).
7. V. Baran, M. Colonna, V. Greco and M. Di Toro, Phys. Rep. (2005) in press.
8. I. Bombaci, Chapter 2 in ref.[5].
9. M.B. Tsang et al., Phys. Rev. Lett. **92**, 062701 (2004).
10. L.W. Chen, C.M. Ko and B.A. Li, Phys. Rev. Lett. **94**, 32701 (2005).
11. B.A. Li, C.B. Das, S. Das Gupta and C. Gale, Phys. Rev. **C69**, 011603 (2004); Nucl. Phys. **A735**, 563 (2004).
12. C.B. Das, S. Das Gupta, C. Gale and B.A. Li, Phys. Rev. **C67**, 034611 (2003).
13. P.E. Hodgson, The Nucleon Optical Model, 1994 (World Scientific).
14. G.W. Hoffmann and W.R. Coker, Phys. Rev. Lett. **29**, 227 (1972).
15. J. Rizzo, M. Colonna, M. Di Toro and V. Greco, Nucl. Phys. **A732**, 202 (2004).
16. B.A. Li, Phys. Rev. **C69**, 064602 (2004).
17. M. Di Toro et al., contribution to this volume.
18. V.R. Pandharipande and S.C. Pieper, Phys. Rev. **C45**, 791 (1991).
19. M. Kohno, M. Higashi, Y. Watanabe, and M. Kawai, Phys. Rev. **C57**, 3495 (1998).
20. D. Persram and C. Gale, Phys. Rev. **C65**, 064611 (2002).
21. F. Sammrruca, private communications and contribution to this volume.
22. B.A. Li, P. Danielewicz and W.G. Lynch, nucl-th/0503038, Phys. Rev. C (2005) in press.
23. F. Rami et al., Phys. Rev. Lett. **84**, 1120 (2000).
24. B.A. Li, Phys. Rev. Lett. 88, 192701 (2002); Nucl. Phys. **A708**, 365 (2003); Phys. Rev. **C67**, 017601 (2003).
25. B.A. Li, G.C. Yong and W. Zuo, Phys. Rev. **C71**, 014608 (2005).
26. G.F. Bertsch, Nature **283**, 280 (1980); A. Bonasera and G.F. Bertsch, Phys. Let. **B195**, 521 (1987).
27. B.A. Li, C.M. Ko, and Z.Z. Ren, Phys. Rev. Let. **78**, 1644 (1997); B. A. Li, Phys. Rev. **C69**, 034614 (2004).
28. L.W. Chen, C.M. Ko and B.A. Li, Phys. Rev. **C68**, 017601 (2003); Nucl. Phys. **A729**, 809 (2003).
29. L.W. Chen, C.M. Ko and B.A. Li, Phys. Rev. **C69**, 054606 (2004).
30. B.A. Li, G.C. Yong and W. Zuo, Phys. Rev. **C71**, 044604 (2005).
31. B.A. Li, Phys. Rev. Let. **85**, 4221 (2000).

No-Core Shell Model and Reactions

Petr Navrátil*, W. Erich Ormand*, Etienne Caurier† and Carlos Bertulani**

*Lawrence Livermore National Laboratory, L-414, P.O. Box 808, Livermore, CA 94551, USA
†Institut de Recherches Subatomiques, IN2P3-CNRS, Universite Louis Pasteur, F-67037 Strasbourg, France
**Department of Physics, University of Arizona, Tucson, AZ 85721

Abstract. There has been a significant progress in *ab initio* approaches to the structure of light nuclei. Starting from realistic two- and three-nucleon interactions the *ab initio* no-core shell model (NCSM) can predict low-lying levels in *p*-shell nuclei. It is a challenging task to extend *ab initio* methods to describe nuclear reactions. In this contribution, we present a brief overview of the NCSM with examples of recent applications as well as the first steps taken toward nuclear reaction applications. In particular, we discuss cross section calculations of p+^{6}Li and ^{6}He+p scattering as well as a calculation of the astrophysically important ^{7}Be(p,γ)^{8}B S-factor.

Keywords: Shell model; forces in hadronic systems and effective interactions; direct reactions; spectroscopic factors; radiative capture
PACS: 21.60.Cs, 21.30.Fe, 24.50.+g, 21.10.Jx, 25.40.Lw

1. INTRODUCTION

Various methods can be used to solve systems of three or four nucleons interacting by realistic interactions [1]. For $A > 4$ systems, a prominent approach has been the Green's function Monte Carlo (GFMC) method [2]. An alternative, and complementary, approach is the no-core shell model (NCSM) [3, 4, 5, 6, 7, 8, 9, 10]. It considers light nuclei as systems of A nucleons interacting by realistic inter-nucleon forces. The calculations are performed using a large but finite harmonic-oscillator (HO) basis. Due to the basis truncation, it is necessary to derive an effective interaction from the underlying inter-nucleon interaction. The effective interaction contains, in general, up to A-body components even if the underlying interaction had, e.g. only two-body terms. In practice, the effective interaction is derived in a sub-cluster approximation retaining just two- or three-body terms. A crucial feature of the method is its convergence to exact solution with the basis size increase and/or the effective interaction clustering increase.

Among successes of the NCSM approach was the first published result of the binding energy of ^{4}He with the CD-Bonn NN potential [6], and the first observation of incorrect ground-state spin in ^{10}B when realistic nucleon-nucleon interactions are employed [9, 11]. This last result is a new example, in addition to the under binding problem, for the need of realistic three-nucleon forces.

In this paper, we give a short overview of the NCSM theory in Sect. 2 with examples of recent results. In particular, we emphasize recent efforts to apply the nuclear structure information obtained within the NCSM to describe nuclear reactions. A derivation of the translationally invariant density from the NCSM wave functions is discussed in Sect. 3. The obtained density then serves as an input for the folding approaches to optical potentials used in the direct reaction coupled channel calculations. As examples of the application, the ^{6}Li and ^{6}He scattering on protons is investigated. Calculation of cluster form factors and spectroscopic factors is discussed in Sect. 4. In Sect. 5, we present preliminary results of the ^{7}Be(p,γ)^{8}B S-factor calculation from NCSM wave functions. Conclusions are given in Sect. 6.

2. *AB INITIO* NO-CORE SHELL MODEL

We consider a system of A point-like nonrelativistic nucleons that interact by realistic two- or two- plus three-nucleon interactions. As the simpler case when just the two-nucleon interaction is considered was discussed in many papers, see e.g. Ref. [8], we focus here on the more general case when both two- and three-nucleon interactions (TNI) are included. The starting Hamiltonian is then

$$H_A = \frac{1}{A}\sum_{i<j}\frac{(\vec{p}_i - \vec{p}_j)^2}{2m} + \sum_{i<j}^{A} V_{\text{NN},ij} + \sum_{i<j<k}^{A} V_{\text{NNN},ijk}, \tag{1}$$

where m is the nucleon mass, $V_{\mathrm{NN},ij}$ the NN interaction, $V_{\mathrm{NNN},ijk}$ the three-nucleon interaction. In the NCSM, we employ a large but finite HO basis. Due to properties of the realistic nuclear interaction in Eq. (1), we must derive an effective interaction appropriate for the basis truncation. To facilitate the derivation of the effective interaction, we modify the Hamiltonian (1) by adding to it the center-of-mass (CM) HO Hamiltonian $H_{\mathrm{CM}} = T_{\mathrm{CM}} + U_{\mathrm{CM}}$, where $U_{\mathrm{CM}} = \frac{1}{2} A m \Omega^2 \vec{R}^2$, $\vec{R} = \frac{1}{A} \sum_{i=1}^{A} \vec{r}_i$. The effect of the HO CM Hamiltonian will later be subtracted out in the final many-body calculation. Due to the translational invariance of the Hamiltonian (1) the HO CM Hamiltonian has in fact no effect on the intrinsic properties of the system in the infinite basis space. The modified Hamiltonian can be cast into the form

$$H_A^\Omega = H_A + H_{\mathrm{CM}} = \sum_{i=1}^{A} h_i + \sum_{i<j}^{A} V_{ij}^{\Omega,A} + \sum_{i<j<k}^{A} V_{\mathrm{NNN},ijk} = \sum_{i=1}^{A} \left[\frac{\vec{p}_i^2}{2m} + \frac{1}{2} m \Omega^2 \vec{r}_i^2 \right]$$
$$+ \sum_{i<j}^{A} \left[V_{\mathrm{NN},ij} - \frac{m\Omega^2}{2A} (\vec{r}_i - \vec{r}_j)^2 \right] + \sum_{i<j<k}^{A} V_{\mathrm{NNN},ijk}. \quad (2)$$

Next we divide the A-nucleon infinite HO basis space into the finite active space (P) comprising of all states of up to $N_{\max}$ HO excitations above the unperturbed ground state and the excluded space ($Q = 1 - P$). The basic idea of the NCSM approach is to apply a unitary transformation on the Hamiltonian (2), $e^{-S} H_A^\Omega e^S$ such that $Q e^{-S} H_A^\Omega e^S P = 0$. If such a transformation is found, the effective Hamiltonian that exactly reproduces a subset of eigenstates of the full space Hamiltonian is given by $H_{\mathrm{eff}} = P e^{-S} H_A^\Omega e^S P$. This effective Hamiltonian contains up to A-body terms and to construct it is essentially as difficult as to solve the full problem. Therefore, we apply this basic idea on a sub-cluster level. When a genuine TNI is considered, the simplest approximation is a three-body effective interaction approximation. The NCSM calculation is then performed with the following four steps:

(i) We solve a three-nucleon system for all possible three-nucleon channels with the Hamiltonian H_A^Ω, i.e., using $h_1 + h_2 + h_3 + V_{12}^{\Omega,A} + V_{13}^{\Omega,A} + V_{23}^{\Omega,A} + V_{\mathrm{NNN},123}$. It is necessary to separate the three-body effective interaction contributions from the TNI and from the two-nucleon interaction. Therefore, we need to find three-nucleon solutions for the Hamiltonian with and without the $V_{\mathrm{NNN},123}$ TNI term. The three-nucleon solutions are obtained by procedures described in Refs. [7] (without TNI) and [12] (with TNI).

(ii) We construct the unitary transformation corresponding to the choice of the active basis space P from the three-nucleon solutions using the Lee-Suzuki procedure [13, 14]. The three-body effective interaction is then obtained as $V_{3\mathrm{eff},123}^{\mathrm{NN+NNN}} = P[e^{-S_{\mathrm{NN+NNN}}}(h_1 + h_2 + h_3 + V_{12}^{\Omega,A} + V_{13}^{\Omega,A} + V_{23}^{\Omega,A} + V_{\mathrm{NNN},123}) e^{S_{\mathrm{NN+NNN}}} - (h_1 + h_2 + h_3)]P$ and $V_{3\mathrm{eff},123}^{\mathrm{NN}} = P[e^{-S_{\mathrm{NN}}}(h_1 + h_2 + h_3 + V_{12}^{\Omega,A} + V_{13}^{\Omega,A} + V_{23}^{\Omega,A}) e^{S_{\mathrm{NN}}} - (h_1 + h_2 + h_3)]P$. The three-body effective interaction contribution from the TNI is then defined as $V_{3\mathrm{eff},123}^{\mathrm{NNN}} \equiv V_{3\mathrm{eff},123}^{\mathrm{NN+NNN}} - V_{3\mathrm{eff},123}^{\mathrm{NN}}$.

(iii) As the three-body effective interactions are derived in the Jacobi-coordinate HO basis but the A-nucleon calculations will be performed in a Cartesian-coordinate single-particle Slater-determinant m-scheme basis, we need to perform a suitable transformation of the interactions. This transformation is a generalization of the well-known transformation on the two-body level that depends on HO Brody-Moshinsky brackets.

(iv) We solve the Schrödinger equation for the A nucleon system using the Hamiltonian $H_{A,\mathrm{eff}}^\Omega = \sum_{i=1}^{A} h_i + \frac{1}{A-2} \sum_{i<j<k}^{A} V_{3\mathrm{eff},ijk}^{\mathrm{NN}} + \sum_{i<j<k}^{A} V_{3\mathrm{eff},ijk}^{\mathrm{NNN}}$, where the $\frac{1}{A-2}$ factor takes care of overcounting the contribution from the two-nucleon interaction. At this point we also subtract the H_{CM}. The A nucleon calculation is then performed using a shell model code generalized to handle three-body interactions.

An interesting example that demonstrates the importance of the TNI is the ground-state spin inversion in ^{10}B. The ground state of ^{10}B is $J^\pi T = 3^+ 0$. Calculations with the high-quality NN potenatials, however, predict a $1^+ 0$ ground state. [2, 9, 11] By including the Tucson-Melbourne TM' TNI, the problem is resolved, see Fig. 1. In the figure, three parameter sets denoted as 81, 93 and 99 are considered for the TM' TNI. All give similar results, dramatically different compared to the calculation with only the two-nucleon potential.

3. TRANSLATIONALLY INVARIANT DENSITY AND JLM OPTICAL POTENTIAL

In general, it is a challenging task to extend the *ab initio* methods to describe nuclear reactions. Concerning direct reactions, in particular the nucleon-nucleus elastic and inelastic scattering, a first and straightforward application of the NCSM is a semi-microscopic approach, e.g. the Jeukenne-Lejeune-Mahaux (JLM) [15], to construct optical potentials from the nuclear densities obtained in the NCSM. Eventually, these optical potentials can be used in coupled channels

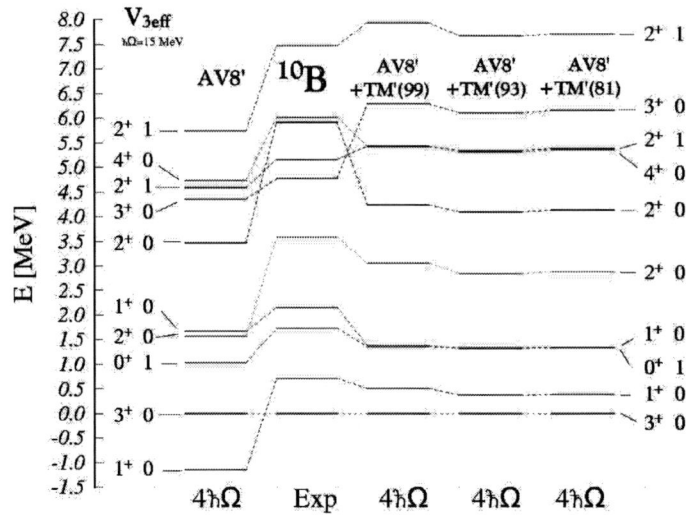

FIGURE 1. Excitation spectra of ^{10}B obtained using the AV8′ NN interactions and AV8′+TM′ interactions, respectively, are compared to experiment. Three parameter sets denoted as 81, 93 and 99 are considered for the TM′ TNI. The $4\hbar\Omega$ basis space and the $\hbar\Omega = 15$ MeV HO frequency were employed.

calculations by employing standard codes, e.g. Fresco [16]. To fully utilize NCSM nuclear structure for this purpose, the spurious CM contribution must be removed from the density.

The nuclear density operator is defined as [17]

$$\rho_{op}(\vec{r}) = \sum_{i=1}^{A} \delta(\vec{r} - \vec{r}_i) = \sum_{i=1}^{A} \frac{\delta(r - r_i)}{rr_i} \sum_m Y_{lm}(\hat{r}_i) Y_{lm}^*(\hat{r}) \ . \tag{3}$$

The physical density should depend on the coordinate measured from the CM of the nucleus, $\vec{r} - \vec{R}$. Using the transformation properties of the HO wave functions, it is possible to relate the physical density to the standard one-body density matrix elements (OBDME) computed in shell model codes from eigenstates obtained in the Slater determinant basis. The final expression is [18]

$$\langle A\lambda_f J_f M_f | \rho_{op}(\vec{r} - \vec{R}) | A\lambda_i J_i M_i \rangle = (\tfrac{A}{A-1})^{3/2} \tfrac{1}{\hat{J}_f} \sum (J_i M_i K k | J_f M_f) Y_{Kk}^*(\widehat{\vec{r} - \vec{R}})$$

$$\times \ R_{nl}(\sqrt{\tfrac{A}{A-1}}|\vec{r} - \vec{R}|) R_{n'l'}(\sqrt{\tfrac{A}{A-1}}|\vec{r} - \vec{R}|)(-1)^K \frac{\hat{l}\hat{l}'(l0l'0|K0)}{\hat{l}_1 \hat{l}_2 (l_1 0 l_2 0|K0)} (M^K)^{-1}_{nln'l', n_1 l_1 n_2 l_2}$$

$$\times \ \langle l_1 \tfrac{1}{2} j_1 \| Y_K \| l_2 \tfrac{1}{2} j_2 \rangle \tfrac{-1}{\hat{K}} \ {}_{SD}\langle A\lambda_f J_f \| (a^\dagger_{n_1 l_1 j_1} \tilde{a}_{n_2 l_2 j_2})^{(K)} \| A\lambda_i J_i \rangle_{SD} \ , \tag{4}$$

where the sum is restricted to both $l + l' + K$ and $l_1 + l_2 + K$ even. The λ_i and λ_f are the additional quantum numbers that classify the initial and final state, respectively. The matrix M^K is defined as

$$(M^K)_{n_1 l_1 n_2 l_2, nln'l'} = \sum_{N_1 L_1} (-1)^{l+l'+K+L_1} \begin{Bmatrix} l_1 & L_1 & l \\ l' & K & l_2 \end{Bmatrix} \hat{l}\hat{l}'$$

$$\times \ \langle nl 00 l | N_1 L_1 n_1 l_1 l \rangle_{\tfrac{1}{A-1}} \langle n'l' 00 l' | N_1 L_1 n_2 l_2 l' \rangle_{\tfrac{1}{A-1}} \ . \tag{5}$$

As an illustration of the significance of the spurious CM removal, we calculated the ^{6}He physical (4) and the shell-model densities $_{SD}\langle A\lambda_f J_f M_f | \rho_{op}(\vec{r}) | A\lambda_i J_i M_i \rangle_{SD}$ using wave functions obtained in Ref. [19]. We note that the relationship between the Jacobi coordinate eigenstates appearing in Eq. (4) and the SD eigenstates is

$$\langle \vec{r}_1 \ldots \vec{r}_A | A\lambda J M \rangle_{SD} = \langle \vec{\xi}_1 \ldots \vec{\xi}_{A-1} | A\lambda J M \rangle \varphi_{000}(\sqrt{A}\vec{R}) \ , \tag{6}$$

In Fig. 2, the proton and the neutron monopole ground state densities are shown. A $10\hbar\Omega$ basis space and the HO frequency of $\hbar\Omega = 13$ MeV was used. The two-body effective interaction was derived from the CD-Bonn NN potential.

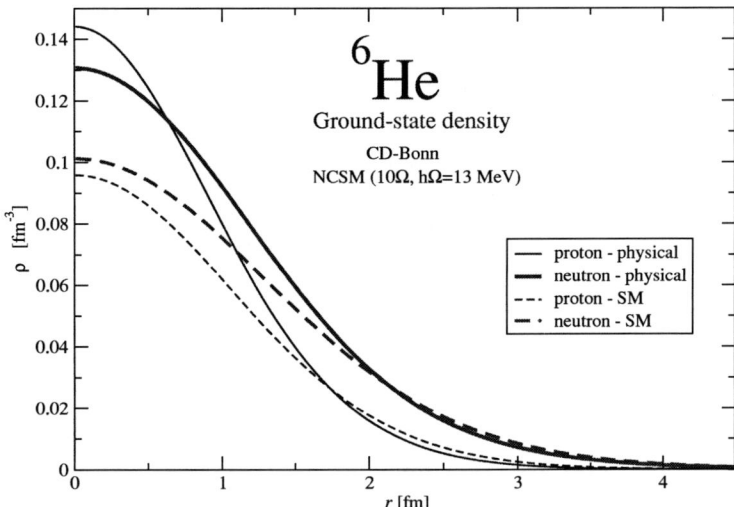

FIGURE 2. ^{6}He proton and neutron monopole ground state densities obtained in the $10\hbar\Omega$ basis space and the HO frequency of $\hbar\Omega = 13$ MeV. The NCSM two-body effective interaction was derived from the CD-Bonn NN potential. The full lines correspond to the physical densities calculated according to Eq. (4) while the dashed lines correspond to the shell-model densities that contain the spurious center-of-mass contribution.

The full lines correspond to the physical densities calculated according to Eq. (4) while the dashed lines correspond to the shell-model densities that contain the spurious CM contribution. Obviously, the same OBDME were employed in both calculations. The normalization of the densities in Fig. 2 is $4\pi \int dr r^2 \rho_{K=0,p(n)}(r) = Z(N)$ with p, n refers to the proton and neutron, respectively, and $\rho_{K=0}(r) = \frac{1}{4\pi} \int d\hat{r} \langle A\lambda JM | \rho_{op}(\vec{r}) | A\lambda JM \rangle$. Not surprisingly, a particularly significant impact of the exact removal of the spurious CM motion is then found for the spin-orbit part of the optical potential proportional to the derivative of the nuclear density.

By multiplying the physical monopole densities by r^2 and integrating we obtain the point-proton and point-neutron rms radii 1.763 fm and 2.361 fm, respectively, identical to those calculated in Ref. [19] in a different way. Performing the same integral using the shell-model densities gives incorrect, larger radii 1.976 fm and 2.524 fm, respectively. The difference between the squares of the two sets of radii is equal to the mean value of the CM $\vec{R}^2$. It should be noted that a recent high-precision measurement of the ^{6}He proton radius reported a point-proton radius of 1.912(18) fm. [20] is larger than that we calculted in the $10\hbar\Omega$ model space with the HO frequency $\hbar\Omega = 13$ MeV. A more detailed investigation of the radius convergence is needed. For example, a new calculation in a larger space, $10\hbar\Omega$ and a different HO frequency $\hbar\Omega = 11$ MeV gives the point-proton radius 1.818 fm.

We performed differential cross sections calculations for the reactions p+^{6}Li at 72 MeV [21] and ^{6}He+p at 71 MeV/A [22]. Starting from the translationally invariant ground-state monopole densities obtained from $12\hbar\Omega$ NCSM calculations using the CD-Bonn NN potential, we constructed the JLM optical potential and the spin-orbit potential proportional to the derivative of the density. We employed the JLM parametrization from Ref. [23] with the local density approximation using the prescription with the mid-point interaction evaluation. Only the real part of the spin-orbit interaction was retained. A simultaneous χ-square fit of the strength parameters of the real and imaginary central potential and the spin-orbit potential was performed for the two reactions. Our results are compared to experimental data in Fig. 3. The values of the fitted parameters are $\lambda_V = 0.88, \lambda_W = 0.92$ and $\lambda_{so} = 0.80$.

4. CHANNEL CLUSTER FORM FACTOR CALCULATION

Detailed knowledge of nuclear structure is important for the description of low-energy nuclear reactions. As the first step in the application of the NCSM to low-energy nuclear reactions, one needs to understand the cluster structure of the eigenstates. That is, one needs to calculate the channel cluster form factors. Those can then, e.g., be integrated to obtain the spectroscopic factors. Let's consider a composite system of A nucleons, a projectile of a nucleons and a target of $A-a$ nucleons. All the nuclei are assumed to be described by eigenstates of the NCSM effective Hamiltonians expanded in the HO basis with identical HO frequency and the same (for the eigenstates of the same parity) or differing

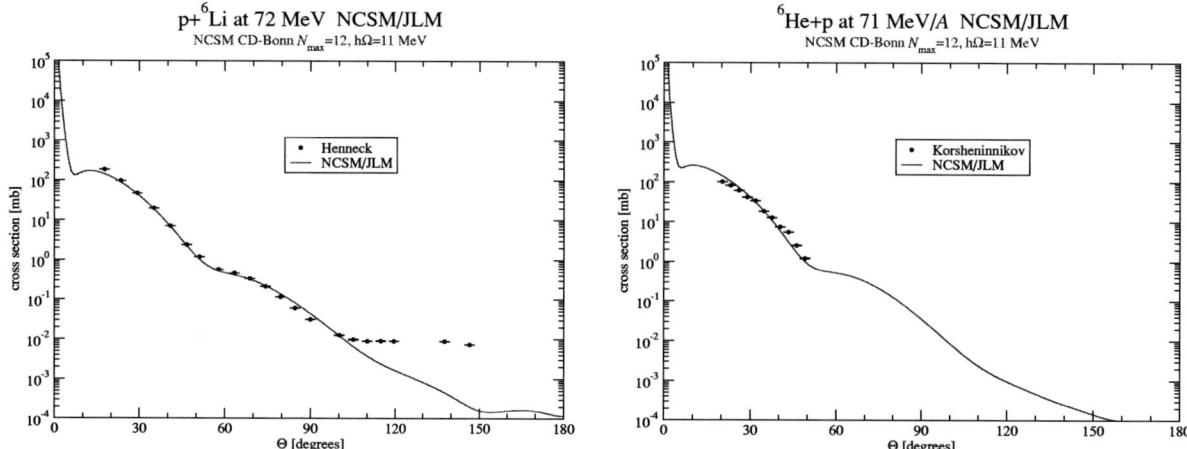

FIGURE 3. Differential cross section for p+^{6}Li at 72 MeV (left) and for ^{6}He+p at 71 MeV/A (right). The NCSM densities obtained in a $12\hbar\Omega$ calculation were used to generate the JLM optical potentials. Experimental data are from Refs. [21] and [22], respectively.

by one unit of the HO excitation (for the eigenstates of opposite parity) definitions of the model space. We limit ourselves to $a \leq 4$ projectiles. In such a case, the projectiles can be efficiently described by a Jacobi-coordinate HO wave functions. The target and the composite system is assumed to be described by Slater determinant single-particle HO basis wave functions which is in general more efficient for $A > 4$. Let us introduce a projectile-target wave function

$$\langle \vec{\xi}_1 \ldots \vec{\xi}_{A-a-1} \eta'_{A-a} \hat{\eta}_{A-a} \vec{\vartheta}_{A-a+1} \ldots \vec{\vartheta}_{A-1} | \Phi^{(A-a,a)JM}_{\alpha I_1, \beta I_2; sl}; \delta_{\eta_{A-a}} \rangle$$
$$= \sum (I_1 M_1 I_2 M_2 | s m_s)(s m_s l m_l | JM) \frac{\delta(\eta_{A-a} - \eta'_{A-a})}{\eta_{A-a} \eta'_{A-a}} Y_{lm_l}(\hat{\eta}_{A-a})$$
$$\times \langle \vec{\xi}_1 \ldots \vec{\xi}_{A-a-1} | A-a \alpha I_1 M_1 \rangle \langle \vec{\vartheta}_{A-a+1} \ldots \vec{\vartheta}_{A-1} | a \beta I_2 M_2 \rangle, \qquad (7)$$

The calculation of the cluster form factor

$$g^{A\lambda JT}_{A-a\alpha I_1, a\beta I_2; sl}(\eta_{A-a}) = \langle A\lambda J | \mathscr{A} \Phi^{(A-a,a)J}_{\alpha I_1, \beta I_2; sl}; \delta_{\eta_{A-a}} \rangle$$
$$= \sqrt{\frac{A!}{(A-a)!a!}} \sum_n R_{nl}(\eta_{A-a}) \langle A\lambda J | \Phi^{(A-a,a)J}_{\alpha I_1, \beta I_2; sl}; nl \rangle, \qquad (8)$$

can then be done in two steps. First, using the relation (6) for both the composite and the target eigenstate and the HO wave function transformations we obtain

$$_{\mathrm{SD}}\langle A\lambda J | \mathscr{A} \Phi^{(A-a,a)J}_{\alpha I_1, \beta I_2; sl}; nl \rangle_{\mathrm{SD}} = \langle nl00l|00nll\rangle_{\frac{a}{A-a}} \langle A\lambda J | \mathscr{A} \Phi^{(A-a,a)J}_{\alpha I_1, \beta I_2; sl}; nl \rangle, \qquad (9)$$

with a general HO bracket due to the CM motion. The nl in (8) and (9) refers to a replacement of $\delta_{\eta_{A-a}}$ by the HO $R_{nl}(\eta_{A-a})$ radial wave function. Second, we relate the SD overlap to a linear combination of matrix elements of a creation operators between the target and the composite eigenstates $_{\mathrm{SD}}\langle A\lambda J | a^\dagger_{n_1 l_1 j_1} \ldots a^\dagger_{n_a l_a j_a} | A-a\alpha I_1 \rangle_{\mathrm{SD}}$. Such matrix elements are easily calculated by shell model codes. To obtain the channel cluster form factor we use the second equality in Eq. (8). The spectroscopic factor is obtained by integrating the square of the form factor.

As an example, we present results for the ^{7}Li$\rightarrow^4$He+t channel cluster form factors in Fig. 4. Apart from the large overlap integrals and spectroscopic factors for the bound $\frac{3}{2}^-_1$ and $\frac{1}{2}^-_1$ states, we find these quantities to be large also for the first excited $\frac{7}{2}^-_1$ and the first excited $\frac{5}{2}^-_1$ state. Both these states appear as resonances in the ^{4}He+t cross section. [24] The present results can be compared to the three-nucleon transfer calculations of Ref. [25] obtained using the phenomenological Cohen-Kurath interaction. [26] The agreement for the lowest four states is quite good. For the second excited $\frac{5}{2}^-_2$ state, however, our spectroscopic factor is significantly smaller than the one obtained in Ref.

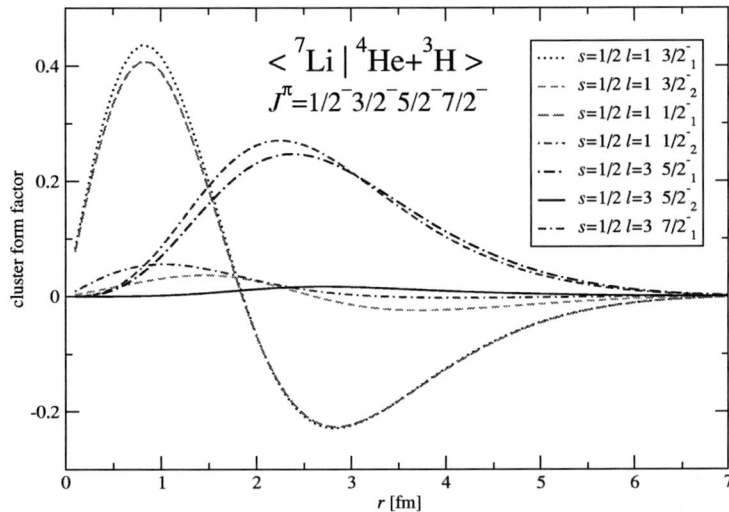

FIGURE 4. Overlap integral, g(r), of the ^{7}Li low-lying $J = \frac{1}{2}^-, \frac{3}{2}^-, \frac{5}{2}^-, \frac{7}{2}^-$ states with the ^{4}He+^{3}H as a function of separation between the ^{4}He and the triton. The CD-Bonn 2000 NN potential, the basis size of $N_{\max} = 8$ (for ^{7}Li), $N_{\max} = 10$ (for ^{4}He and ^{3}H) and the HO frequency of $\hbar\Omega = 13$ MeV were used.

[25]. The other system involving ^{7}Li as the composite nucleus that we investigated is ^{6}Li+n. As in the ^{7}Li→^{4}He+t case, we observe large overlap integrals and spectroscopic factors for the two bound states $\frac{3}{2}^-_1$ and $\frac{1}{2}^-_1$. Contrary to the ^{7}Li→^{4}He+t case, however, we find a large overlap integral and the spectroscopic factor for the $\frac{5}{2}^-_2$ state. The lowest $\frac{7}{2}^-_1$ and $\frac{5}{2}^-_1$ states have negligible overlap integrals for the ^{6}Li+n system. The large overlap integral and the spectroscopic factor for the $\frac{5}{2}^-_2$ state is consistent with the observed resonance in the ^{6}Li+n cross section.

5. ^{7}BE(P,γ)^{8}B S-FACTOR FROM NCSM WAVE FUNCTIONS

The ^{7}Be(p,γ)^{8}B capture reaction serves as an important input for understanding the solar neutrino flux [27]. S-factor of this reaction needs to be known with a precision of better than 9%. Current experimental uncertainties are around 20%. Many theoretical investigations were devoted to this reaction. Here we present a calculation for the S-factor starting from the *ab initio* NCSM wave functions for ^{8}B and ^{7}Be. The NCSM calculations were performed using the CD-Bonn NN potential in the model spaces up to $10\hbar\Omega$ for both nuclei. The harmonic oscillator frequency $\hbar\Omega = 12$ MeV, which produces the ground-state energy minimum in the $10\hbar\Omega$ space, was selected as the starting point also for the present S-factor calculation. The ground-state wave functions for ^{8}B and ^{7}Be were utilized to calculate the cluster form factors or overlap integrals that serve as the S-factor calculation input. The two most important channels are the p-waves, $l = 1$, with the proton in the $j = 3/2$ and $j = 1/2$ states, $\vec{j} = \vec{l} + \vec{s}, s = 1/2$. In these channels, we obtain the spectroscopic factors of 0.96 and 0.10, respectively. The dominant $j = 3/2$ overlap integral is presented in the left of Fig. 5 by the full line. Despite the fact, that a very large basis was employed in the present calculation, it is apparent that the overlap integral is nearly zero at about 10 fm. This is a consequence of the HO basis asymptotics. The proton capture on ^{7}Be to the weakly bound ground state of ^{8}B associated dominantly by the $E1$ radiation is a peripheral process. Consequently, the overlap integral with an incorrect asymptotic behavior cannot be used to calculate the S-factor.

It is our expectation, however, that the interior part of the overlap integral as obtained from our large-basis NCSM calculation is realistic. It is straightforward then to correct the asymptotic behavior of the overlap integral. We performed a least-square fit of a Woods-Saxon potential solution to the interior of the NCSM overlap in the range of $0-4$ fm. The Woods-Saxon potential parameters were varied in the fit under the constrain that the experimental separation energy of ^{7}Be+p was reproduced. In this way we obtain a perfect fit to the interior of the overlap integral and a correct asymptotic behavior at the same time. The resulting overlap is presented in Fig. 5 by the dashed line. Eventually, we rescale the overlap to preserve the original NCSM spectroscopic factor. The same procedure is applied

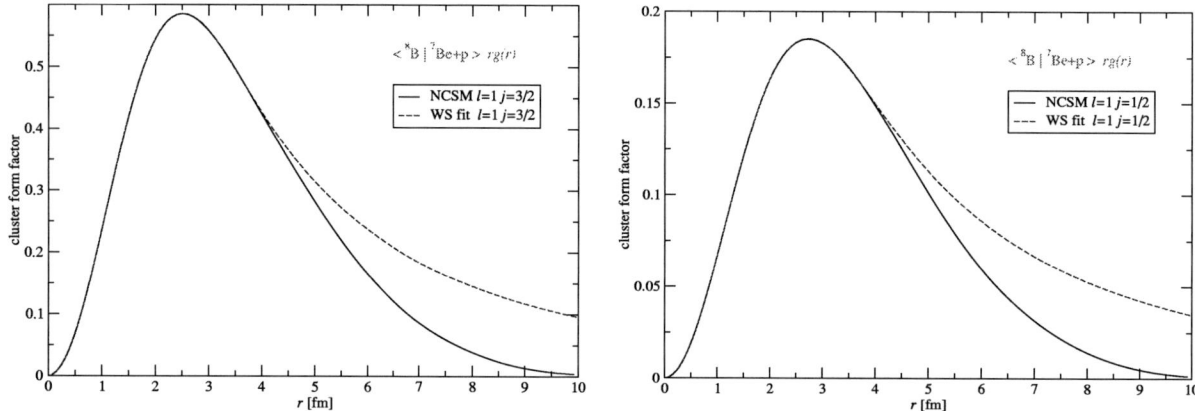

FIGURE 5. Overlap integral, $rg(r)$, for the ground state of ^{8}B with the ground state of ^{7}Be plus proton as a function of separation between the ^{7}Be and the proton. Left (right), the p-wave channel with $j = 3/2$ ($j = 1/2$) is shown. The full line represents the NCSM result, while the dashed line represents a renormalized overlap obtained from a Woods-Saxon potential whose parameters were fit to the NCSM overlap up to 4.0 fm under the constraint to reproduce the experimental separation energy.

to the other p-wave channel (right panel in Fig. 5). Obviously, the Woods-Saxon parameters obtained for the two channels are different. The corrected overlap integrals then serve as the input for the S-factor calculation performed as described in Ref. [28]. The scattering ^{7}Be+p s- and d-wave states are obtained within the potential model approach using a Woods-Saxon potential used in Ref. [29].

Our obtained S-factor is presented in Figs. 6 where contribution from the two partial waves are shown together with the total result (left figure). It is interesting to note a good agreement of our calculated S-factor with the recent Seattle direct measurement [30]. The sensitivity of the S-factor to the size of the NCSM basis is also presented in Figs. 6 (right figure). The overlap integrals were obtained in 6, 8 and 10$\hbar\Omega$ calculations and independently corrected to insure the proper asymptotic behavior. The same scattering states were used in all three cases. It is apparent that the sensitivity to the basis change is rather moderate. We observe a small oscillation at this frequency ($\hbar\Omega = 12$ MeV). More detailed inverstigation of the basis size and the HO frequency sensitivity is under way.

6. CONCLUSION

Substantial progress has been made towards an exact description of nuclear structure. In this work, we described the *ab initio* no-core shell model and recent results. In particular, we find that realistic NN interactions by themselves are inadequate and that three-nucleon forces play an important role in determining nuclear properties. We are also in the process of extending the no-core shell model into a formalism capable of providing a description of nuclear reactions. Overall, the prospects are good that exact results for both structure and reactions for nuclei up to Oxygen utilizing the fundamental forces between nucleons can be achieved in the near future.

ACKNOWLEDGMENTS

We thank I. Thompson for assistance and help with the direct reaction code Fresco as well as with the code for constructing the JLM optical potential. This work was performed under the auspices of the U. S. Department of Energy by the University of California, Lawrence Livermore National Laboratory under contract No. W-7405-Eng-48. Support from the LDRD contract No. 04-ERD-058 is acknowledged.

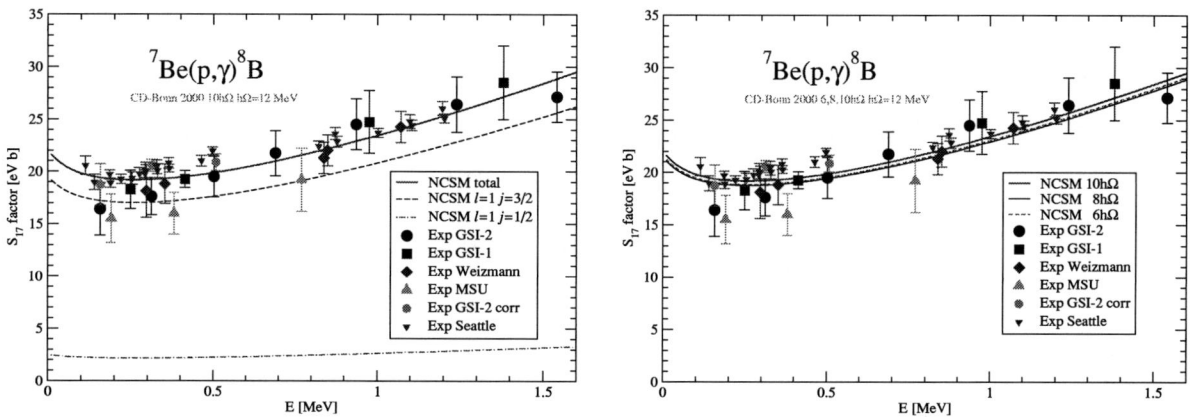

FIGURE 6. The ^{7}Be(p,γ)^{8}B S-factor obtained using the NCSM cluster form factors with corrected asymptotics as described in the text. Left, the dashed and dashed-dotted lines show the contribution due to the $l=1$, $j=3/2$ and $j=1/2$ partial waves, respectively. Right, the dependence on the size of the basis form $6\hbar\Omega$ to $10\hbar\Omega$ is shown. Experimental values are from Refs. [30, 31].

REFERENCES

1. J. L. Friar et al., Phys. Lett. **B 311**, 4 (1993); W.Glöckle and H. Kamada, *Phys. Rev. Lett.* **71**, 971 (1993); M. Viviani, A. Kievsky, and S. Rosati, *Few-Body Systems* **18**, 25 (1995).
2. B. S. Pudliner et al., Phys. Rev. C **56** 1720, (1997); R. B. Wiringa, *Nucl. Phys.* **A 631**, 70c (1998); R. B. Wiringa et al., Phys. Rev. C **62**, 014001 (2000); S. C. Pieper et al., Phys. Rev. C **64**, 014001 (2001).
3. D. C. Zheng, B. R. Barrett, L. Jaqua, J. P. Vary and R. J. McCarthy, Phys. Rev. C **48**, 1083 (1993); D. C. Zheng, J. P. Vary, and B. R. Barrett, Phys. Rev. C **50**, 2841 (1994); D. C. Zheng, B. R. Barrett, J. P. Vary, W. C. Haxton and C.-L. Song, Phys. Rev. C **52**, 2488 (1995).
4. P. Navrátil and B. R. Barrett, Phys. Rev. C **54**, 2986 (1996).
5. P. Navrátil and B. R. Barrett, Phys. Rev. C **57**, 562 (1998).
6. P. Navrátil and B. R. Barrett, Phys. Rev. C **59**, 1906 (1999).
7. P. Navrátil, G. P. Kamuntavičius and B. R. Barrett, Phys. Rev. C **61**, 044001 (2000).
8. P. Navrátil, J. P. Vary and B. R. Barrett, Phys. Rev. Lett. **84**, 5728 (2000); Phys. Rev. C **62**, 054311 (2000).
9. P. Navrátil and W. E. Ormand, Phys. Rev. Lett. **88**, 152502 (2002).
10. P. Navrátil and W. E. Ormand, Phys. Rev. C **68**, 034305 (2003).
11. E. Caurier, P. Navrátil, W. E. Ormand and J. P. Vary, Phys. Rev. C **66**, 024314 (2002).
12. D. C. J. Marsden, P. Navrátil, S. A. Coon and B. R. Barrett, Phys. Rev. C **66**, 044007 (2002).
13. K. Suzuki and S. Y. Lee, Prog. Theor. Phys. **64**, 2091 (1980).
14. K. Suzuki and R. Okamoto, Prog. Theor. Phys. **92**, 1045 (1994).
15. J.-P. Jeukenne, A. Lejeune and C. Mahaux, Phys. Rev. C **16**, 80 (1977).
16. I. J. Thompson, *Computer Physics Reports* **7**, 167 (1988).
17. G. R. Satchler, *Direct Nuclear Reactions*, Oxford University Press 1983.
18. P. Navrátil, Phys. Rev. C **70**, 014317 (2004).
19. P. Navrátil, J. P. Vary, W. E. Ormand and B. R. Barrett, Phys. Rev. Lett. 87, 172502 (2001).
20. L.-B. Wang et al. Phys. Rev. Lett. **93**, 142501 (2004).
21. R. Henneck et al., *Nucl. Phys.* **A571**, 541 (1994).
22. A. A. Korsheninnikov et al., *Nucl. Phys.* **A 617**, 45 (1997).
23. E. Bauge, J. P. Delaroche and M. Girod, Phys. Rev. C **58**, 1118 (1998).
24. D. R. Tilley, C. M. Cheves, J. L. Godwin, G. M. Hale, H. M. Hofmann, J. H. Kelley, C. G. Sheu and H. R. Weller, *Nucl. Phys.* **A 708**, 3 (2002).
25. D. Kurath and D. J. Millener, Nucl. Phys. **A238** 269, (1975).
26. S. Cohen and D. Kurath, *Nucl. Phys.* **73** 1, (1965).
27. E. Adelberger it et al., *Rev. Mod. Phys.* **70**, 1265 (1998).
28. C. Bertulani, *Comp. Phys. Comm.* **156**, 123 (2003).
29. H. Esbensen and G. F. Bertsch, *Nucl. Phys.* **A 600**, 37 (1996).
30. A. R. Junghans, E. C. Mohrmann, K. A. Snover, T. D. Steiger, E. G. Adelberger, J. M. Casandjian, H. E. Swanson, L. Buchmann, S. H. Park, A. Zyuzin, and A. M. Laird, Phys. Rev. C **68**, 065803 (2003).
31. N. Iwasa et al., Phys. Rev. Lett. **83**, 2910 (1999); B. Davids et al., Phys. Rev. Lett. **86**, 2750 (2001); L.T. Baby et al., Phys. Rev. Lett. **90**, 022501 (2003); F. Schuemann et al., Phys. Rev. Lett. **90**, 232501 (2003).

Combined method to extract the spectroscopic factors from transfer reactions

A. M. Mukhamedzhanov* and F.M. Nunes[†]

Cyclotron Institute, Texas A&M University, College Station, TX 77843, USA
[†]*N.S.C.L. and Department of Physics and Astronomy, Michigan State University, MI 48824, USA*

Abstract. We revise the standard method for extracting spectroscopic factors from transfer reactions and discuss a new combined method which is based on including the asymptotic normalization coefficient (ANC) of the overlap functions into the transfer analysis. This new method determines the SF from fitting the internal part of the reaction amplitude, while the contribution of the external part, typically dominant, is fixed. This is in contrast with the standard approach in which the SF is extracted at the expense of the arbitrary variation of the external part contribution. As examples, we show results for (d,p) reactions on a variety of targets: ^{208}Pb, ^{7}Li ^{12}C, and ^{84}Se. The modified method has the potential of improving the reliability and accuracy of the structure information. Ultimately, it can be used as a check of DWBA by simultaneously introducing the information on the ANC and SF into the analysis.

Keywords: Spectroscopic factor, asymptotic normalization coefficient, transfer reaction.
PACS: 21.10.Jx, 24.10.-i, 24.50.+g, 25.40.Hs

INTRODUCTION

In [1], we discuss the standard method of extracting spectroscopic information from transfer reactions and introduce a combined method that includes the asymptotic normalization coefficient (ANC), determined from an independent measurement, into the reaction description. Here, we make a more detailed formulation and further the discussion on the combined method.

Spectroscopic factors (SF) from transfer reactions have made a come back due to the interest on the properties of nuclei on the driplines. SF were introduced by the shell model formalism and are typically related to the shell occupancy of a state n in one nucleus relative to a state m in a nearby nucleus [2]. At present ab-initio calculations are improving the accuracy of the calculated spectroscopic factors and one can, in some cases, find surprises and disagreements, especially when moving toward the driplines. It is timely to think of an accurate probe that will test the predictions of these models and will disentangle the relevant elements of the NN force that are still missing (e.g. [3]).

For conventional nuclei there are many experiments available providing SFs, which are often lower than those predicted by shell model [2]. Electron-induced knockout or electron scattering is supposed to provide a better accuracy in extracting SFs than transfer [4, 5]. However, for exotic nuclei near or on the driplines, transfer reactions are a unique tool and, hence, can have a large impact in the programs of the new generation rare isotope laboratories. Given the experimental difficulties faced with measurements on the driplines, it is crucial to have a reliable method for analyzing and extracting useful information from each single data set.

For more than forty years since the dawn of nuclear physics, direct transfer reactions, such as (d,p), (d,t), $(^3\text{He},d)$ and $(^3\text{He},\alpha)$, have been the central tool to determine SFs [6, 7, 8]. Analyzes are usually based on a single transfer data set. Angular distributions of transfer reactions are analyzed within the framework of the distorted-wave Born approximation (DWBA). The SF determined by normalizing the calculated DWBA differential cross section to the experimental one in the main peak of the angular distribution (e.g. [9, 10]) are compared with the SF predicted by shell model. Inaccuracies are typically due to: i) optical potentials ambiguity, ii) the inadequacy of the DWBA reaction theory, or iii) the dependence on the single-particle potential parameters. Recently, a systematic analysis including many data sets was performed [11] to address the first point. Reduced error bars could be expected and were obtained. The second point needs to be addressed case by case, and examples of improved reaction models are the coupled channel Born approximation (e.g. [12]) or the continuum discretized coupled channel method (e.g. [13]).

It is important to note from the start that, in the combined method, one needs to use two independent data sets, two transfer reactions that probe different regions of the overlap function with the purpose of extracting different bits of information: a more peripheral reaction for extraction of the ANC and a reaction which contains an interior contribution and will allow to determined the SF. Also, the combined method assumes DWBA provides a good description of the transfer angular distribution at forward angle and that the changes in the single particle parameters needed to scan the ANCs do not modify the angular distribution significantly. Finally, as in the combined method, the exterior part is fixed, it requires a significantly higher accuracy from DWBA in the calculation of the nuclear interior, than that required before. So, it also offers a crucial test of the existing theories of nuclear transfer reactions widely used in nuclear physics.

MODIFIED APPROACH VERSUS STANDARD APPROACH

Here we review the derivations in [1] with additional comments. We consider $A(d,p)B$ reaction and disregard spins (naturally these are included in the applications). The DWBA amplitude for this reaction is given by:

$$M = <\psi_f^{(-)} I_{An}^B |\Delta V| \varphi_{pn} \psi_i^{(+)}>, \qquad (1)$$

where $\Delta V = V_{pn} + V_{pA} - U_{pB}$ is the transition operator in the post-form, V_{ij} is the interaction potential between i and j, U_{pB} is the optical potential in the final-state. The distorted waves in the initial and final states are $\psi_i^{(+)}$ and $\psi_f^{(-)}$, φ_{pn} is the deuteron bound-state wave function and $I_{An}^B(\mathbf{r})$ is the overlap function of the bound-states of nuclei B and A which depends on $\mathbf{r}$, the radius-vector connecting the center of mass of A with n. The overlap function is not an eigenfunction of an Hermitian Hamiltonian and is not normalized to unity [14]. The square norm of the overlap function gives a model-independent definition of the SF:

$$S = N <I_{An}^B | I_{An}^B>. \qquad (2)$$

Here, N is the antisymmetrization factor in the isospin formalism (N will be included in the overlap function from now on).

The leading asymptotic term of the radial overlap function (for $B = A+n$) is

$$I_{An(lj)}^B(r) \stackrel{r \geq R}{\approx} C_{lj} i \kappa h_l(i\kappa r), \qquad (3)$$

where $h_l(i\kappa r)$ is the spherical Bessel function, $\kappa = \sqrt{2\mu_{An}\varepsilon_{An}}$, ε_{An} is the binding energy for $B \to A+n$, and μ_{An} is the reduced mass of A and n. Similarly, the asymptotics of the neutron single-particle wave function is $\varphi_{An(n_r l j)}(r) \stackrel{r \geq R}{\approx} b_{n_r l j} i\kappa h_l(i\kappa r)$, where n_r is the principle quantum number. The asymptotic behaviour is valid beyond R, the channel radius. It is clear that, in the asymptotic region, the overlap function is proportional to the single particle wave function. The normalization C_{lj} introduced in Eq.(3) is the ANC which relates to the single-particle ANC $b_{n_r l j}$ by $C_{lj} = K_{n_r l j} b_{n_r l j}$, where $K_{n_r l j}$ is an asymptotic proportionality coefficient. It is standard practice to assume that the proportionality between the overlap function and the single particle function extends to all r values

$$I_{An(lj)}^B(r) = K_{n_r l j} \varphi_{An(n_r l j)}(r). \qquad (4)$$

Since $\varphi_{An(n_r l j)}(r)$ is normalized to unity, this approximation (Eq. 4) implies that $S_{lj} = K_{n_r l j}^2$. We have to emphasize, however, that the overlap function is a many-body object and its radial dependence in the interior is nontrivial and may well differ from the single particle wavefunction, since $I_{An(lj)}^B(r)$ includes (on the shell model language) the effects of the mean field and of residual interactions. The later may have a strong influence on the form of the overlap function near the nuclear surface region, which gives the most important contribution to nucleon transfer reactions. Thus, the multiparticle character of the overlap function reveals itself in the deviation of its norm unity and in its radial shape. In the DWBA we use only the SF definition given by Eq. (2). Approximating the radial dependence of the overlap function as described above leads to the DWBA amplitude $M = K_{n_r l j} <\psi_f^{(-)} \varphi_{An(n_r l j)} |\Delta V| \varphi_{pn} \psi_i^{(+)}>$. Normalizing the calculated DWBA cross section,

$$\sigma^{DW} = |<\varphi_{An(n_r l j)} |\Delta V| \varphi_{pn} \psi_i^{(+)}>|^2, \qquad (5)$$

to the experimental data provides the phenomenological SF $S_{lj} = K_{n_r l j}^2$. Assuming that Eq.(4) is valid for all r, we can infer from Eq.(2) that the main contribution to the norm of the overlap function comes from the nuclear interior.

In order to make the dependence on the single-particle more explicit, we split the reaction amplitude into an interior part and an exterior part:

$$M = K_{n_r l j} \tilde{M}_{int}[b] + K_{n_r l j} b_{n_r l j} \tilde{M}_{ext}, \tag{6}$$

where the internal part of the matrix element $\tilde{M}_{int}[b_{n_r l j}] = <\psi_f^{(-)} \varphi_{An(n_r l j)}|\Delta V|\varphi_{pn} \psi_i^{(+)}>_{r<R}$ depends on $b_{n_r l j}$ through the bound state wavefunction $\varphi_{An(n_r l j)}$, while the external part $\tilde{M}_{ext} = <\psi_f^{(-)} i\kappa h_l(i\kappa r)|\Delta V|\varphi_{pn} \psi_i^{(+)}>_{r>R}$ does not depend on $b_{n_r l j}$. Here, R is the channel radius taken so that for $r > R$ the overlap function can be approximated by its asymptotic form Eq.(3) (R is only used to illustrate the method as in the end this separation is not required). The contribution from the nuclear exterior is fixed by the ANC, whereas the SF determines the normalization of the internal part of the radial matrix element. Since transfer reactions are dominantly peripheral, SFs can only be extracted from transfer reactions due to a small contribution from the nuclear interior. We now introduce the ANC into the DWBA cross section:

$$\frac{d\sigma^{DW}}{d\Omega} = C_{lj}^2 \frac{\sigma^{DW}}{b_{n_r l j}^2}. \tag{7}$$

Introducing Eq. (6) into Eq. (7) and dividing by C_{lj}^2, we arrive at a function $R^{DW}(b)$

$$R^{DW}(b_{n_r l j}) = |\frac{\tilde{M}_{int}[b]}{b_{n_r l j}} + \tilde{M}_{ext}|^2. \tag{8}$$

Note that the single-particle ANC $b_{n_r l j}$ itself is a function of the geometrical parameters of the bound state $n - A$ nuclear potential (r_0, a) which are, a priori, not known. If the ANC and the cross section for the (d,p) reaction have been measured, the experimental counterpart of R^{DW}, $R^{exp} = \frac{d\sigma^{exp}}{d\Omega}/C_{lj}^2$ can be experimentally fixed. Then, imposing the equality

$$R^{exp} = R^{DW}(b_{n_r l j}), \tag{9}$$

will provide the correct b_{nlj} and consequently the SF $S_{lj} = C_{lj}^2/b_{nlj}^2$.

At this stage, a few points should be made clear. First of all, for specific optical potentials, Eq. (5) depends on two independent parameters, S_{lj} and b_{nlj}. In the standard approach, to evaluate this cross section, the second parameter is fixed by arbitrarily choosing the bound state $n-A$ potential geometry. Thus, the extracted product $S_{lj} b_{n_r l j}^2$ does not coincide necessarily with the correct ANC. Since the ANC determines the normalization of the external part of the DWBA amplitude, in the standard approach the SF is determined by an unrealistic variation of the external contribution. In the modified method here discussed, since the contribution of the external part is fixed through the correct ANC, the whole DWBA procedure loses this artificial degree of freedom.

Secondly, if the reaction is peripheral, i. e. the first term in Eq. (6) is negligible,

$$M \approx K_{n_r l j} b_{n_r l j} \tilde{M}_{ext} = C_{lj} \tilde{M}_{ext}, \tag{10}$$

one can determine the ANC.

So, the modified approach makes use of two experiments: the first to fix the ANC, the second to determine the SF consistent with that ANC. In present experiments and with the new generation of rare isotope facilities, ANCs can be determined with 5% accuracy. Since the determination of the SF comes from the internal region, the second experiment needs to be performed at a beam energy for which the contribution from the interior is significant. The higher the contribution of the internal region, the stronger the dependence on $b_{n_r l j}$ in $R^{DW}(b_{n_r l j})$ and the smaller the uncertainty of the extracted SF, although a balance needs to be found since large interior contributions may not be well describe by DWBA. The DWBA differential cross section near the main peak of the angular distribution and, correspondingly, $R^{DW}(b_{n_r l j})$ are the functionals of the single-particle ANC $b_{n_r l j}$. Note that one given $b_{n_r l j}$ can be produced by an infinite number of single-particle potentials, local and non-local. However, the dependence of $d\sigma^{DW}/d\Omega$ or $R^{DW}(b_{n_r l j})$ on the shape of the single-particle potential for a fixed $b_{n_r l j}$ is minor. To explain this very important result we present in Fig. 1 the dependence of the neutron single particle $1p_{1/2}$ bound state wave of the ^{13}C(gr. st.) on the geometrical parameters of the bound state Woods-Saxon potential by fixing the single particle ANC $b_{1p1/2} = 1.78$ fm$^{-1/2}$ and the spin-orbit term $V_{so} = 6.39$ MeV. A similar conclusion can also be drawn from Fig. 2, where the dependence of the neutron single particle $2d_{1/2}$ bound state wave of the ^{83}Ge(gr. st.) on the geometrical

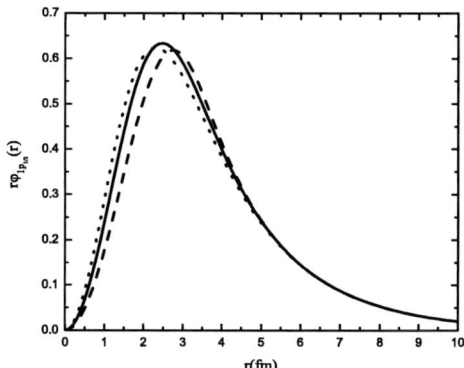

FIGURE 1. The neutron single particle $1p_{1/2}$ bound state wave function ^{13}C(gr. st.) for a fixed single particle ANC $b_{1p1/2} = 1.78$ fm$^{-1/2}$: $r_0 = 1.20$ fm, $a = 0.60$ fm- solid line; $r_0 = 1.478$ fm, $a = 0.30$ fm- dashed line; $r_0 = 0.80$ fm, $a = 0.853$ fm- dotted line.

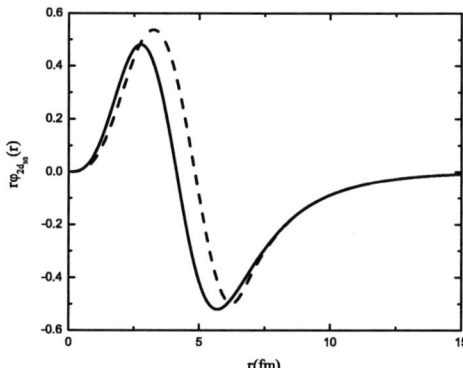

FIGURE 2. The neutron single particle $2d_{5/2}$ bound state wave function of ^{83}Ge(gr. st.) for a fixed single particle ANC $b_{2d5/2} = -3.1$ fm$^{-1/2}$: $r_0 = 1.30$ fm, $a = 0.65$ fm- solid line; $r_0 = 1.537$ fm, $a = 0.15$ fm- dashed line.

parameters of the bound state Woods-Saxon potential at fixed single particle ANC $b_{2d5/2} = -3.1$ fm$^{-1/2}$ and the spin-orbit term $V_{so} = 2.636$ MeV is shown. As we can see for $b_{n_r l j} = const$ and varying geometrical parameters r_0 and a of the bound state Woods-Saxon potential the bound state wave function changes only in the region of small r, which does not contribute to the differential cross section of the transfer reaction in the region of the main peak. It explains why the differential cross section of the transfer reaction is eventually a functional of the single particle ANC $b_{n_r l j}$ rather than the individual geometrical parameters r_0 and a of the bound state potential. However, if condition $,b_{n_r l j} = const$ is not satisfied, then the absolute value and the angular dependence of the transfer differential cross section change significantly with variation of r_0 and a. In DWBA, the differential cross section near the main peak of the angular distribution is essentially a functional of the single-particle ANC $b_{n_r l j}$, thus the extracted SF in the modified method has a very weak dependence on the details of the single-particle potential.

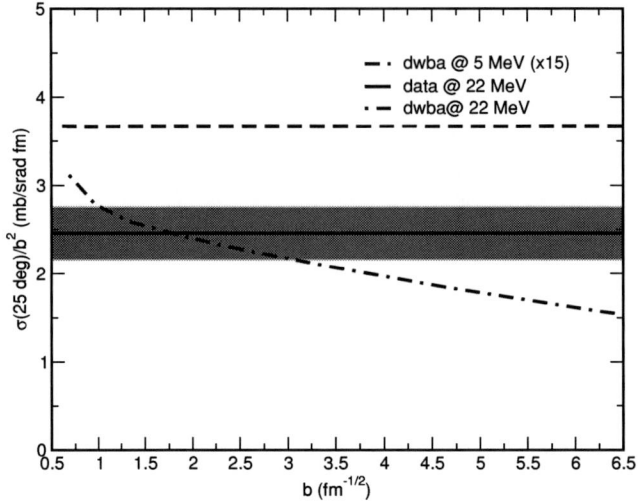

FIGURE 3. $R(b) = \sigma(25°)/b^2$ function for the ^{208}Pb(d,p)^{209}Pb(g.s.) reaction at 22 MeV: experimental value (solid line), experimental error bar (shaded area) and the DWBA prediction (dot-dashed).

APPLICATIONS

We illustrate the method presenting four different applications: i) ^{209}Pb, ii) ^{8}Li, iii) ^{13}C and iv) ^{85}Se. We will drop the subscripts for b for simplicity.

$$^{208}Pb(d,p)^{209}Pb$$

Let us consider the reaction ^{208}Pb(d,p)^{209}Pb from [13]. Although the ANC for $<^{209}$Pb$|^{208}$Pb$>$ is not published, it can be determined from the sub-Coulomb reaction [15] ^{208}Pb(^{13}C,^{12}C)^{209}Pb as the other vertex $<^{13}C|^{12}C>$ is well known [16]. Sub-Coulomb reactions are extremely peripheral and insensitive to details of the optical potentials. For this reason they present an excellent probe for extracting the ANC accurately. From [15] we obtain an ANC $C^2_{g9/2} = 2.15(0.16)$ fm^{-1} for ^{209}Pb. Then using ^{208}Pb(d,p)^{209}Pb data at $E_d = 22$ MeV [13] we obtain $R^{exp} = 2.46(0.31)$ fm mb/srad, where the error bar is calculated based on both, the ANC and the cross sections errors, taken as independent. The experimental data in [13] has 10% uncertainty but is taken only down to $\theta_{cm} = 35°$ whereas the peak of the DWBA distribution is at $\theta_{cm} = 25°$. We extrapolate the data based on the shape predicted by DWBA and include a 10% error in the cross section to account for this difference. Measurements at $25°$ could improve the error bar in R^{exp} considerably. We next perform a series of finite range DWBA calculations for ^{208}Pb(d,p)^{209}Pb ($E_d = 22$ MeV), using the optical potentials from [17]. The adiabatic prescription [18] was used to take into account deuteron breakup which is important for this reaction. The Reid-soft-core potential was used for the deuteron wavefunction, as well as in all other examples. For illustration purposes, we use a Woods Saxon well to generate the ^{208}Pb$+n$ single-particle wavefunctions and obtain a range of the single-particle ANCs b by varying the single particle parameters (r_0, a) and adjusting the depth to reproduce the correct binding for the $2g_{9/2}$ in each case. We use the same s.o. strength as that in [15] although the s.o. strength does not affect the final result.

The results of our calculations R^{DW} (dot-dashed line) and the experimental value R^{exp} (solid line and shaded area) are presented, as a function of b, in Fig. 3. From R^{exp} one finds $b = 1.82$ fm$^{-1/2}$ and $S = 0.65$. It is worth noting that in the standard approach standard parameters $(r_0, a) = (1.2, 0.6)$ fm produce $b = 1.34$ fm$^{-1/2}$. The direct comparison of the DWBA cross section using $(r_0, a) = (1.2, 0.6)$ fm, with the data, give S=0.866 and consequently, $C^2 = 1.56$ fm^{-1}, beyond the experimental range. As pointed out before, in the standard approach the SF is determined at the cost of an artificial ANC. Since the ANC determines the normalization of the external (peripheral) part of the transfer cross section, in the standard approach the SF is determined on the expense of the 38% decrease of the external part contribution.

The beam energy of 22 MeV is above the Coulomb barrier, thus the reaction is not peripheral. This can be seen in

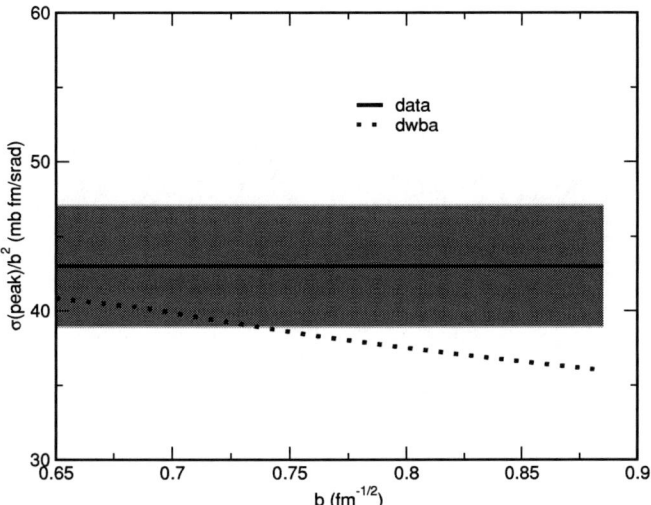

FIGURE 4. $R(b) = \sigma(peak)/b^2$ function for the ^{7}Li$(d,p)^8$Li(g.s.) reaction at 12 MeV: experimental value (solid line), experimental error bar (shaded area) and the DWBA prediction (dot-dashed).

Fig. 3 through the slope of the dot-dashed curve. In fact for this particular energy, the interior contribution is around 10%. The uncertainty in $b \in [1.1, 3.1]$ fm$^{-1/2}$ propagates into a large uncertainty in $S \in [0.35, 1.0]$. This is due to the fact that at higher b the relative contribution of the nuclear exterior increases. The smaller the contribution from the interior, the smaller the accuracy with which the SF can be determined.

Also in Fig. 3 we show the results for R^{DW} corresponding to the calculation at $E_d = 5$ MeV (dashed line). This is to illustrate that, at sub-Coulomb energies, the reaction becomes completely peripheral and the dependence on b disappears. Measurements at these energies could provide $C^2_{g9/2}$ with accuracy $< 5\%$. In addition, measurements at higher energy (> 30 MeV) would increase the slope of $R^{DW}(b)$ and decrease further the error on the extracted SF.

$^7Li(d,p)^8Li$

In Fig. 4 we show an identical analysis for ^{7}Li$(d,p)^8$Li. A series of finite range DWBA ^{7}Li$(d,p)^8$Li calculations were performed for $E_d = 12$ MeV, varying the single particle parameters (r_0, a) to span a range of b values. We neglect the spin orbit of the ^{7}Li$-n$ and calculate the differential cross section at the peak of the distribution as a function of the single particle ANC b_{1p}. Optical potentials are taken from [9]. We compare the theoretical predictions for $R^{DW} = \frac{\sigma_{peak}(b_{1p})}{b_{1p}^2}$ (dashed line) with $R^{exp} = \frac{\sigma(peak)}{C^2_{1,3/2}+C^2_{1,1/2}}$. For R^{exp} we use the cross section data from [9] at the most forward angle, and the ANC from [19]. We obtain $R^{exp} = 43.0(4.1)$ fm mb/srad (represented in Fig. 4). Taking the average value for R^{exp} would be consistent with $b_{1p} \approx 0.65$ fm$^{-1/2}$ and would imply $S_{g.s.} \approx 1.0$. The standard DWBA analysis in [9] provides $S_{g.s.} = 0.87$ both consistent with the predictions from VMC [21] $S_1 = S_{p1/2} + S_{p3/2} = 0.923$. However, given the error bar for R_{exp}, the uncertainty is amplified due the weak dependence on b_{1p} and the only safe conclusion is that $S_{g.s.} > 0.8$. In addition, the dependence on optical potentials and any higher order effects would need to be considered.

$^{12}C(d,p)^{13}C$

Another standard case is the ^{12}C$(d,p)^{13}$C reaction, for which many data sets are conveniently compiled in a recent publication [11]. We studied three cases (8.9 MeV, 30 MeV and 51 MeV), using the same JLM optical potentials as [11]. We perform a series of finite range DWBA calculations varying the $1p_{1/2}$ ^{12}C$-n$ single particle parameters, in order to obtain $R^{DW}(b)$ as described before. Results for the less peripheral case (51 MeV) are plotted in Fig. 5.

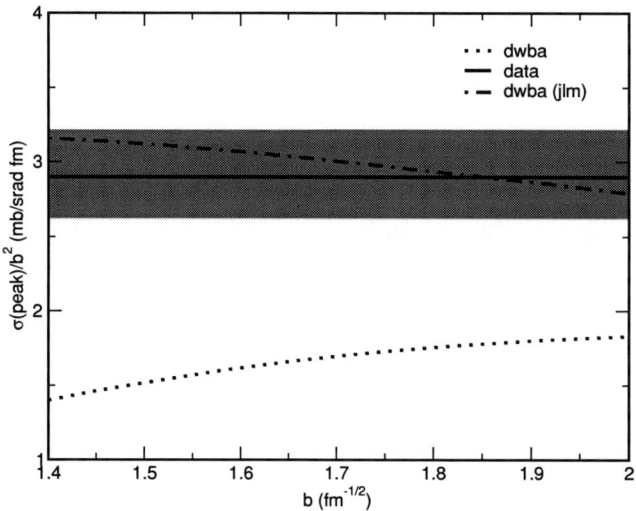

FIGURE 5. $R(b) = \sigma(peak)/b^2$ function for the ^{12}C(d,p)^{13}C(g.s.) reaction at 51 MeV: experimental value (solid line), experimental error bar (shaded area) and the DWBA prediction (adiabatic deuteron-^{12}C optical potential- dot-dashed line, deuteron-^{12}C optical potential from the elastic scattering fit- dotted line).

We take the data from [11] and the ANC from [16], to obtain $R^{exp} = \frac{\sigma(2.5°)}{C^2_{1,1/2}} = 2.92(0.35)$ fm mb/srad. An $S = 0.66$ (shell model) would require $b = 1.89$ which is contained in our results. However, such a conclusion is misleading. Fig. 5 shows that even for this relatively large energy, the dependence of R^{DW} on b is weak. Consequently, it is not possible to extract a SF.

It was pointed out in [11] that the deuteron breakup is important for this reaction and should be taken into account. To emphasize this fact, we compare our results using the adiabatic deuteron potential [18] from [11] (dot-dashed line in Fig. 5) with those obtained using an optical potential fitted to the deuteron elastic scattering (dotted line in Fig. 5). The disagreement is very large. Interestingly, the method here described is also able to detect inadequate optical potential parameterizations.

$$^{84}Se(d,p)^{85}Se$$

Oak-Ridge has developed a program to measure a series of inverse kinematics (d,p) reactions for nuclei on the neutron dripline [20]. As one of the nuclei in the program is ^{85}Se, we have performed exploratory calculations for ^{84}Se(d,p)^{85}Se. We take global parameterizations for the optical potentials [17] and perform a series of calculations varying the single particle parameters. We compare the dependence of R^{DW} on b for $E_d = 4, 15, 20$, and 50 MeV. We verify that, expectedly, the dependence on b increases with beam energy. We find that Oak-Ridge energies (10 MeV/A) are adequate to determine ANCs but not SFs. However, a facility that allows for the production of ^{84}Se at $E > 25$ MeV/A (such as NSCL-MSU, GANIL or RIKEN) could provide accurate spectroscopic information.

SUMMARY

An alternative method to extract SFs, taking into account the sensitivity of the transfer data to the interior part of the overlap function and combining that information with the ANC, is presented. This contribution focuses on one, and one only, source of uncertainty in the analysis, which up to now has been the most elusive part of the analysis: the single particle parameters. There are still: the optical model dependence and higher order effects. These need to be considered in addition to the single particle dependence here discussed. Also, if the single particle parameters consistent with the ANC are very different from the standard values, the transfer angular distribution may change, and the chosen optical potential may no longer provide a good description. These issues need to be addressed and work along these lines is underway. The combined method still assumes that the interior part has a Woods Saxon

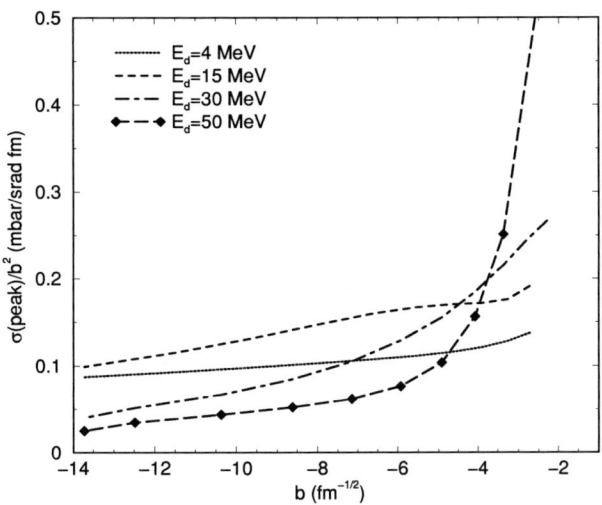

FIGURE 6. The DWBA predicition for the $R(b) = \sigma(peak)/b^2$ function for the ^{84}Se(d,p)^{85}Se(g.s.) reaction at several beam energies.

single particle wavefunction shape. This has been corroborated by recent Green's Function Monte Carlo calculations on light nuclei [21]. The transfer differential cross section is a functional essentially of the single particle ANC (b). This b itself is a function of many parameters (geometrical parameters of the local potential, and other parameters of the non-local part). The condition b=constant strongly limits the variation of the transfer cross section, making them practically indistinguishable for different single particle potentials, both local and non-local.

Transfer data can only become useful within this method if it has a significant contribution from the interior, and is well described through a one-step DWBA formalism. The balance between these two conditions is not a trivial one. By reducing the error bars in both the measured transfer cross section and the ANC, this prescription determines the single particle asymptotics and from it, a SF with reduced uncertainty. The ANC needs to be determined independently; it can be pinned down accurately with the same transfer reaction at sub-Coulomb energies or using heavy-ion induced reactions, both safely peripheral.

Introducing the ANC into the analysis sets the problem into its correct physical ground: i) It guarantees the correct normalization of the external reaction amplitude, which typically gives the dominant contribution to the transfer reaction. Only then is the analysis well rooted; ii) It enables the spectroscopic factor to be extracted from the contribution of the nuclear interior, as it should be. The great weakness of the conventional analysis is that the spectroscopic factor is often determined from the external part.

As we have shown, the new method determines the geometrical parameters of the bound-state potential (in fact, the single particle ANC). Then, disagreements can no longer be fixed by arbitrarily changing these parameters. In this way: i) it provides a critical test of the standard approach ii) and it is a check on other reaction theory uncertainties (DWBA method and optical potentials).

We have shown four examples. The first is a standard single particle case (Pb209), the second (Li8) is a case for which a very accurate ANC is available as well as ab-initio SF, the third case (C13) was focus of a recent study, and finally in the fourth example we make projections that are relevant for experimental programs in various world wide RIB facilities (Se85). Even though we show how it works with the present data, in the end this is a suggestion for future experimental work and the optimum application of the method will require specific measurements that include backward angles for sub-barrier and forward angles for above-barrier transfer cross sections, as well as a reliable determination of optical potentials.

The new method suggests a way to check the consistency of the existing direct reaction theory based on the DWBA. The introduction of the ANC into the analysis fixes the contribution of the peripheral transfer reaction amplitude. This is not done in the standard analysis where the peripheral contribution is artificially varied. The introduction of the ANC immediately decreases the degrees of freedom and demands a higher accuracy of calculation of the internal part of the reaction amplitude and the adopted optical potential. It might well be that DWBA will fail when the new approach is applied. As a test of DWBA one can consider a case where the ANC and the SF are well determined.

Incorporating both quantities into the transfer analysis will demonstrate the consistency of the theory. In particular, its ability to reproduce the experimental angular distributions and the absolute value. It will also be a crucial test for the optical model.

This same method can equally be used for transfer to excited states. In addition, these same ideas can be extended to other reactions, in particular breakup reactions which also have an impact on Astrophysics.

ACKNOWLEDGMENTS

We thank Xiandong Liu for providing the JLM potentials and Prof. Goldberg for useful comments. This work has been partially supported by the NSCL at Michigan State University, U. S. DOE under Grant No. DE-FG03-93ER40773, by NSF Award No. PHY-0140343 and by Fundação para a Ciência e a Tecnologia (F.C.T.) of Portugal, under the grant POCTIC/36282/99.

REFERENCES

1. A.M. Mukhamedzhanov and F.M. Nunes, *Phys. Rev. C in press*.
2. B.A. Brown *et al.*, *Phys. Rev. C*, **65**, 061601-061604(R) (2002).
3. L. Lapikas, J. Wesseling, R.B. Wiringa, Phys. Rev. Lett. 82 (1999) 4404.
4. A. Navin *et al.*, *Phys. Rev. Lett.*, **81**, 5089-5092 (1998).
5. G.J. Kramer, H.P. Blok and L. Lapikas, *Nucl. Phys. A*, **679**, 267-286 (2001).
6. S. A. Goncharov *et al.*, *Yad. Fiz.*, **35**, 662-674 (1982) (*Sov. J. Nucl. Phys.*, **35**, 383-390 (1982)).
7. N. Austern, Direct Nuclear Reaction Theories (Wiley, New York, 1970).
8. J. Al-Khalili and F. Nunes, *J. Phys. G: Nucl. Part. Phys.*, **29**, R89-132 29 (2003) R89. (2003).
9. J. P. Schiffer *et al.*, *Phys. Rev.*, **164**, 1274-1284 (1967).
10. C. Iliadis, M. Wiescher, *Phys. Rev. C* **69**, 064305-064317 (2004).
11. X.D. Liu *et al.*, *Phys. Rev. C* **69**, 064313-064317 (2004).
12. H. Ohnuma *et al.*, *Nucl. Phys. A* **448**, 205-220 (1986).
13. K. Hirota *et al.*, *Nucl. Phys. A* **628**, 547-579 (1998).
14. L. D. Blokhintsev, I. Borbely, E. I. Dolinskii, *Sov. J. Part. Nucl.* **8**, 485 (1977).
15. M.A Franey, J.S. Liley, W.R. Phillips, *Nucl. Phys. A* **324**, 193-220 (1979).
16. A. M. Mukhamedzhanov and N. K. Timofeyuk, *Yad. Fiz.***51**, 679-685 (1990) (*Sov. J. Nucl. Phys.* **51**, 431-437 (1990)).
17. C. M. Perey and F. G. Perey, *At. Data Nucl. Data Tables* **17**, 1 (1976).
18. R.C. Johnson and P.J.R. Soper, *Phys. Rev. C* **1**, 976-990 (1970).
19. L. Trache et al., *Phys. Rev. C* **67**, 062801-062805(R) (2003).
20. K.L. Jones, private communication 2004.
21. B. Wiringa, private communication, Argonne April 2004.

Ground state neutron spectroscopic factors for Z=3-24 isotopes from transfer reactions

M.B. Tsang[1], H.C. Lee[1,2]

[1]National Superconducting Cyclotron Laboratory and Department of Physics and Astronomy, Michigan State University, East Lansing, MI 48824
[2]Physics Department, Chinese University of Hong Kong, Shatin, Hong Kong, China

Abstract

We have extracted the ground state to ground state neutron spectroscopic factors for 79 nuclei ranging in Z from 3 to 24 by analyzing the past angular distribution measurements from (d,p) and (p,d) reactions in a systematic and consistent manner using a standard parameter set. For the Ca isotopes from ^{40}Ca to ^{48}Ca, the spectroscopic factors follow the independent particle model predictions. For the 60 nuclei where modern shell model calculations are available, the experimental spectroscopic factors for most nuclei agree with the calculations to within 20%.

As discussed in this workshop, spectroscopic factors measured with (e,e'p) on a few nuclei from ^{7}Li to ^{208}Pb near the closed shells show that the measured values are lower than predictions from independent particle models by about 40% [1]. Recent measurements of spectroscopic factors from single-nucleon "knock-out" reactions with radioactive beams and stable nuclei show increasing quenching of the spectroscopic factor values with nucleon separation energy [2]. However, there is no systematic analysis of the transfer reaction data over a wide range of nuclei to quantify any trends even though abundant amount of transfer reaction data has been collected in the past four decades.

We analyze angular distribution measurements from the (d, p) and (p,d) reactions found in the literatures on targets ranging from Li to Cr isotopes. The experimental spectroscopic factor is defined to be the ratio of the measured differential cross-section divided by the calculated cross section using a reaction model. To describe the transfer reaction, a fast one-step process using the Distorted Born Wave Approximation (DWBA) is assumed [3]. The transfer cross-sections are calculated within the Johnson-Soper adiabatic approximation to the neutron, proton, and target three-body system, thus including the effects of deuteron breakup in the field of the target [4]. Both the entrance and exit channels are dominated by elastic scatterings, which are described by optical model potentials. The phenomenological nucleon nucleus optical model potential, Chapel-Hill 89 [5] is adopted for proton and neutron. The deuteron optical model potentials are obtained from the folding of the n and p potentials. All calculations include the local energy approximation (LEA) for finite range effects using the zero-range strength (D_o^2=150006.25) and range (β=0.7457) parameters of the Reid soft-core $^3S_1$-$^3D_1$ neutron-proton interaction. Nonlocality corrections with range parameters of 0.85 and 0.54 are included in the proton and deuteron channels, respectively. For the neutron bound wave function, the neutron potential is assumed to have a Woods-Saxon shape with fixed radius parameter r_o=1.25 fm and diffuseness parameters a_o=0.65 fm. The potential depth is adjusted to the experimental binding energy. For the present work, we use the Surrey version of TWOFNR, a direct reaction model code, which contains all the input options discussed above [6]. We fit mainly the first peak and only one set of parameters as described above is used throughout the analysis of many nuclei. The analysis in the present work follow closely ref. [7] in which the extracted spectroscopic factors of the $p_{1/2}$ neutron coupled to the ^{12}C are consistent to within 15% for deuteron energy ranging from 12-60 MeV.

In ref [3], a simple relationship between the number of valence nucleons (n) and the spectroscopic factors (S) is derived based on the independent particle model [IPM] with pairing.

$$S = n \text{ for n=even;} \quad S = 1 - \frac{n-1}{2j+1} \text{ for n=odd} \tag{1}$$

The spectroscopic factors for j=7/2 which corresponds to the $f_{7/2}$ orbits of the $^{41-48}$Ca isotopes are given on Pg. 291 in Austern's book [3]. A textbook example for these spectroscopic factors is shown in Figure 1. The IPM predictions shown as thin blue bars as a function of the mass number A for the Ca isotopes are compared directly with the extracted neutron spectroscopic factors as represented by the red stars with error bars. The agreement is remarkable except for the ^{49}Ca isotope where the $2p_{3/2}$ neutron SF value is about 30% lower than the prediction. The excellent agreement suggests that in the simple shell model, the double magic ^{40}Ca isotope with closed proton and neutron shells can be well described as a spherical core and that the valence neutron orbitals in the $f_{7/2}$ shell for the other isotopes are good single particle states.

Despite the success of the IPM in the description of the Ca isotopes, most nuclei cannot be described by such simple model, which assumes no interactions between the nucleons and the core except pairing. Residual interactions between nucleons are included in the large basis shell model [8]. However, most shell model calculations are limited by the computation time and the size of the model space. Using one of the most up-to-date shell model programs, Oxbash [9], the ground state neutron spectroscopic factors for 60 nuclei, most of which with Z≤20 have been calculated. The uncertainties in the calculations are about 10-20% [8]. For the Ca isotopes, there is close agreement between shell model calculations (thick green bars) and IPM values (thin blue bars) as shown in Figure 1.

Figure 2 compares the experimental extracted spectroscopic factors (y axis) and the predicted shell model SF values (x axis) for 51 nuclei. In this figure, we exclude the deformed nuclei ^{24}Mg, Li, Ne and F isotopes. This latter group also includes nuclei with small calculated or experimental SF values. In general small spectroscopic factors, an indication of substantial configuration mixing, are more susceptible to uncertainties in the calculations. Furthermore, large experimental uncertainties arise from small cross-sections. Each element in Fig. 2 is represented by one symbol. Open symbols represent odd Z elements and the closed symbols represent even Z elements. The three different colors of green ($3 \leq Z \leq 8$), blue ($9 \leq Z \leq 18$) and red ($19 \leq Z \leq 22$) represent approximately the p-shell, sd-shell and sdf shell nuclei, respectively. The solid line indicates perfect agreement. For most nuclei, the agreements between data and shell model predictions are within 20% as indicated by the two dashed lines. Within the experimental and calculation uncertainties, we do not see any systematic dependence of quenching of the spectroscopic factors. For ^{40}Ca, the ground state spectroscopic factors of the valence protons and neutrons should be similar. If the spectroscopic factors we obtain from the transfer reactions are absolute values, then our result is at odds with the quenching of the ground state proton spectroscopic factor by 35% as observed in (e,e'p) reactions.

Figure 3 shows the extracted spectroscopic factor normalized to the shell model values as a function of neutron separation energy. Over the range of 0.5 to 19 MeV, we do not see increasing quenching of the spectroscopic factors with n-separation energy as observed in knockout reactions [2]. Rather, there seems to be an opposite trend. There are indications that the spectroscopic factors for the neutron rich nuclei with n-separation energy less than 5 MeV are suppressed.

For discussion purpose, we raise the following questions:

1. The simple analysis procedure employed in the present work yields spectroscopic factors that are consistent with the shell model. Is this a result from the "tuning" of the input parameters in the past to the expected sum rule of one? As the phenomenological interactions used in shell models explore the same nuclear structure physics as transfer reactions, it maybe no surprise that the predicted shell model spectroscopic factors are the same as the experimental spectroscopic factors extracted using the transfer reaction models.

2. Both the transfer reaction and the knockout reactions are more sensitive to the surface of the wavefunctions, it is rather peculiar that they do not yield the same

spectroscopic factor values for the ^{16}O and ^{12}C nuclei. Are they measuring the same quantities? In addition, the dependence of the spectroscopic factor on separation energy is also different for both types of reactions. What is the theoretical basis for the dependence on separation energy?

3. The electron probes the interior of the wave function and therefore sensitive to the nucleon-nucleon correlations. However the knockout reactions probe the tail of the wavefunctions, it is rather puzzling that both reactions yield the same quenching of the spectroscopic factors. Within experimental uncertainties, there is no separation energy dependence of the quenching factors in the (e,e'p) reactions.

In summary, our analysis shows that spectroscopic factors extracted over a wide range of nuclei agree with modern day shell model predictions. As the spectroscopic factor values depend on the input potentials used in the reaction model calculations [9,10], it is possible to find other potentials or parameter sets which will give an overall quenching of the spectroscopic factors so that the result agrees with (e,e'p) analysis. The challenge is how to choose such potentials, preferably, in an ab initio manner to address the short range nucleon-nucleon interactions in the nuclei.

This work is supported by the National Science Foundation under Grant No. PHY-01-10253 and the Summer Undergraduate Research Experience (SURE) program sponsored by the Hong Kong Chinese University.

References:
1. GJ Kramer, HP Blok and L. **Lapikas**, Nucl. Phys. A. 679, 267 (2001) and references therein
2. A. Gade et. al., Phys. Rev. Lett. **93**, 042501 (2004), and references therein.
[3] N. Austern, Direct Nuclear Reaction Theories, John Wiley & Sons, New York, 1970.
[4] R.C. Johnson and P.J.R. Soper, Phys. Rev. **C 1,** 976 (1970).
[5] R.L.Varner, W.J. Thompson, T.L. McAbee, E.J. Ludwig and T.B. Clegg, Phys. Rep. **201,** 57 (1991).
[6] M. Igarashi, M. Toyoma and N. Kishida, Computer Program TWOFNR (Surrey University version).
[7] X. D. Liu et. al., Phys. Rev. C69, 064313 (2004).

[8] Progress in Particle and Nuclear Physics, 47, 517 (2001).

[9] B. A. Brown et al., Computer program, http://www.nscl.msu.edu/~brown/resources/oxbash-augsut-2004.pdf

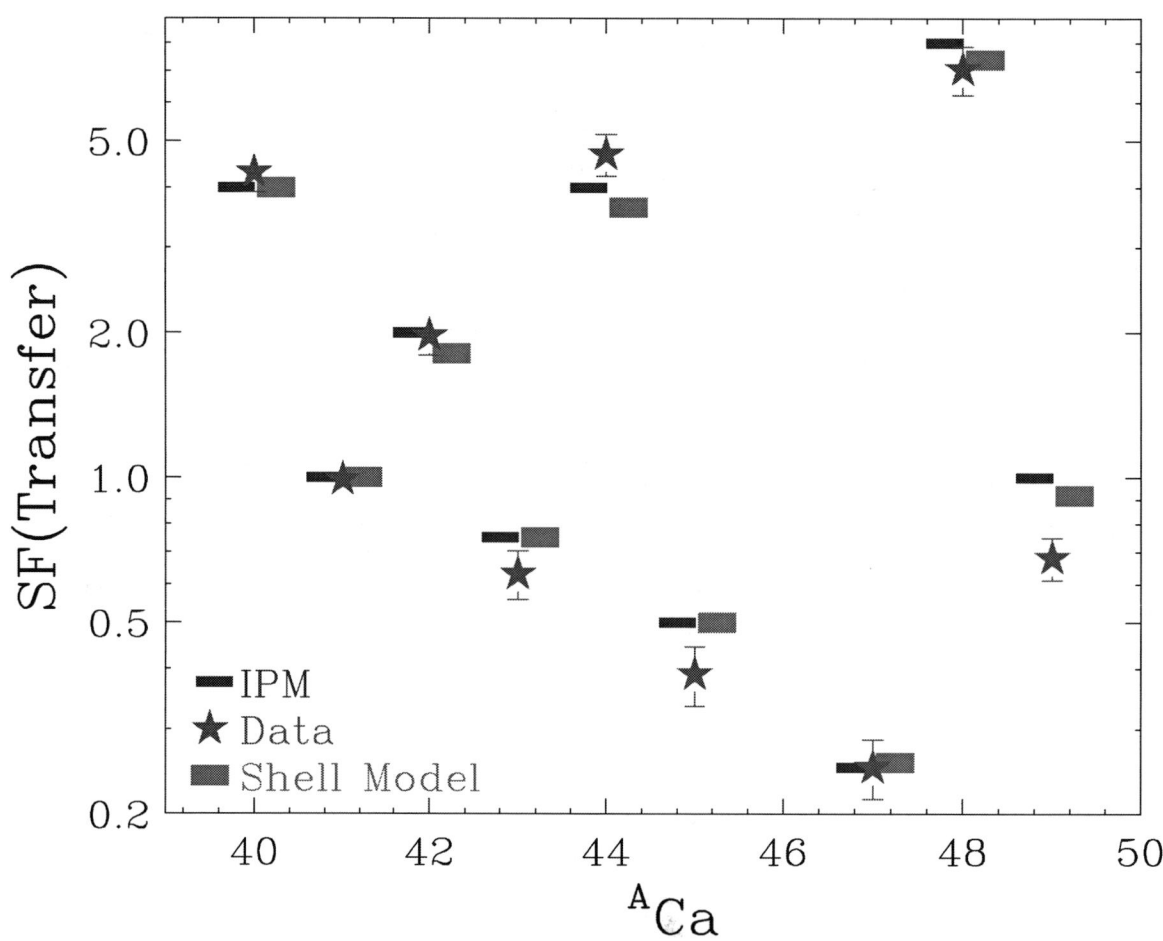

Figure 1 : (Color online) Ground state neutron spectroscopic factors for Calcium isotopes from ^{40}Ca to ^{49}Ca, star symbols represent values extracted from present analysis. Blue lines are IPM values and thick green bars represent predictions from Oxbash.

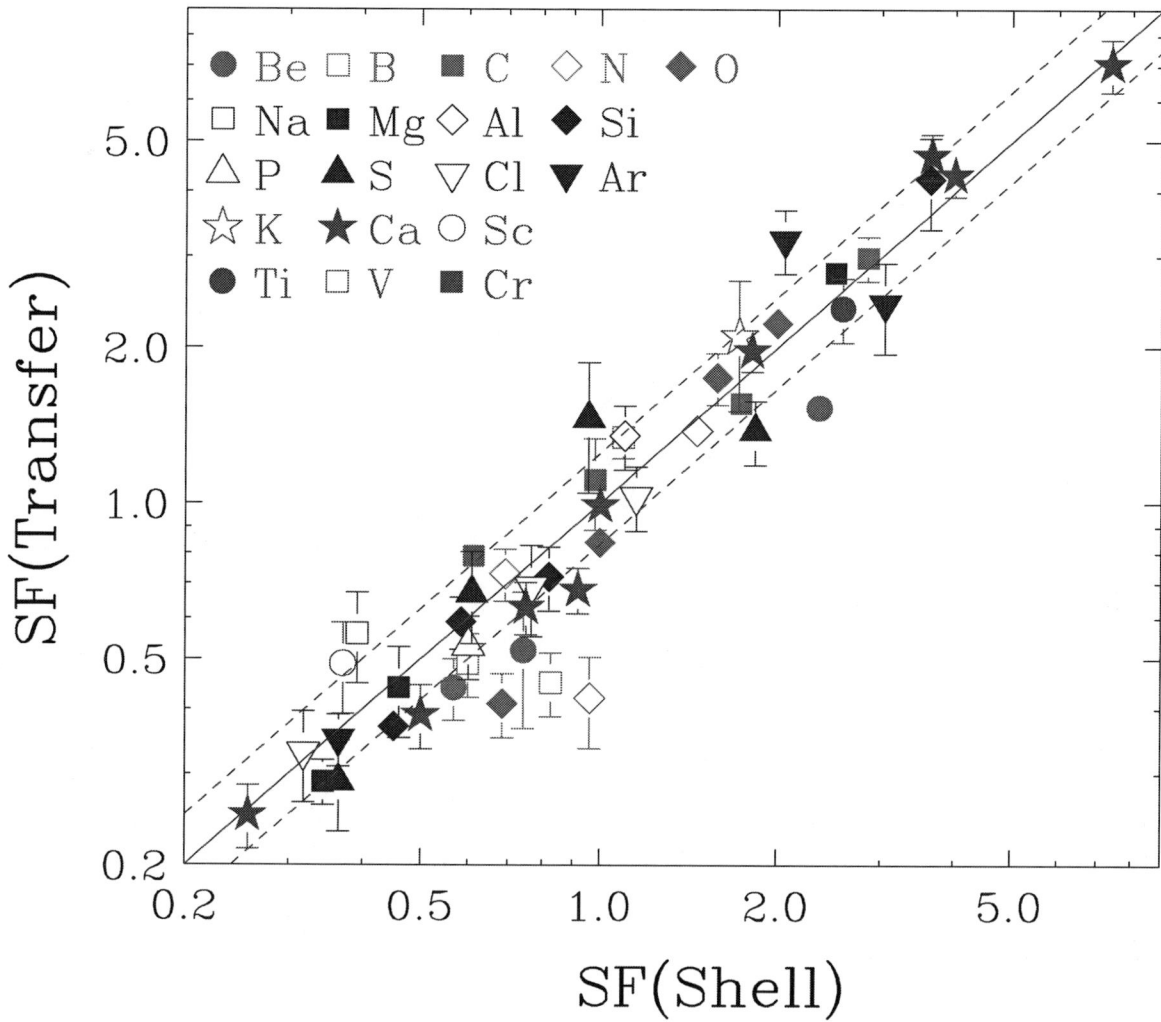

Figure 2: (Color online) Comparison of experimental spectroscopic factors to predictions from the shell model program Oxbach. The solid line indicates perfect agreement between the two values. Solid symbols represent even Z elements and open symbols depict odd Z elements. The green color show elements from Z=3 to 8, mostly p-shell nuclei. The blue symbols represent Z=9 to 18, mostly sd shell nuclei. Finally the red symbols represent Z=19 to 24 isotopes, mostly sdf shell nuclei.

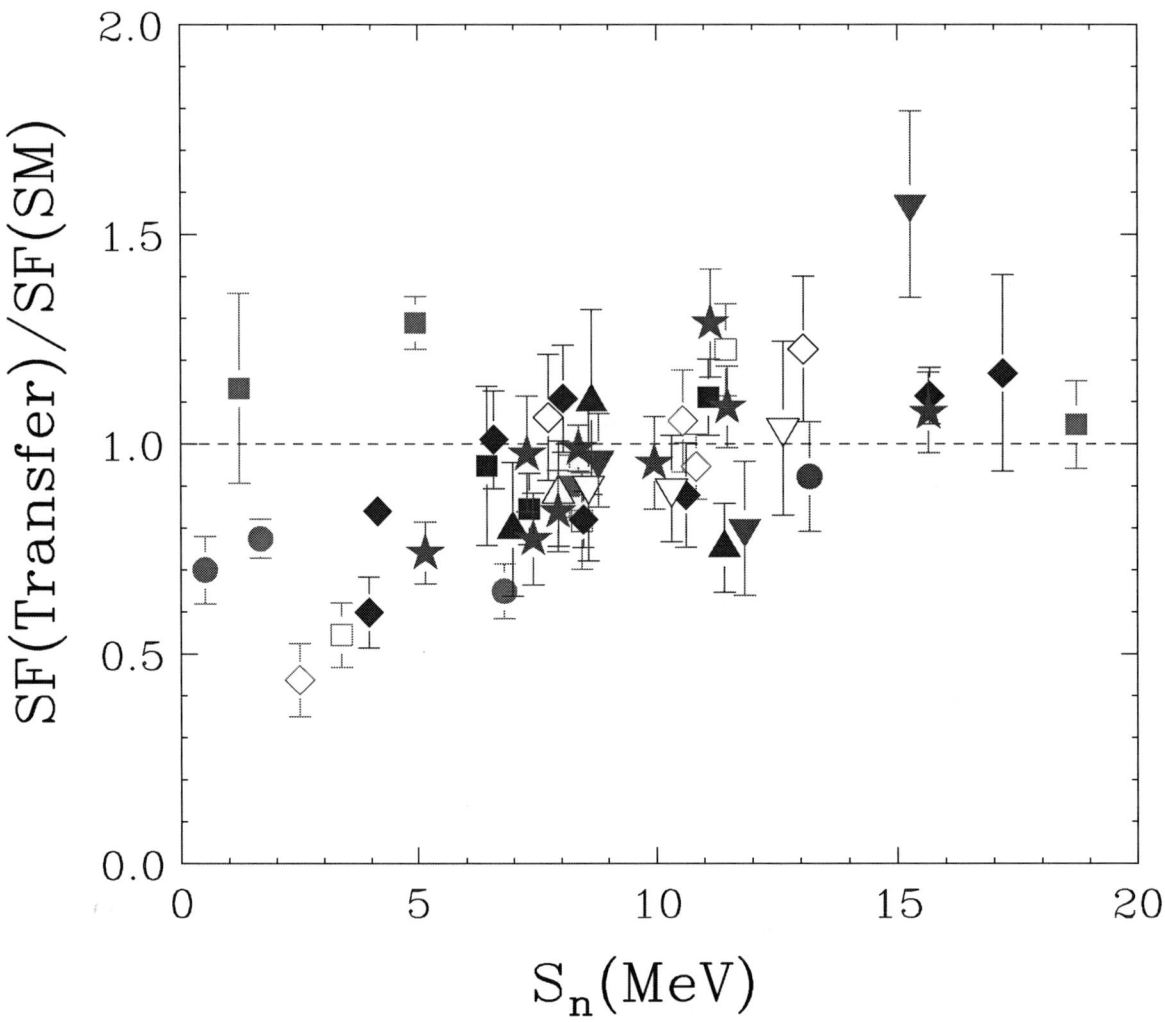

Figure 3 : (Color online) Neutron separation dependence of the ratios of the spectroscopic factors to shell model values. See Figure 2 for explanation of symbols.

Contribution to the Wednesday afternoon discussion on spectroscopic factors

C. Barbieri

TRIUMF, 4004 Wesbrook Mall, Vancouver, British Columbia, Canada V6T 2A3

Abstract. This part of the discussion would like to review the concept of spectroscopic factors and how they relate to measured cross sections and nuclear correlations. A profound knowledge of how correlations affect the spectral function can help to better understand transfer reactions. Nowadays, we have a fairly complete picture for protons in stable nuclei but a lot remain to be learned regarding exotic species.

Keywords: Spectroscopic factors, spectral function, nucleon transfer, nuclear correlations
PACS:

ON THE MEANING AND USEFULNESS OF SPECTROSCOPIC FACTORS

Several contributions to the present meeting emphasized that experimental cross sections (for transfer of a nucleon) are on average lower than what is predicted by the independent particle model (IPM). Whence, the need for introducing spectroscopic factors that account for (and provide a measure of) the occupation of the relevant orbitals. However, the attempts to extract experimentally such information have produced discordant results—and a quite a bit of confusion—over the years. Still, as not all of the relevant astrophysical reactions can be determined from measurement, it is imperative to understand the mechanisms of such quenching and how to provide reliable predictions of it.

From the theoretical point of view, there exist a precise definition of spectroscopic factors as the norm of the one-body overlap functions,

$$Z \equiv \int d^3r \, |\psi_n(r)|^2 \, , \qquad \text{where} \qquad \psi_n(r) = \langle \Psi_n^{A-1} | a_\mathbf{r} | \Psi^A \rangle \qquad (1)$$

and $a_\mathbf{r}$ is the operator destroying a particle at position $\mathbf{r}$. If $|\Psi_n^{A-1}\rangle$ and $|\Psi^A\rangle$ were Slater determinants, Z would be equal to 1 and $\psi_n(r)$ would be the mean field orbital occupied only in $|\Psi^A\rangle$. However, in Eqs. (1) they represent *exact* eigenstates of the nuclear Hamiltonian in the complete (A-1)- and A-body Hilbert spaces. In this situation $\psi_n(r)$ is the solution of the microscopic optical potential derived from Feshbach theory and the corresponding orbital is depleted to $Z < Z_{IPM}$.

We note that once the nuclear Hamiltonian has been chosen, Eqs. (1) uniquely define the spectroscopic factors. However, one must remember that different models of the NN interaction have different off-shell behaviors while they reproduce the same scattering data. This poses an issue of consistency between the treatment of the initial state (hence the spectroscopic factor), the final state interactions and the interactions with the probe: all these contributions come together to generate the observed cross section and should, in principle, be derived from the same microscopic Hamiltonian [1]—also using consistent electromagnetic currents, in case of electron scattering. This is a formidable problem in many-body nuclear physics, hence one is often forced to model scattering and bound states in terms of Wood-Saxon potentials. In general, a fit to a few Woods-Saxon parameters does not guarantee consistency and the uncertainty increases when the reaction mechanism is only sensitive to the nuclear surface (see, for example, contribution by P. Capel). There are two types of uncertainties here: one due to our incapability of solving exactly the many-body problem and one intrinsic to the unknown off-shell behavior of the nuclear force. Still, for a given realistic NN interaction, a proper *microscopic* theory should be able to describe different transfer reactions on a nucleus using the same occupation numbers. That this is possible was shown, for example, by Kramers et al. [2], where $(e,e'p)$ and $(d,^3\text{He})$ data are reconciled by analyzing both reactions with the same overlap functions. Recently, Ref. [3] achieved a coherent description of initial and final states for $^{16}\text{O}(e,e'p)$, showing that data at very different kinematics can be explained with the same spectroscopic factors—and in particular independently of the Q^2 transferred by the electron.

We note that the spectroscopic factors extracted from $(e,e'p)$ experiments with different parameterizations of initial and final states tend to agree with each other to within a 10% of the IPM value. This is possible due to the sensitivity

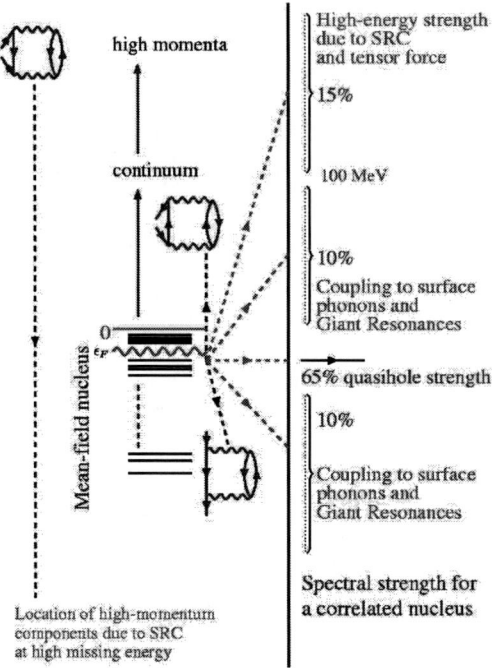

FIGURE 1. Distribution of single particle strength for a nucleus like ^{208}Pb. This applies to an orbital near to the Fermi level.

of the cross section to the overlap function in the interior of the nucleus and to the lower uncertainty on the optical potential for protons emitted at energies of 90 MeV (or larger). A particularly well studied reaction is $^{16}O(e,e'p)$, for which data exists in different kinematical regimes [4, 5] and several theoretical studies have been done, including relativistic effects. The fact that these studies give similar spectroscopic factors is an indication that the uncertainty intrinsic to the nuclear Hamiltonian is contained and that the quenching observed for the quasiparticle peaks hint at actual physics happening in the nucleus.

Nuclear correlations and spectral function

If one is to make predictions for the quenching of the occupation of a given orbital, he/she will be facing the questions of where the missing strength goes and what are the possible mechanisms (or correlations) that induce its distribution. A general picture of the correlations in nuclei can be achieved by considering the one-body spectral function [6]

$$S(r,\omega) = S^h(r,\omega) + S^p(r,\omega) = \sum_n |\langle \Psi_n^{A-1}|a_{\mathbf{r}}|\Psi^A\rangle|^2 \delta(\omega - E^A + E_n^{A-1}) + \sum_n |\langle \Psi_n^{A+1}|a_{\mathbf{r}}^\dagger|\Psi^A\rangle|^2 \delta(\omega - E_n^{A+1} + E^A) \quad (2)$$

which, for transitions to discrete states, contains the same information of Eqs. (1). The hole part of the spectral function, $S^h(r,\omega)$, is interpreted as the probability of finding a nucleon with energy ω at position **r**. Analogously, $S^p(r,\omega)$ is the probability of adding a nucleon. In spite these two quantities refer to different processes (knock out and capture), it must be remembered that they are intimately related through the Pauli principle, namely, by the sum rule

$$\int_{-\infty}^{\infty} d\omega\, S(r,\omega) = 1 \,. \quad (3)$$

This imply that small probabilities of extracting a nucleon (low occupation) leave room to large cross sections for capturing one to the same orbital.

The characteristics of the nuclear spectral function of protons have been extensively investigated in recent years for nuclei along the stability line [7]. Electron scattering experiments have shown that in closed shell mean field orbits are

TABLE 1. Dependence of theoretical spectroscopic factors (as a fraction of the IPM value) on the inclusion of various short- and long-range correlations (LRC) effects.

Spect. factors of ^{16}O	$Z_{p1/2}$	$Z_{p3/2}$
SRC only [10, 13]	∼0.90	∼0.90
SRC + LRC(RPA) [14]	0.74	0.72
SRC + LRC(RPA + sp dressing) [14]	0.77	0.72
experiment [3, 15]	0.64-0.71	0.54-0.61

indeed well defined near to the Fermi level, however, these have spectroscopic factors of the order of 60% of the IPM value [8]. Orbitals at larger energies are spread over a wider range of energies due to the coupling to 2h1p (2p1h) states and to the continuum. Here, occupation numbers are no longer a well defined concept. Still, one can approximately think in terms of occupation by integrating the strength function, Eq. (2), over a proper range of energies.

Short-range correlations (SRC) are induced by the repulsive core at small distances and by the tensor operator present in the NN potential. Different theoretical evaluations of SCR suggest that these move about 10-15% of the total strength to very high momenta and energies [9, 10]. The experimental observation of such effects has been a challenging issue for many years and could be achieved only recently at Jefferson Lab [11]. In this experiment, the high energy of the electron and the emitted proton allow for a semiclassical analysis of the reaction mechanism [12], which supports the evidence for the presence of high momentum components in the nuclear wave function.

Figure 1, gives a qualitative sketch of the strength distribution of Eq. (2) for an orbit just below the Fermi energy, together with the self-energy diagrams responsible for spreading it in the particle and hole domains. The IPM is depicted in the central column. The contributions to the reduction of spectroscopic factors are summarized in table 1 for the p shell orbitals of ^{16}O. A large part of the quenching can be described as coupling of single particle motion to collective surface phonons [14]. However, the dressing of quasiparticles partially screens these correlations. This particular nucleus has so far defeated a complete theoretical description. Ref. [14] suggested that this may be due to the poor description of the excitation spectrum in random phase approximation (RPA). Further work is being pursued to clarify this point [16].

The above experimental and theoretical works identify the global properties of protons in nuclei along the stability line[7]. However it is not clear how these features can be extrapolated toward the drip lines. Recent nuclear break up experiments at NSCL have begun providing first information for nucleons (including neutrons) in isotopes near the stability lines[17]. These point, for example, to larger occupation for halo states (see contribution by J. Tostevin). A proper understanding of this information will be crucial to push forward our knowledge of exotic nuclei.

ACKNOWLEDGMENTS

One of us (C.B.) would like to acknowledge helpful discussions with W. H. Dickhoff, L. Lapikás, B. Jennings, M. Radici, D. Rohe, I. Sick and E. Vogt. The work of C.B. is supported by the Natural Sciences and Engineering Research Council of Canada (NSERC).

REFERENCES

1. S. Boffi, C. Giusti, F. D. Pacati, and M. Radici, *Electromagnetic Response of Atomic Nuclei*, Oxford Studies in Nuclear Physics (Clarendon Press, Oxford, 1996); S. Boffi, C. Giusti, and F. D. Pacati, Phys. Rep. **226**, 1 (1993).
2. G. J. Kramers, H. P. Blok, and L. Lapikás, Nucl. Phys. **A679**, 267 (2001).
3. M. Radici, W. H. Dickhoff, and E. Roth Stoddard, Phys. Rev. C **66**, 014613 (2002).
4. M. Leuschner *et al.*, Phys. Rev. C **49**, 955 (1994).
5. K. G. Fissum *et al.* [Jefferson Lab Hall A Collaboration], Phys. Rev. C **70**, 034606 (2004)
6. A. L. Fetter and J. D. Walecka, *Quantum Theory of Many-Particle Physics* (McGraw-Hill, New York, 1971).
7. W. H. Dickhoff and C. Barbieri, Prog. Part. Nucl. Phys. **52**, 377 (2004).
8. L. Lapikás, Nucl. Phys. **A553**, 297c (1993).
9. O. Benhar, A. Fabrocini, S. Fantoni, and I. Sick, Nucl. Phys. **A579** 493 (1994).
10. H. Müther, A. Polls, and W. H. Dickhoff, Phys. Rev. C **51**, 3040 (1995).
11. D. Rohe, *et al.*, Phys. Rev. Lett. **93**, 182501 (2004).

12. C. Barbieri and L. Lapikás, Phys. Rev. C **70**, 054612 (2004).
13. D. Van Neck, M.Waroquier, A. E. L. Dieperink, S C. Pieper, and V. R. Pandharipande, Phys. Rev. C **57**, 2308 (1998).
14. C. Barbieri and W. H. Dickhoff, Phys. Rev. C **65**, 064313 (2002).
15. M. Radici, A. Meucci and W. H. Dickhoff, Eur. Phys.. J. **A 17**, 65 (2003).
16. C. Barbieri, in preparation.
17. P. G. Hansen and J. A. Tostevin, Ann. Rev. Nucl. Part. Sci. **53**, 219 (2003).

Direct Reactions with Exotic Nuclei

G. Baur* and S. Typel[†]

Institut für Kernphysik, Forschungszentrum Jülich, D-52425 Jülich, Germany
[†]*Gesellschaft für Schwerionenforschung mbH (GSI), Planckstraße 1, D-64291 Darmstadt, Germany*

Abstract. We discuss recent work on Coulomb dissociation and an effective-range theory of low-lying electromagnetic strength of halo nuclei. We propose to study Coulomb dissociation of a halo nucleus bound by a zero-range potential as a homework problem. We study the transition from stripping to bound and unbound states and point out in this context that the Trojan-Horse method is a suitable tool to investigate subthreshold resonances.

Keywords: Direct reactions, exotic nuclei, Coulomb dissociation, halo nuclei, Trojan-horse method, transition from bound to unbound states, subthreshold resonances
PACS: 24.10.-i, 24.50.+g, 25.20.Lj, 25.60.-t

1. INTRODUCTION AND OVERVIEW

With the exotic beam facilities all over the world - and more are to come - direct reaction theories are experiencing a renaissance. We report on recent work - just finished and in progress - on Coulomb dissociation of halo nuclei [1, 2] and on transfer reactions to bound and scattering states. We hope to report on further progress at the next workshop at MSU/ANL/INT/JINA/RIA or elsewhere.

Electromagnetic strength functions of halo nuclei exhibit universal features that can be described in terms of characteristic scale parameters. For a nucleus with nucleon+core structure the reduced transition probability, as determined, e.g., by Coulomb dissociation experiments (for a review see [3, 4]), shows a typical shape that depends on the nucleon separation energy and the orbital momenta in the initial and final states. The sensitivity to the final-state interaction (FSI) between the nucleon and the core can be studied systematically by varying the strength of the interaction in the continuum. In the case of neutron+core nuclei analytical results for the reduced transition probabilities are obtained by introducing an effective-range expansion. The scaling with the relevant parameters is found explicitly. General trends are observed by studying several examples of neutron+core and proton+core nuclei in a single-particle model assuming Woods-Saxon potentials. Many important features of the neutron halo case can be obtained already from a square-well model. Rather simple analytical formulae are found. The nucleon-core interaction in the continuum affects the determination of astrophysical S factors at zero energy in the method of asymptotic normalisation coefficients (ANC). It is also relevant for the extrapolation of radiative capture cross sections to low energies.

Coulomb dissociation of a neutron halo nucleus in the limit of a zero-range neutron-core interaction in the Coulomb field of a target nucleus can be studied in various limits of the parameter space and rather simple analytical solutions can be found. We propose to solve the scattering problem for this model Hamiltonian by means of the various advanced numerical methods that are available nowadays. In this way their range of applicability can be studied by comparison to the analytical benchmark solutions.

The Trojan-Horse Method [5, 6] is a particular case of transfer reactions to the continuum under quasi-free scattering conditions. Special attention is paid to the transition from reactions to bound and unbound states and the role of subthreshold resonances. Since the binding energies of nuclei close to the drip line tend to be small, this is expected to be an important general feature in exotic nuclei.

2. EFFECTIVE RANGE THEORY OF HALO NUCLEI

At low energies the effect of the nuclear potential is conveniently described by the effective-range expansion [7]. An effective-range approach for the electromagnetic strength distribution in neutron halo nuclei was introduced in [1] and applied to the single neutron halo nucleus ^{11}Be. Recently, the same method was applied to the description of electromagnetic dipole strength in ^{23}O [8]. A systematic study sheds light on the sensitivity of the electromagnetic

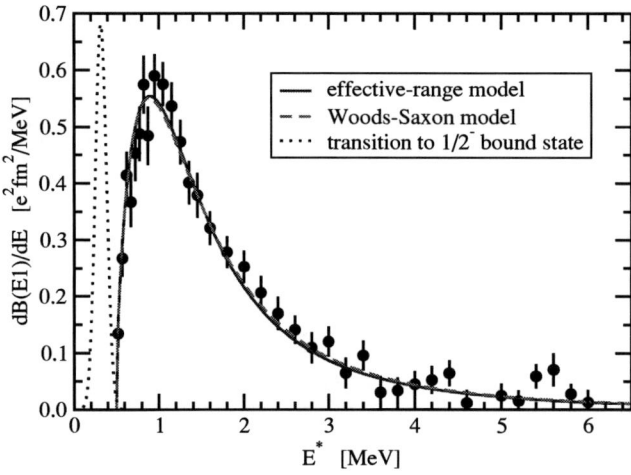

FIGURE 1. Reduced probability for dipole transitions as a function of the excitation energy $E^* = E + S_n$ in comparison to experimental data extracted from Coulomb dissociation of ^{11}Be [14].

strength distribution to the interaction in the continuum. We expose the dependence on the binding energy of the nucleon and on the angular momentum quantum numbers. Our approach extends the familiar textbook case of the deuteron, that can be considered as the prime example of a halo nucleus, to arbitrary nucleon+core systems, for related work see [9, 10, 11]. We also investigate in detail the square-well potential model. It has great merits: it can be solved analytically, it shows the main characteristic features and leads to rather simple and transparent formulae. As far as we know, some of these formulae have not been published before. These explicit results can be compared to our general theory for low energies (effective-range approach) and also to more realistic Woods-Saxon models. Due to shape independence, the results of these various approaches will not differ for low energies. It will be interesting to delineate the range of validity of the simple models.

Our effective-range approach is closely related to effective field theories that are nowadays used for the description of the nucleon-nucleon system and halo nuclei [12]. The characteristic low-energy parameters are linked to QCD in systematic expansions. Similar methods are also used in the study of exotic atoms ($\pi^- A$, $\pi^+\pi^-$, $\pi^- p$, ...) in terms of effective-range parameters. The close relation of effective field theory to the effective-range approach for hadronic atoms was discussed in Ref. [13].

In Fig. 1 we show the application of the method to the electromagnetic dipole strength in ^{11}Be. The reduced transition probability was deduced from high-energy ^{11}Be Coulomb dissociation at GSI [14]. Using a cutoff radius of $R = 2.78$ fm and an inverse bound-state decay length of $q = 0.1486$ fm^{-1} as input parameters we extract an ANC of $C_0 = 0.724(8)$ fm$^{-1/2}$ from the fit to the experimental data. The ANC can be converted to a spectroscopic factor of $C^2 S = 0.704(15)$ that is consistent with results from other methods. In the lowest order of the effective-range expansion the phase shift δ_l^j in the partial wave with orbital angular momentum l and total angular momentum j is written as $\tan\delta_l^j = -(xc_l^j\gamma)^{2l+1}$, where $\gamma = qR = 0.4132 < 1$ is the halo expansion parameter and $x = k/q = \sqrt{E/S_n}$ with the neutron separation energy S_n. The parameter c_l^j corresponds to the scattering length $a_l^j = (c_l^j R)^{2l+1}$. We obtain $c_1^{3/2} = -0.41(86, -20)$ and $c_1^{1/2} = 2.77(13, -14)$. The latter is unnaturally large because of the existence of a bound $\frac{1}{2}^-$ state close to the neutron breakup threshold in ^{11}Be. For a further discussion we refer to [1].

3. HOMEWORK PROBLEM

We consider a three-body system consisting of a neutron n, a core c and an (infinitely heavy) target nucleus with charge Ze. The Hamiltonian is given by

$$H = T + V_{cZ} + V_{nc} \tag{1}$$

where T is the kinetic energy. The Coulomb interaction between the core and the target is given by $V_{cZ} = ZZ_c e^2/r_c$ and V_{nc} is a zero-range interaction between c and n. The s-wave bound state of the $a = (c+n)$ system is given by the wave function $\Phi_0 = \sqrt{q/(2\pi)}\exp(-qr)/r$, where q is related to the binding energy E_b by $E_b = \hbar^2 q^2/(2\mu)$ and the reduced mass of the $c + n$ system is denoted by μ. We refer to [3] (see especially Ch. 4 there) for details. (The present homework problem is simpler than the one assigned by I. Thompson: in his case there is a p-wave bound state in ^{8}Li, and, in addition, the interactions between the target and the projectile are much more complicated.)

One can study elastic scattering (influence of the polarisation potential) as well as breakup of the halo nucleus a in the Coulomb field of the target nucleus Z. Although the Coulomb dissociation of this zero-range halo nucleus is governed by a rather simple Hamiltonian, the solution of this problem is nontrivial, as is often the case in physics. This model is also relevant for the Oppenheimer-Phillips process (polarisation of a deuteron in the Coulomb field of a nucleus) [15], see also [16] for a criticism. The parameter space is given by the charge Ze of the target and $Z_c e$ of the core c, the binding energy E_b of the $(c+n)$ system, the neutron and core masses m_n and m_c respectively.

In this model one can study elastic scattering as well as breakup. The beam momentum is denoted by $\vec{q}_a$ (the beam velocity is denoted by v), the momenta of the outgoing fragments c and n are $\vec{q}_c$ and $\vec{q}_n$, respectively (or $\vec{q}'_a$ in the case of elastic scattering). In the case of elastic scattering, the influence of the polarisation potential can be studied [17]. The polarizability of a zero-range neutron halo nucleus is given by

$$\alpha_{pol} = \frac{\hbar c}{2\pi^2}\sigma_{-2} = \frac{(Z_c m_n e \hbar)^2}{6\mu(m_n+m_c)^2 E_b^2}. \tag{2}$$

For a small binding energy E_b this can be a large effect. In 1982 the electric dipole polarizability of the deuteron was determined by measuring elastic scattering of deuterons on ^{208}Pb at energies from 3.0 to 7.0 MeV [18]. (By the way, two of the authors of this paper were participating in this workshop.) The measured value of the electric polarizability $\alpha_{pol} = (0.70 \pm 0.05)$ fm^3 is in fair agreement with eq. 2, if the necessary finite range corrections are applied see, e.g., [19].

The kinematics of the breakup process is given by $\vec{q}_a \to \vec{q}_{cm} + \vec{q}_{rel}$ where $\vec{q}_{cm}$ and $\vec{q}_{rel}$ are directly related to $\vec{q}_c$ and $\vec{q}_n$, respectively. Analytic results are known for the plane-wave limit, the Coulomb-wave Born approximation (CWBA, "Bremsstrahlung integral") and the adiabatic approximation (Ron Johnson, this workshop and [20]). A first derivation of the "Bremsstrahlung formula" was given by Landau and Lifshitz [21], it was improved by Breit in [22]; an early review is given in [23].

In the plane-wave limit the result does not depend on q_a itself but only on the "Coulomb push" $\vec{q}_{coul} = \vec{q}_a - \vec{q}_{cm}$.

In the semiclassical high energy straight-line and electric dipole limit, first and second order analytical results are available, as well as for the sudden limit. E.g., in the straight-line dipole approximation a shape parameter $x = k/q$ and a strength parameter $y = m_n \eta/[(m_n+m_c)bq]$ determine the breakup probability (in the sudden limit). The impact parameter is denoted by b and the Coulomb parameter is $\eta = ZZ_c e^2/(\hbar v)$. In [24] it was found that the breakup probability is given in leading order by

$$\frac{dP_{LO}}{dk} = \frac{16}{3\pi q} y^2 \frac{x^4}{(1+x^2)^4} \tag{3}$$

and in next-to-leading order by

$$\frac{dP_{NLO}}{dk} = \frac{16}{3\pi q} y^4 \frac{x^2(5-55x^2+28x^4)}{15(1+x^2)^6}. \tag{4}$$

Another important scaling parameter, in addition to x and y, is $\xi = \omega b/v$, where $\hbar\omega$ is the excitation energy of the $(c+n)$ system. In the sudden approximation we have $\xi = 0$ and there is an analytical solution [26].

This homework problem can be studied, e.g., in the CDCC method, which was widely discussed at the workshop. It would be very interesting to see how well this method works in various limits of the parameter space. An especially interesting limit is the limit of low beam energies, where the CWBA is very appropriate. We would expect that higher-order effects are very important under these conditions and it would be good to see that the CDCC method converges. We refer to [3], especially Sect. 4.2, for further details and references on experimental and theoretical work on $E_d = 12$ MeV deuteron breakup on ^{197}Au.

It would also be extremely interesting to apply the three-body methods of [25] to the homework problem. In this work, the so-called post-decay acceleration of the fragments is studied and genuine three-particle wave functions for the final state are used. In their case there are three charged particles in the final state, but the problem is non-trivial even for only two (out of three) charged particles in the final state.

A related problem, the Coulomb breakup of antideuterons bound in an orbit with quantum numbers n, l, m [27, 28] can also be studied with this Hamiltonian: in this case the charge of the core nucleus c is negative, $Z_c = -|Z_c|$. In [27] the adiabatic method is used: the antideuteron c.m. motion is assumed to be slow compared to the internal $\bar{p}$ and $\bar{n}$-motion and the authors calculate the antideuteron tunnel probability through the Coulomb barrier which is provided by the nucleus Z.

4. TRANSFER REACTIONS

Exotic nuclei have low thresholds for particle emission. It is expected that in transfer reactions one will often meet a situation where the transferred particle is in a state close to the particle threshold. In "normal" nuclei, the neutron threshold is around an excitation energy of about 8 MeV, and the pure single particle picture is not directly applicable. Much is known from stripping treactions like (d, p) and thermal neutron scattering, see, e.g., [29]. The single particle strength is fragmented over many more complicated compound states. The interesting quantity is the strength function which is proportional to Γ/D where Γ is the width and D the level spacing. This ratio is $\ll 1$, as can be estimated from a square well model (see, e.g., [29]). For $l = 0$ there are no sharp resonances, since $\Gamma > E$ around threshold. Due to the angular momentum (and/or) Coulomb barrier, one has $\Gamma/E \ll 1$ at threshold for all the other cases.

For neutron rich (halo) nuclei the neutron threshold is much lower, of the order of one MeV. In this case the single-particle properties are dominant and the ideas developed in the following can become relevant, see also [30]. The level density is also much lower. In normal nuclei the level density at particle threshold is generally so high that the single particle structure is very much dissolved. This can be quite different in exotic nuclei which can show a very pronounced single particle structure.

4.1. Trojan-Horse Method

A similarity between cross sections for two-body and closely related three-body reactions under certain kinematical conditions [31] led to the introduction of the Trojan-Horse method [32, 33, 34, 5]. In this indirect approach a two-body reaction

$$A + x \to C + c \tag{5}$$

that is relevant to nuclear astrophysics is replaced by a reaction

$$A + a \to C + c + b \tag{6}$$

with three particles in the final state. One assumes that the Trojan horse a is composed predominantly of clusters x and b, i.e. $a = (x + b)$. This reaction can be considered as a special case of a transfer reaction to the continuum. It is studied experimentally under quasi-free scattering conditions, i.e. when the momentum transfer to the spectator b is small. The method was primarily applied to the extraction of the low-energy cross section of reaction (5) that is relevant for astrophysics. However, the method can also be applied to the study of single-particle states in exotic nuclei around the particle threshold.

4.2. Continuous Transition from Bound to Unbound State Stripping

Motivated by this we look again at the relation between transfer to bound and unbound states. Our notation is as follows: we have the reaction

$$A + a \to B + b \tag{7}$$

where $a = (b + x)$ and B denotes the final $B = (A + x)$ system. It can be a bound state B with binding energy $E_{bind} = -E_{Ax}(>0)$, the open channel $A + x$, with $E_{Ax} > 0$, or another channel $C + c$ of the system $B = (A + x)$. In particular, the reaction $x + A \to C + c$ can have a positive Q value and the energy E_{Ax} can be negative as well as positive. As an example we quote the recently studied Trojan horse reaction $d + ^6\text{Li}$ [35] applied to the $^6\text{Li}(p,\alpha)^3\text{He}$ two-body reaction (the neutron being the spectator). In this case there are two charged particles in the initial state ($^6\text{Li} + p$). Another example with a neutral particle x would be $^{10}\text{Be} + d \to p + ^{11}\text{Be} + \gamma$. The general question which we want to answer here is how the two regions $E_{Ax} > 0$ and $E_{Ax} < 0$ are related to each other. E.g., in Fig. 7 of [35] the coincidence yield is plotted as a function of the ^{6}Li-p relative energy. It is nonzero at zero relative energy. How does the theory [5] (and the experiment) continue to negative relative energies? With this method, subthreshold resonances can be investigated rather directly. We treat two cases separately, one where system B is always in the $(A+x)$ channel, with a real potential V_{Ax} between A and x. In the other case, there are also other channels $C+c$, at positive and negative energies E_{Ax}.

4.2.1. One Channel Case

We imagine the following situation: The potential V_{Ax} gives rise to a bound state with angular momentum l close to threshold. Now we decrease the potential so that the bound state disappears and reappears as a resonance in the continuum. For $l > 0$ there are sharp resonances and we can define a cross section for stripping to a resonance by integrating over the resonance line (over an energy range which is several times larger than the width) and they join smoothly to the stripping to the bound states, see [36].

Due to the absence of the angular momentum barrier for $l = 0$ there are some peculiarites which we study now. Stripping to bound states is determined by the asymptotic normalization constant B (see eqs. (A54) - (A56) of [2]) of the bound-state wave function and the function $h_l(iqr)$ where q is related to the binding energy. Since

$$B \sim q^{3/2} \quad \text{for} \quad l = 0 \tag{8}$$

and

$$B \sim q^{l+1} R^{l-1/2} \quad \text{for} \quad l > 0 \tag{9}$$

the stripping cross section (see, e.g., eq. (17) of [36]) to a (halo) state with $l = 0$ tends to zero for q going to zero, while it stays finite for $l > 0$. We note that the presence of a bound state close to zero energy leads to a large scattering length in the $A+x$ system which leads to an enhancement of the elastic breakup cross section. The double differential cross section at threshold is proportional to

$$\frac{d^2\sigma}{d\Omega dE} \sim \frac{\sin^2 \delta_0}{k}. \tag{10}$$

The quantity $\sin^2 \delta_0$ is given by $k^2/(q^2+k^2)$ for a bound and virtual state. Thus the double differential cross section tends to zero like $k \sim \sqrt{E_n}$ for $l = 0$.

When the strength of the potential is decreased, the bound state becomes a virtual state, which again leads to a very large scattering length, see also [30]. In this context it seems interesting to note that about 30 years ago a new type of threshold effects was predicted in [37] (what is now called a halo state was referred to as a puffy state in those days). Related to this is the qualitative difference of $l = 0$ and $l > 0$ in the location of the poles of the S matrix in the complex plane [38, 39]. Only for $l > 0$ there are poles of the S matrix close to the real axis.

4.2.2. Absorption at Zero Energy, Multichannel Case

We follow the work of Ichimura, Austern, Vincent, and Kasano [40, 41] who have studied the case $E_{Ax} > 0$ and we now extend it to the case of $E_{Ax} < 0$. The exclusive case can be also studied by generalizing, e.g., eq. (61) of [5].

For positive energies E_{Ax} the inclusive cross section for $A+a \rightarrow b+X$ where X is any state of the system $B = (A+x)$ consists of an elastic and inelastic component, see eq. (2.20) of [40] or eq. (8) of [41]. For negative energies E_{Ax} the elastic breakup component is zero, and only the inelastic component remains. For positive energies this inclusive inelastic breakup cross section is written as [41]

$$\frac{d\sigma_{inel}}{d^3k_b} = \frac{(2\pi)^4}{\pi \hbar v_a} \int d^3r W(r) \left| \int G_x(\vec{r},\vec{r}')\rho(\vec{r}')d^3r' \right|^2. \tag{11}$$

The "source term" ρ can be calculated from the distorted waves in the incident and final channel and is given by eq. (3) of [41]. The Green's function in the $x+A$ channel is given by G_x and $W = -\text{Im}U(r)$ where U is the optical potential (assumed to be local) in the $x+A$ channel.

It is now our aim to give a meaning to W and G_x for negative energies E_{Ax} and show that the cross section behaves smoothly when going from positive energies to negative energies.

In [41] the Green's function is expanded in partial waves as

$$G_x(\vec{r},\vec{r}') = -\frac{2m}{\hbar^2 k_x} \sum_{lm} \frac{f_l(r_<) h_l(r_>)}{rr'} Y_{lm}(\hat{r}) Y_{lm}^*(\hat{r}') \tag{12}$$

where f_l and h_l are regular and outgoing radial wave functions in the potential U.

The imaginary part $-W$ of the optical model potential is related to the partial wave reaction cross section σ_l of $x+A$ scattering by (this is eq. (26) of [41])

$$\int_0^\infty W(x)|f_l(r)|^2 dr = \frac{\hbar^2 k_x}{2m} \sigma_l. \tag{13}$$

The total reaction cross section σ_{reac} is given by $\sigma_{reac} = \sum_l (2l+1)\sigma_l$ and σ_l is related to the imaginary part of the phase shift by $\sigma_l = \pi[1-\exp(-4\mathrm{Im}\delta_l)]/k^2$. We now derive this equation and generalize it to the case of negative energies E_{Ax}. According to (A.20) in [2] we normalize the regular scattering wave function g_l as (our normalization differs from the one of Ref. [41] by a factor of k)

$$g_l \to \frac{1}{2i}\left[\exp(2i\delta_l) u_l^{(+)} - u_l^{(-)}\right] \tag{14}$$

valid for r ouside the range of the potential. The ingoing and outgoing wave functions $u_l^{(\pm)}$ are given by

$$u_l^{(\pm)} = x(-y_l \pm i j_l) \tag{15}$$

for neutrons and

$$u_l^{(\pm)} = \exp(\mp i\sigma_l)(G_l \pm i F_l) \tag{16}$$

for charged particles, respectively. The asymptotic behaviour is $u_l^{(\pm)} \to \exp[\pm i(x - \eta \ln(2x) - l\pi/2)]$. For positive energies $E_{Ax} > 0$ we have $x = kr$. By the usual procedure we obtain

$$-2i\frac{2m}{\hbar^2}\int_0^\infty W(r)|g_l|^2 dr = \left(g_l^* \frac{dg_l}{dr} - g_l \frac{dg_l^*}{dr}\right)\bigg|_{r=\infty}. \tag{17}$$

From the Wronskian relation $G\frac{dF}{dx} - F\frac{dG}{dx} = 1$ we obtain $u_l^{(+)}\frac{du_l^{(-)}}{dx} - u_l^{(-)}\frac{du_l^{(+)}}{dx} = -2i$. Using this we can evaluate the RHS. For positive energies we have $[u_l^{(\pm)}]^* = u_l^{(\mp)}$ and the RHS is given by

$$RHS = \frac{k}{2i}\left[1 - \exp(-4\mathrm{Im}\delta_l)\right]. \tag{18}$$

This quantity is directly related to the partial wave reaction cross section σ_l and eq. (13) is established. For low energies $E_{Ax} > 0$ the phase shift is small and we can expand

$$RHS = -2ik\mathrm{Im}\delta_l. \tag{19}$$

For negative energies $E_{Ax} < 0$ we put $x = iqr$. The functions $u^{(\pm)}$ are exponentially decreasing and increasing respectively. (A bound state corresponds to a pole of $S_l = \exp(2i\delta_l)$.) They are given asymptotically by (disregarding the logarithmic Coulomb phase)

$$u^{(\pm)} = i^{\pm l} \exp(\mp qr). \tag{20}$$

Using these properties we can evaluate the Wronskians and the RHS is found to be

$$RHS = \frac{q}{2}(-1)^l \left[\exp(2i\delta_l) - \exp(-2i\delta_l^*)\right]. \tag{21}$$

Close to the threshold δ_l is small and we have

$$RHS = iq(-1)^l(\delta_l + \delta_l^*) = 2iq(-1)^l \mathrm{Re}\delta_l. \tag{22}$$

We can assume that the interior logarithmic derivative L_i is smooth when E_{Ax} goes from positive to negative values. Now we can relate the value of δ_l to this logarithmic derivative and show in this way that the transition from positive to negative values of E_{Ax} is smooth. In the presence of an imaginary part W the LHS is non-vanishing. The logarithmic derivative L_i is complex. This means that for $E_{Ax} > 0$ δ_l acquires an imaginary part, for $E < 0$ the "phase shift" δ_l acquires a real part.

Let us deal with neutral particles. For low (positive) energies we can express the phase shift in terms of the scattering length a_l by $\tan(\delta_l) = -a_l k^{2l+1}$ where the scattering length is related to the interior logarithmic derivative L_i by eq. (A.31) of [2]

$$a_l = a_l^{hs}\left(1 - \frac{2l+1}{L_i + l}\right) \tag{23}$$

where the hard sphere scattering length is given by $a_l^{hs} = R^{2l+1}/[(2l+1)!!(2l-1)!!]$. In order to obtain this result, the expansion of the Bessel and Neumann functions for small values of kr was used: $j_l = (kr)^l/(2l+1)!!$ and $n_l = -(2l-1)!!/(kr)^{l+1}$. We can write

$$\delta_l = -k^{2l+1} a_l^{hs}\left(1 - \frac{2l+1}{L_i + l}\right). \tag{24}$$

Thus the Wronskian can be expressed in terms of L_i. For $E_{Ax} > 0$ we find $RHS = -2ik^{2l+2}a_l^{hs}\text{Im}[(2l+1)/(L_i+l)]$. For negative energies we put $k = -iq$. Carrying through the corresponding steps as for the positive energy case we obtain

$$\delta_l = i(-1)^l q^{2l+1} a_l^{hs}\left(1 - \frac{2l+1}{L_i + l}\right). \tag{25}$$

This leads to $RHS = -2iq^{2l+2}a_l^{hs}\text{Im}[(2l+1)/(L_i+l)]$. In our approach we have used the surface approximation, see eqs. (24) and (25) of [41]. This means that the r-coordinate in eq. (11) is associated with the $r_<$-coordinate in eq. (12) and r'' with $r_>$. The k^{2l+1} and q^{2l+1} factors which enter in eqs. (24) and (25) are cancelled by the term coming from h_l, see eqs. (11), (12) and eq. (25) of [41]. Thus there is a continuous transition in the stripping from bound to unbound states.

Quite similarly, one can relate the logarithmic derivative L_i to the phase shift for charged particles and establish the smooth transition from positive to negative energies. We do not give the details here.

4.2.3. Imaginary part of the optical model potential and solution of a toy model

A formal expression for the optical potential is given in Eq. (2.16) of [40] by the Feshbach projector formalism. In a schematic two-state model we want to illustrate the smooth transition from positive to negative energies. We assume two channels with $l = 0$, the coupled radial equations are

$$\left(\frac{d^2}{dr^2} - u_1(r) + k_1^2\right) f_1(r) = u_{12}(r) f_2(r) \tag{26}$$

and

$$\left(\frac{d^2}{dr^2} - u_2(r) + k_2^2\right) f_2(r) = u_{21}(r) f_1(r). \tag{27}$$

We have $k_2^2 = k_1^2 + Q(>0)$ and the channel 2 is open for $k_1^2 = 0$ down to $k_1^2 > -Q$. Introducing the Green's function $G_2(r,r')$ we can express f_2 as $f_2(r) = \int G_2(r,r') u_{21} f_1(r') dr'$. Inserting this into eq. (26) we obtain an equation for f_1 in an optical potential. This optical potential has a real and an imaginary part. We are especially interested here in the imaginary part which can be found as follows: We can express the Green's function as $G_2 = \int dE\, \chi_E(r)\chi_E(r')/(E^+ - E)$. Using $\lim \frac{1}{x-x_0\pm i\varepsilon} = PP\frac{1}{x-x_0} \mp i\pi\delta(x-x_0)$ we obtain $\text{Im} G_2 = -i\pi \chi_E(r)\chi_E(r')$ where $\chi_E(r)$ is the regular solution of the homogeneous part of eq. (26) (with the coupling potential $u_{21} = 0$). This leads to a nonlocal, separable imaginary part given by $W(r,r') = -\pi V_{12}(r)\chi_E(r)\chi_E(r')V_{21}(r')$.

It is instructive to solve eqs. (26) and (27) analytically for a square-well model with delta-function coupling. We take $u_1 = -|u_1|, u_2 = -|u_2|$ for $r < R$ and zero otherwise and $u_{12} = u_{21} = u\delta(r - R)$. This leads to a Sprungbedingung in the logarithmic derivatives. According to eqs. (22) ff. of [5] we have the following asymptotic behaviour of the (s-wave) radial wave functions:

$$f_1(r) \to \frac{i}{2}\left[S_{12}^* \exp(-ik_1 r)\right] \tag{28}$$

and

$$f_2 \to \frac{i}{2}\sqrt{\frac{v_2}{v_1}}\left[S_{22}^* \exp(-ik_2 r) - \exp(ik_2 r)\right]. \tag{29}$$

The two logarithmic matching conditions determine $z_1 = \sqrt{k_2/k_1}S_{12}^*$ and $z_2 = S_{22}^*$. The interior logarithmic derivatives L_1 and L_2 are real (somewhat differently from the previous subsection they are defined here as $L_i = f_i'/f_i, i = 1, 2$). Introducing $\tilde{L} = L_2 - u^2/(L_1 + ik_1)$ one can express $z_2 = \exp(2ik_2 R)(\tilde{L} - ik_2)/(\tilde{L} + ik_2)$ and $z_1 = 2ik_2 u \exp[i(k_1 + k_2)R]/[(L_1 + ik_1)(\tilde{L} + ik_2)]$. From these expressions one can derive the unitarity of the S matrix (2 by 2 for $k_1^2 > 0$). The S-matrix element ($k_1^2 > 0$) S_{12} has the threshold behaviour $S_{12} \propto \sqrt{k_1}$ which is characteristic for the s wave. It should be straightforward to generalize to $l > 0$ and to Coulomb interactions.

For $k_1^2 < 0$ there is only one open channel (channel 2) and the S matrix consists only of one S-matrix element S_{22}. We put $k_1 = -iq$ ($|E_n| = \hbar^2 q^2/(2m)$). One sees that $\tilde{L}$ is real (rather than complex for the 2 channel case) and z_2 is unitary (modulus is one). The quantity z_1 tends to a well defined number, of interest for the THM method. For $E_n = 0$ it is given by $z_1 = 2ik_2 u \exp(ik_2 R)/[L_1(\tilde{L} + ik_2)]$. Since channel 1 is closed, S_{12} is not an S-matrix element, but it can still be used as an input in eqs. (64), (65) of [5]. The quantitiy $J_l^{(+)}$ there can also be defined for imaginary values of k_{Ax} (closed channel case).

5. CONCLUSION

While the foundations of direct reaction theory have been laid several decades ago, the new possibilites which have opened up with the rare isotope beams are an invitation to revisit this field. The general frame is set by nonrelativistic many-body quantum scattering theory, however, the increasing level of precision demands a good understanding of relativistic effects notably in intermediate energy Coulomb excitation, see the talk by Carlos Bertulani at this workshop.

The properties of halo nuclei depend very sensitively on the binding energy and despite the ever increasing precision of microscopic approaches using realistic NN forces it will not be possible, say, to predict the binding energies of nuclei to a level of about 100 keV. Thus halo nuclei ask for new approaches in terms of some effective low-energy constants. Such a treatment was provided in Ch. 2 and an example to the one-neutron halo nucleus ^{11}Be was given. With RIA one will be able to study also neutron halo nuclei for intermediate mass nuclei. This is expected to be relevant also for the astrophysical r process. It is a great challenge to extend the present approach for one-nucleon halo nuclei to more complicated cases, like two-neutron halo nuclei.

The treatment of the continuum is a general problem, which becomes more and more urgent when the dripline is approached. In the present proceedings we studied the transition from bound to unbound states as a typical example.

ACKNOWLEDGMENTS

We wish to thank Carlos Bertulani, Kai Hencken, Radhey Shyam and Dirk Trautmann for their collaboration on various topics in this field.

REFERENCES

1. S. Typel and G. Baur, *Phys. Rev. Lett.*, **93**, 142502 (2004).
2. S. Typel and G. Baur, nucl-th/0411069.
3. G. Baur, K. Hencken, and D. Trautmann, *Progress in Particle and Nuclear Physics*, **51**, 487 (2003).
4. G. Baur et al., in *Proceedings of Hirschegg04*, nucl-th/0402012.
5. S. Typel and G. Baur, *Ann. Phys.*, **305**, 228 (2003).
6. G. Baur, and S. Typel, *Progress of Theoretical Physics Supplement*, **154**, 333 (2004).
7. H. A. Bethe, *Phys. Rev.*, **76**, 38 (1949).
8. C. Nociforo et al., *Phys. Lett. B*, **605**, 75 (2005).

9. D. M. Kalassa and G. Baur, *J. Phys. G*, **22**, 115 (1996).
10. C. A. Bertulani and A. Sustich, *Phys. Rev. C*, **46**, 2340 (1992).
11. C. A. Bertulani, nucl-th/0503053.
12. C. A. Bertulani, H.-W. Hammer, and U. van Kolck, *Nucl. Phys. A*, **721**, 37 (2002).
13. B. R. Holstein, *Phys. Rev. D*, **60**, 114030 (1999).
14. R. Palit et al., *Phys. Rev. C*, **68**, 034318 (2003).
15. J. R. Oppenheimer and M. Phillips, *Phys. Rev.*, **48**, 500 (1935).
16. Gy. Bencze and C. Chandler, *Phys. Rev. C*, **53**, 880 (1996).
17. G. Baur, F. Rösel and D. Trautmann, *Nuclear Physics A*, **288**, 113 (1977).
18. N. L. Rodning, L. D. Knutson, W. G. Lynch and M. B. Tsang, *Phys. Rev. Lett.* **49**, 909 (1982)
19. T. Y. Wu and T. Ohmura, *Quantum Scattering Theory*, Prentice-Hall, Englewood Cliffs, NJ, 1962.
20. J. A. Tostevin, S. Rugmai, and R. C. Johnson, *Phys. Rev. C* **57**, 3225 (1998).
21. L. Landau and E. Lifshitz, *JETP*, **18**, 750 (1948). (English translation in Collected Papers of L. Landau.)
22. G. Breit, in *Handbuch der Physik, Vol.* **41**. edited by S. Flügge, Springer-Verlag, Berlin, 1959, pp. 304–320.
23. G. Baur, D. Trautmann, *Physics Reports*, **25C**, 293 (1976).
24. S. Typel and G. Baur, *Phys. Rev. C*, **64**, 024601 (2001).
25. E. O. Alt, B. F. Irgaziev and A. M. Mukhamedzhanov, *Phys. Rev. C*, **71**, 024605 (2005).
26. S. Typel and G. Baur, *Nucl. Phys. A* **573**, 486 (1994).
27. T. E. O. Ericson and P. Osland, *Nuclear Physics A*, **249**, 445 (1975).
28. G. Baur, *Phys. Lett.*, **60B**, 137 (1976).
29. A. G. Bohr and B. Mottelson, *Nuclear Structure*, Vol. II, Benjamin, Reading, MA, 1975.
30. G. Blanchon, A. Bonaccorso, and N. Vinh Mau, *Nucl. Phys. A*, **739**, 259 (2004).
31. H. Fuchs et al., *Phys. Lett. B*, **37**, 285 (1971).
32. G. Baur, F. Rösel, D. Trautmann, and R. Shyam, *Phys. Rep.*, **111**, 333 (1984).
33. G. Baur, *Phys. Lett. B*, **178**, 135 (1986).
34. S. Typel and H. H. Wolter, *Few Body Systems*, **29**, 75 (2000).
35. A. Tumino et al., *Phys. Rev. C*, **67**, 065803 (2003).
36. G. Baur and D. Trautmann, *Z. Phys.*, **267**, 103 (1974).
37. P. v. Brentano, B. Gyarmati, and J. Zimanyi, *Phys. Lett.*, **46B**, 177 (1973).
38. H. M. Nussenzveig, *Nucl. Phys.*, **11**, 499 (1959).
39. C. Mahaux and H. A. Weidenmüller, *Shell-model approach to nuclear reactions*, North-Holland, Amsterdam, 1969.
40. M. Ichimura, N. Austern, and C. M. Vincent, *Phys. Rev. C*, **32**, 431 (1985).
41. A. Kasano and M. Ichimura, *Phys. Lett.*, **115B**, 81 (1982).

On the splitting of nucleon effective masses at high isospin density: reaction observables

M.Di Toro*, M.Colonna* and J.Rizzo*

Laboratori Nazionali del Sud INFN, via S.Sofia 62, I-95123 Catania, Italy and Dipartimento di Fisica e Astronomia, Universita' di Catania

Abstract. We review the present status of the nucleon effective mass splitting *puzzle* in asymmetric matter, with controversial predictions within both non-relativistic *and* relativistic approaches to the effective in medium interactions. Based on microscopic transport rimulations we suggest some rather sensitive observables in collisions of asymmetric (unstable) ions at intermediate (*RIA*) energies: i) Energy systematics of Lane Potentials; ii) Isospin content of fast emitted nucleons; iii) Differential Collective Flows. Similar measurements for light isobars (like $^3H - ^3He$) could be also important.

Keywords: Asymmetric nuclear matter, neutron/proton effective mass splitting, Lane potential, relativistic heavy ion collisions, particle emission, collective flows

PACS: 21.30.Fe, 21.65.+f, 24.10.Jv, 25.75.-q

ISOSPIN MOMENTUM DEPENDENCE IN SKYRME FORCES

We start from a discussion on the isospin dependence of these widely used non-relativistic effective interactions [1]. In a Skyrme-like parametrization the symmetry term has the form:

$$\varepsilon_{sym} \equiv \frac{E_{sym}}{A}(\rho) = \frac{\varepsilon_F(\rho)}{3} + \frac{C(\rho)}{2}\frac{\rho}{\rho_0} \qquad (1)$$

with the function $C(\rho)$, in the potential part, given by:

$$\frac{C(\rho)}{\rho_0} = -\frac{1}{4}\left[t_0(1+2x_0) + \frac{t_3}{6}(1+2x_3)\rho^\alpha\right]$$

$$+\frac{1}{12}\left[t_2(4+5x_2) - 3t_1x_1\right]\left(\frac{3\pi^2}{2}\right)^{2/3}\rho^{2/3} \equiv \frac{1}{\rho_0}\left[C_{Loc}(\rho) + C_{NLoc}(\rho)\right] \qquad (2)$$

with $\alpha > 0$ and the usual Skyrme parameters. We remark that the second term is related to isospin effects on the momentum dependence [2].

In the Fig. 1(left) we show the density dependence of the potential symmetry term of various Skyrme interactions, *SIII*, *SGII*, *SKM** (see [3] and refs. therein) and the more recent Skyrme-Lyon forms, *SLya* and *Slyb* (or *Sly*4), see [2, 4]. We also separately present the local and non-local contributions, first and second term of the Eq.(2). We clearly see a sharp change from the earlier Skyrme forces to the Lyon parametrizations, with almost an inversion of the signs of the two contributions, [5]. The important repulsive non-local part of the Lyon forces leads to a completely different behavior of the neutron matter *EOS*, of great relevance for the neutron star properties. Actually this substantially modified parametrization was mainly motivated by a very unpleasant feature in the spin channel of the earlier Skyrme forces, the collapse of polarized neutron matter, see discussion in [8, 2, 4, 6]. In correspondence the predictions on the isospin effects on the momentum dependence of the symmetry term are quite different, see Fig.1 (left). A very important consequence for the reaction dynamics is the expected inversion of the sign of the n/p effective mass splitting.

Effective masses in neutron-rich matter

In asymmetric matter we consistently have a splitting of the neutron/proton effective masses given by:

$$m_q^{*-1} = m^{-1} + g_1\rho + g_2\rho_q, \qquad (3)$$

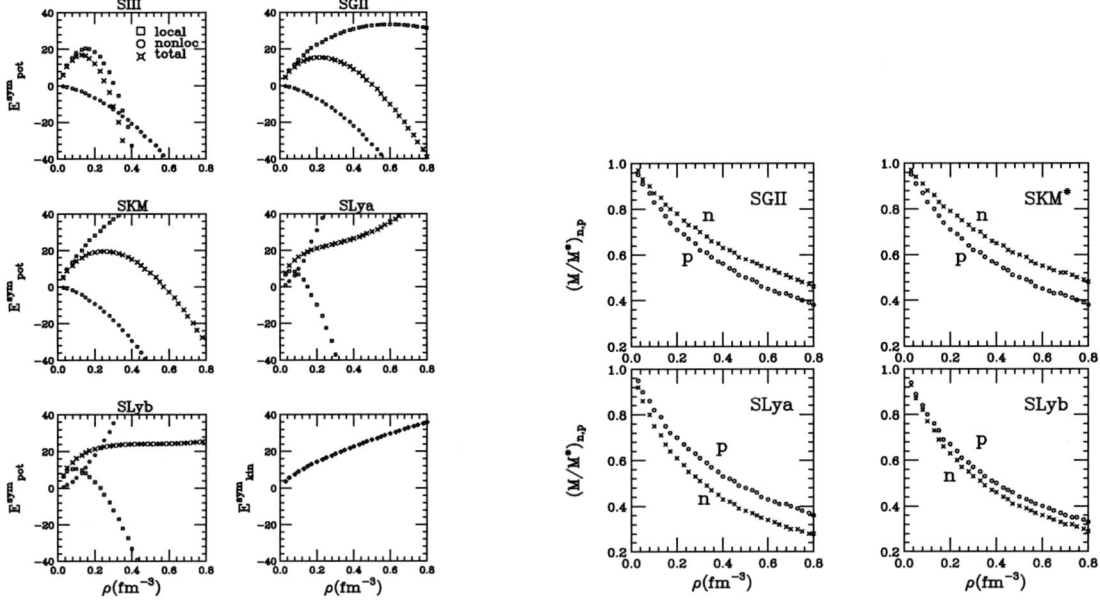

FIGURE 1. *Left: Density dependence of the potential symmetry term for various Skyrme effective forces, see text. The bottom right panel shows the kinetic contribution. Right: Density dependence of the neutron/proton effective mass splitting for various Skyrme effective forces, see text. The asymmetry is fixed at I = 0.2, not very exotic.*

with

$$\rho_{q=n,p} = \frac{1 \pm I}{2}\rho \quad (+,n).$$

The g_1, g_2 coefficients are simply related to the momentum dependent part of the Skyrme forces:

$$g_1 = \frac{1}{4\hbar^2}[t_1(2+x_1) + t_2(2+x_2)]$$

$$g_2 = \frac{1}{4\hbar^2}[t_2(1+2x_2) - t_1(1+2x_1)] \qquad (4)$$

This result derives from a general $q - structure$ of the momentum dependent part of the Skyrme mean field

$$U_{q,MD} = m(g_1\rho + g_2\rho_q)E \qquad (5)$$

where E is the nucleon kinetic energy, while the total field seen by the q-nucleon has the form [7]

$$U_q(\rho,\rho_q,k) = U_{q,MD} + \frac{g_1\hbar^2\tau}{2} + \frac{g_2\hbar^2\tau_q}{2} + \frac{\partial \varepsilon_{Loc}(Pot)}{\partial \rho_q} \qquad (6)$$

where $\hbar^2\tau$ is the kinetic energy density and $\varepsilon_{Loc}(Pot)$ the local part of the potential energy density.

In the Fig. 1(right) we show the density behavior of $m^*_{n,p}$ in neutron rich matter $I = 0.2$ for the same effective interactions. From the Eqs.(3, 4) we see that the sign of the g_2 univocally assigns the sign of the splitting, i.e. $g_2 < 0$ gives larger neutron masses $m^*_n > m^*_p$ while we have the opposite for $g_2 > 0$.

In the Table 1 we report some results obtained with various Skyrme forces for quantities of interest, around saturation, for the present discussion. We show also the $E - slope$ of the corresponding Lane Potential, see later, simply related to the isospin dependent part of Eq.(5). For the effective mass parameters of Eq.(3) we observe that while the g_1 coefficients are always positive, corresponding to a decrease of the nucleon mass in the medium, the isospin dependent part shows different signs. In particular we see that in the Lyon forces the g_2 values are positive, with neutron effective masses below the proton ones for n-rich matter as shown in Fig. 1 (right).

TABLE 1. Properties at saturation

Force	SIII	SGII	SkM*	SLya	SLy4	SLy7
g_1 $(10^{-3})(MeV^{-1}fm^3)$	+3.85	+3.31	+3.53	$+10^{-5}$	+1.67	+1.70
g_2 $(10^{-3})(MeV^{-1}fm^3)$	−3.14	−2.96	−3.50	+5.78	+2.53	+2.76
$\rho_0(fm^{-3})$	0.150	0.1595	0.1603	0.160	0.1595	0.1581
$a_4(MeV)$	28.16	26.83	30.03	31.97	32.01	32.01
$C(\rho_0)(MeV)$	31.72	29.06	35.46	39.40	39.42	39.42
$E-slope(LanePot.)$	−0.22	−0.21	−0.26	+0.43	+0.19	+0.20

We note that the same is predicted from microscopic relativistic Dirac-Brueckner calculations [9, 10, 11] and in general from the introduction of scalar isovector virtual mesons in *RMF* approaches [12, 13]. At variance, non-relativistic Brueckner-Hartree-Fock calculations are leading to opposite conclusions [14, 16]. We remind that a comparison between relativistic effective (*Dirac*) masses and non-relativistic effective masses requires some attention. This point will be carefully discussed later.

Energy dependence of the Lane Potential

The sign of the splitting will directly affect the energy dependence of the Lane Potential, i.e. the difference between (n,p) optical potentials on charge asymmetric targets, normalized by the target asymmetry [17]. From the Eqs.(5,6) we obtain the explicit Skyrme form of the Lane Potential:

$$U_{Lane} \equiv \frac{U_n - U_p}{2I} = C(\rho_0) - \frac{2}{3}m\rho_0\varepsilon_F\left[g_1 + \frac{7}{12}g_2\right] + \frac{m\rho_0}{2}g_2 E \quad (7)$$

where $C(\rho_0)$ gives the potential part of the a_4 parameter in the mass formula. We see that the $E-slope$ has just the sign of the g_2 parameter, and so we have opposite predictions from the various Skyrme forces analysed here. The change in the energy slope is reported in the last row of the Table 1. The difference in the energy dependence of the Lane Potential is quite dramatic.

The second term of the Eq.(7) is also interesting. In the case $g_2 > 0$ (positive slope) decreases the starting zero energy point, while in the case $g_2 < 0$ (negative slope) it represents a small correction to the symmetry energy $C(\rho_0)$. We then expect to see a *crossing* of the two prescriptions at very low energies, i.e. low momentum nucleons will see exactly the same Lane potentials.

An important physical consequence of the negative slopes is that the isospin effects on the optical potentials tend to disappear at energies just above 100 *MeV* (or even change the sign for "old" Skyrme-like forces). Unfortunately results derived from neutron/proton optical potentials at low energies are not conclusive, [17, 18, 19], since the effects appear of the same order of the uncertainty on the determination of the local contribution. Moreover at low energies we expect the crossing discussed before therefore for a wide energy range we cannot see differences. More neutron data are needed at higher energies, in particular a systematics of the energy dependence.

We can expect important effects on transport properties (fast particle emission, collective flows) of the dense and asymmetric *NM* that will be reached in Radioactive Beam collisions at intermediate energies, i.e. in the *RIA* energy range.

SYMMETRY ENERGY IN QUANTUM-HADRO-DYNAMICS

The *QHD* effective field model represents a very successful attempt to describe, in a fully relativistic picture, equilibrium and dynamical properties of nuclear systems at the hadronic level [20, 21, 22]. Here we focus on the dynamical response and static (equilibrium) properties of Asymmetric Nuclear Matter (*ANM*) in a "minimal" effective meson field model, with two meson (scalar and vector) contributions in each isospin channel, (σ,ω) for the isoscalar and (δ,ρ) for the isovector part, see details in [5] and refs. therein. In particular we will discuss the effects of the isovector/scalar δ meson, since this is directly related to the nucleon effective mass splitting of interest here.

The symmetry energy in *ANM* is defined from the expansion of the energy per nucleon $E(\rho_B, I)$ in terms of the asymmetry parameter $I \equiv -\frac{\rho_{B3}}{\rho_B} = \frac{\rho_{Bn} - \rho_{Bp}}{\rho_B} = \frac{N-Z}{A}$. We have

$$E(\rho_B, I) \equiv \frac{\varepsilon(\rho_B, I)}{\rho_B} = E(\rho_B) + E_{sym}(\rho_B)I^2 + O(I^4) + ... \quad (8)$$

and so in general

$$E_{sym} \equiv \frac{1}{2} \frac{\partial^2 E(\rho_B, I)}{\partial I^2}\Big|_{I=0} = \frac{1}{2}\rho_B \frac{\partial^2 \varepsilon}{\partial \rho_{B3}^2}\Big|_{\rho_{B3}=0} \quad (9)$$

In the Hartree case an explicit expression for the symmetry energy is easily derived that can be reduced to a very simple form, with transparent δ-meson effects [12, 13]:

$$E_{sym}(\rho_B) = \frac{1}{6}\frac{k_F^2}{E_F^*} + \frac{1}{2}\left[f_\rho - f_\delta \left(\frac{m^*}{E_F^*}\right)^2\right]\rho_B \quad (10)$$

where k_F is the nucleon Fermi momentum corresponding to ρ_B, $E_F^* \equiv \sqrt{(k_F^2 + m^{*2})}$ and m^* is the effective nucleon mass in symmetric *NM*, $m^* = m + \Sigma_s(\sigma)$, with the isoscalar, scalar self-energy $\Sigma_s(\sigma) \equiv -f_\rho \rho_s$ (ρ_s scalar density). $f_i \equiv (\frac{g_i}{m_i})^2$, $i = \sigma, \omega, \delta, \rho$), are the effective meson coupling constants [24].

We see that, when the δ is included, the empirical a_4 value actually corresponds to the combination $[f_\rho - f_\delta(\frac{M}{E_F})^2]$ of the (ρ, δ) coupling constants. Therefore if $f_\delta \neq 0$ we have to increase correspondingly the ρ-coupling.

Now the symmetry energy at saturation density is actually built from the balance of scalar (attractive) and vector (repulsive) contributions, with the scalar channel becoming weaker with increasing baryon density. This is clearly shown in Fig.2(left). This is indeed the isovector counterpart of the saturation mechanism occurring in the isoscalar channel for symmetric nuclear matter.

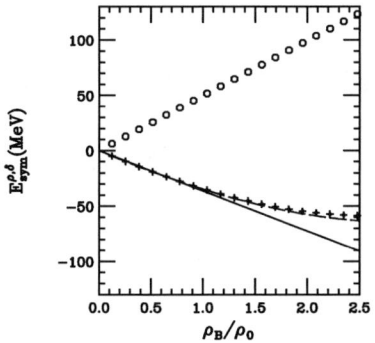
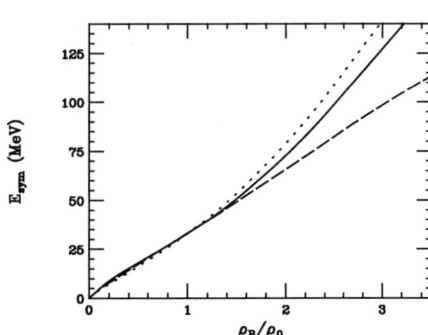

FIGURE 2. *Left: ρ- (open circles) and δ- (crosses) contributions to the potential symmetry energy. The solid line is a linear extrapolation of the low density δ contribution. Right: Total (kinetic + potential) symmetry energy as a function of the baryon density. Dashed line (NLρ). Dotted line (NL$\rho\delta$). Solid line NLHF, only Fock correlations [13].*

In Fig.2(right) we show the total symmetry energy for the different models. At subnuclear densities, $\rho_B < \rho_0$, in both cases, $NL\rho$ and $NL\rho\delta$, from Eq.(10) we have an almost linear dependence of E_{sym} on the baryon density, since $m^* \simeq E_F$ as a good approximation. Around and above ρ_0 we see a steeper increase in the $(\rho + \delta)$ case since m^*/E_F is decreasing.

In conclusion when the δ-channel is included the behaviour of the symmetry energy is stiffer at high baryon density from the relativistic mechanism discussed before. This is in fact due to a larger contribution from the ρ relative to the δ meson. We expect to see these effects more clearly in the relativistic reaction dynamics at intermediate energies, where higher densities are reached.

An important qualitatively new result of the δ-meson coupling is the n/p-effective mass splitting in asymmetric matter, [5] and refs. therein, *Dirac* mass, see later:

$$m_D^*(q) = m + \Sigma_s(\sigma) \pm f_\delta \rho_{S3} , \qquad (11)$$

where + is for neutrons. Since $\rho_{S3} \equiv \rho_{Sp} - \rho_{Sn}$, in n-rich systems we have a neutron *Dirac* effective mass always smaller than the proton one. We note again that a decreasing neutron effective mass in n-rich matter is a direct consequence of the relativistic mechanism for the symmetry energy, i.e. the balance of scalar (attractive) and vector (repulsive) contributions in the isovector channel.

Dirac and Schrödinger Nucleon Effective Masses in Asymmetric Matter

The prediction of a definite $m_D^*(n) < m_D^*(p)$ effective mass splitting in *RMF* approaches, when a scalar $\delta - like$ meson is included, is an important result that requires some further analysis, in particular relative to the controversial predictions of non-relativistic approaches.

Here we are actually discussing the *Dirac* effective masses $m_D^*(n, p)$, i.e. the effective mass of a nucleon in the in-medium Dirac equation with all the meson couplings. The relation to the *Schrödinger* effective masses $m_S^*(n,p)$, i.e. the "k-mass" due to the momentum dependence of the mean field in the non-relativistic in-medium Schrödinger equation is not trivial, see [25, 21, 26, 27]. We will extend the argument of the refs. [26, 27] to the case of asymmetric matter.

We start from the simpler symmetric case without self-interacting terms. The nucleon Dirac equation in the medium contains the scalar self-energy $\Sigma_s = -f_\sigma \rho_S$ and the vector self-energy (fourth component) $\Sigma_0 = f_\omega \rho_B$ and thus the corresponding energy-momentum relation reads:

$$(\varepsilon + m - \Sigma_0)^2 = p^2 + (m + \Sigma_s)^2 = p^2 + m_D^{*2} \qquad (12)$$

i.e. a dispersion relation

$$\varepsilon = -m + \Sigma_0 + \sqrt{p^2 + m_D^{*2}} \qquad (13)$$

From the total single particle energy $E = \varepsilon + m$ expressed in the form $E = \sqrt{k_\infty^2 + m^2}$, where k_∞ is the relativistic asymptotic momentum, using Eq.(12) we can get the relation

$$\frac{k_\infty^2}{2m} = \varepsilon + \frac{\varepsilon^2}{2m} =$$
$$\frac{p^2}{2m} + \Sigma_s + \Sigma_0 + \frac{1}{2m}(\Sigma_s^2 - \Sigma_0^2) + \frac{\Sigma_0}{m}\varepsilon \equiv \frac{p^2}{2m} + U_{eff}(\rho_B, \rho_s, \varepsilon) \qquad (14)$$

i.e. a Schrödinger-type equation with a momentum dependent mean field that with the dispersion relation Eq.(13) is written as

$$U_{eff} = \Sigma_s + \frac{1}{2m}(\Sigma_s^2 + \Sigma_0^2) + \frac{\Sigma_0}{m}\sqrt{(p^2 + m_D^{*2})}$$
$$\simeq \Sigma_s + \frac{\Sigma_0 m_D^*}{m} + \frac{1}{2m}(\Sigma_s^2 + \Sigma_0^2) + \frac{p^2}{2m}\frac{\Sigma_0}{m_D^*} \qquad (15)$$

The relation between Schrödinger and Dirac nucleon effective masses is then

$$m_S^* = \frac{m}{1 + \frac{\Sigma_0}{m_D^*}} = m_D^* \frac{m}{m + \Sigma_s + \Sigma_0} \qquad (16)$$

Since at saturation the two self-energies are roughly compensating each other, $\Sigma_s + \Sigma_0 \simeq -50 MeV$ the two effective masses are not much different, with the $S-mass$ slightly larger than the $D-mass$.

In the case of asymmetric matter, neutron-rich as always considered here, we can have two cases:

- *Only ρ meson coupling*
  Now the scalar part is not modified, we have the same scalar self energies Σ_s for neutrons and protons and so the same Dirac masses. The vector self energies will show an isospin dependence with a new term $\mp f_\rho \rho_{B3}$, repulsive for neutrons ($-$ sign, since we use the definition $\rho_{(B,S)3} \equiv \rho_{(B,S)p} - \rho_{(B,S)n}$). As a consequence we see a splitting at the level of the Schrödinger masses since Eq.(16) becomes

$$m_S^*(n,p) = m_D^* \frac{m}{m + (\Sigma_s + \Sigma_0)_{sym} \mp f_\rho \rho_{B3}} \quad (17)$$

in the direction of $m_S(n)^* < m_S(p)^*$ (here and in the following upper signs are for neutrons).

- *$\rho + \delta$ coupling*
  The above splitting is further enhanced by the direct effect of the scalar isovector coupling, see Eq.(11). Then the Schrödinger masses are

$$m_S^*(n,p) = (m_{D\,sym}^* \pm f_\delta \rho_{S3}) \frac{m}{m + (\Sigma_s + \Sigma_0)_{sym} \mp (f_\rho \rho_{B3} - f_\delta \rho_{S3})} \quad (18)$$

The new term in the denominator will further contribute to the $m_S^*(n) < m_S^*(p)$ splitting since we must have $f_\rho > f_\delta$ in order to get a correct symmetry parameter a_4. The effect is larger at higher baryon densities because of the decrease of the scalar ρ_{S3}, due to the faster $\frac{m_{Dn}^*}{E_{Fn}^*}$ reduction of ρ_{Sn}.

Actually in a non-relativistic limit we can approximate in Eq.(14) directly the energy ε with the asymptotic kinetic energy leading to the much simpler relation:

$$m_S^* = \frac{m}{1 + \frac{\Sigma_0}{m}} \simeq m - \Sigma_0 = m_D^* - (\Sigma_s + \Sigma_0). \quad (19)$$

In the case of asymmetric matter this leads to the more transparent relations

- *Only ρ meson coupling*

$$m_S^*(n,p) = m_D^* - (\Sigma_s + \Sigma_0)_{sym} \pm f_\rho \rho_{B3} \quad (20)$$

- *$\rho + \delta$ coupling*

$$m_S^*(n,p) = m_{D\,sym}^* - (\Sigma_s + \Sigma_0)_{sym} \pm (f_\rho \rho_{B3} - f_\delta \rho_{S3}) \quad (21)$$

In conclusion any *RMF* model will predict the definite isospin splitting of the nucleon effective masses $m_S(n)^* < m_S(p)^*$ in the non-relativistic limit. We expect an increase of the difference of the neutron/proton mean field at high momenta, with important dynamical contributions that will enhance the transport effects of the symmetry energy.

Beyond RMF: k-dependence of the Self-Energies

Correlations beyond the Mean Field picture will lead to a further density and, most important, momentum dependence of the effective couplings and consequently of the Self-Energies. This can be clearly seen from microscopic Dirac-Brueckner (*DBHF*) calculations, see [9, 10, 11] and refs. therein. Even the basic Fock correlations are implying a density dependence of the couplings, see [13]. In fact from the very first relativistic transport calculations a momentum dependence of the vector fields was required in order to reduce the too large repulsion of the otpical potential at high momenta, see [28]. From the above discussion we can expect that, when the δ-meson is included [29], the sign of the *Dirac*-mass splitting will not be modified, i.e. $m_{Dn}^* < m_{Dp}^*$. In fact this is the result of the microscopic *DBHF* calculations in asymmetric matter of refs.[9, 10, 11]. The problem is however that in correspondence the Schrödinger mass splitting can have any sign. Indeed this is the case of the very recent *DBHF* analysis of the Tübingen group [11, 30]. For the S-masses they get an opposite behavior $m_{Sn}^* > m_{Sp}^*$, with large fluctuations around the Fermi momenta, due to the opening of the phase space for intermediate inelastic channels. These results are actually dependent on the way the self-energies (mean fields) are derived from the full correlated theory and different predictions can be obtained, see the recent refs.[31, 32, 33].

All that clearly shows that experimental observables are in any case strongly needed.

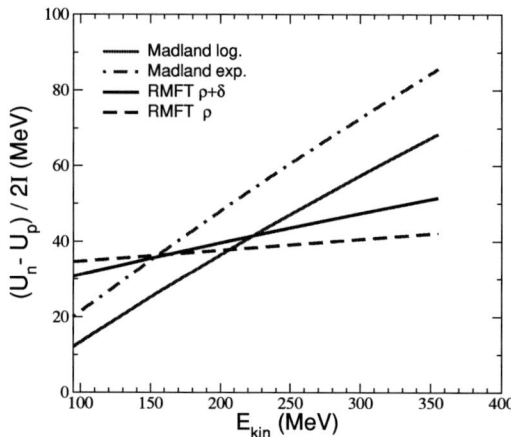

FIGURE 3. *Energy dependence of the Dirac-Lane potential in the RMF picture (solid: NLρ; dashed: NLρδ) and in the phenomenologic Dirac Optical Model of Madland et al. [37, 38], see text.*

The Dirac-Lane Potential

In the non-relativistic limit of the Eq.(14) we can easily extract the neutron-proton mean optical potential using the isospin dependence of the self-energies

$$\Sigma_{0,q} = \Sigma_{0,sym} \mp f_\rho \rho_{B3} ,$$
$$\Sigma_{s,q} = \Sigma_{s,sym} \pm f_\delta \rho_{S3} , \qquad (22)$$

With some algebra we get a compact form of the *Dirac − Lane Potential*

$$U_{Dirac-Lane} \equiv \frac{U_n - U_p}{2I} =$$
$$\rho_0 \left[f_\rho (1 - \frac{\Sigma_{0,sym}}{m}) - f_\delta \frac{\rho_{S3}}{\rho_{B3}}(1 + \frac{\Sigma_{s,sym}}{m}) \right] + f_\rho \frac{\rho_0}{m} \varepsilon \qquad (23)$$

It is interesting to compare with the related discussion presented for the non-relativistic effective forces, mainly of Skyrme-like form. First of all we predict a definite positive $E − slope$ given by the quantity $f_\rho \frac{\rho_0}{m}$, which is actually not large within the simple *RMF* picture described here. This is a obvious consequence of the fact that the "relativistic" mass splitting is always in the direction $m_S(n)^* < m_S(p)^*$. The realistic magnitude of the effect could be different if we take into account that some explicit momentum dependence should be included in the scalar and vector self energies, [28], as also discussed before. An interesting point in this direction comes from the phenomenological Dirac Optical Potential (*Madland − potential*) constructed in the refs. [37, 38], fitting simultaneously proton and neutron (mostly total cross sections) data for collisions with a wide range of nuclei at energies up to 100 *MeV*. Recently this Dirac optical potential has been proven to reproduce very well the new neutron scattering data on ^{208}Pb at 96 *MeV* [39] measured at the Svendberg Laboratory in Uppsala.

The phenomenological *Madland − potential* has different implicit momentum dependences (*exp/log*) in the self-energies, [37, 38]. In Fig.3 we show the corresponding energy dependences for the *Dirac − Lane* potentials, compared to our *NLρ* and *NLρδ* estimations. When we add the δ-field we have a larger slope since the $ρ − coupling$ should be increased. The slopes of the phenomenological potentials are systematically larger, interestingly similar to the ones of the *Skyrme − Lyon* forces. We note that in the *Madland − potential* the Coulomb interaction is included, i.e. an extra repulsive vector contribution for the protons. The first, not energy dependent, term of Eq.(23) is directly related to the symmetry energy at saturation, exactly like in the non-relativistic case, see Eq.(7).

The conclusion is that a good *systematic* measurement of the Lane potential in a wide range of energies, and particularly around/above 100 *MeV*, would answer many fundamental questions in isospin physics.

REACTION OBSERVABLES

In order to directly test the influence of the *S*-effective mass splitting we will present results from reaction dynamics at intermediate energies analysed in a non-relativistic transport approach, of *BNV* type, see [5]. The *Iso − MD* effective interaction is derived via an asymmetric extension of the *GBD* force, [34, 35], constructed in order to have a strict correspondence to the Skyrme forces, see also [36].

The energy density can be parametrized as follows:

$$\varepsilon = \varepsilon + \varepsilon_{kin} + \varepsilon(A', A'') + \varepsilon(B', B'') + \varepsilon(C', C'') \tag{24}$$

where ε_{kin} is the usual kinetic energy density and

$$\varepsilon(A', A'') = (A' + A''\beta^2)\frac{\rho^2}{\rho_0}$$

$$\varepsilon(B', B'') = (B' + B''\beta^2)\left(\frac{\rho}{\rho_0}\right)^\sigma \rho$$

$$\varepsilon(C', C'') = C'(\mathscr{I}_{NN} + \mathscr{I}_{PP}) + C''\mathscr{I}_{NP} \tag{25}$$

Here $\mathscr{I}_{\tau\tau'}$ are integrals of the form:

$$\mathscr{I}_{\tau\tau'} = \int d\vec{p}\, d\vec{p}'\, f_\tau(\vec{r},\vec{p}) f_{\tau'}(\vec{r},\vec{p}') g(\vec{p},\vec{p}')$$

with $g(\vec{p},\vec{p}') = g(\vec{p}-\vec{p}')^2$ This choice of the function $g(\vec{p},\vec{p}')$ corresponds to a Skyrme-like behaviour and it is suitable for *BNV* simulations. In this frame we can easily adjust the parameters in order to have the same density dependence of the symmetry energy *but with two opposite n/p effective mass splittings*. So we can separately study the correspondent dynamical effects, [36]. In Fig. 4 we show the *E*-dependence of the Lane potentials with the parametrizations corresponding to the two choices of the sign of the n/p mass splitting (shown in the insert for the $I = 0.2$ asymmetry). The value of the splitting is exactly the same, only the sign is opposite.

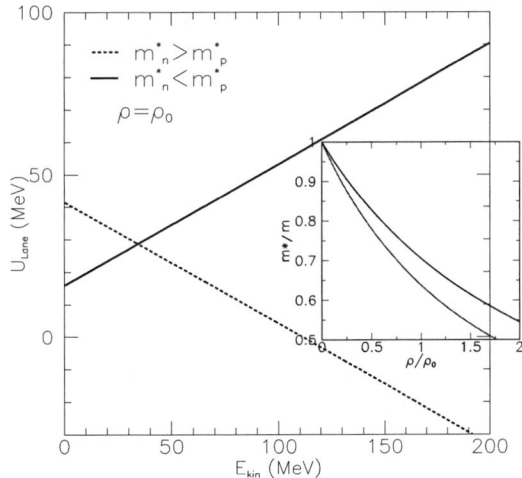

FIGURE 4. *Lane potential for the used parametrizations; small panel: the effective mass splitting, related to the slope of the Lane potential.*

The upper curve well reproduces the Skyrme-Lyon (in particular $SLy4, Sly7$) results. The lower (dashed) the $SIII, SKM^*$ ones. We clearly see the crossing at low energy, as expected from the previous discussion.

From the figure we can immediately derive the expectation of very different symmetry effects for nucleons around $100 MeV$ kinetic energy, enhancement (with larger neutron repulsion) in the $m_n^* < m_p^*$ case vs. a disappearing (and even larger proton repulsion) in the $m_n^* > m_p^*$ choice. So it seems natural to look at observables where neutron/proton mean fields at high momentum are playing an important role. We will show results for fast nucleon emissions and collective flows at intermediate energies, in particular for high transverse momentum selections.

Isotopic content of the fast nucleon emission

We have performed realistic "ab initio" simulations of collisions at intermediate energies of n-rich systems, in particular $^{132}Sn + ^{124}Sn$ at 50 and $100, AMeV$, central selection. Performing a local low density selection of the test particles ($\rho < \rho_0/8$) we can follow the time evolution of nucleon emissions (gas phase) and the corresponding asymmetry. In Fig. 5 we show the evolution of the gas asymmetry ($I(t) \equiv (N-Z)/A$) for the $50AMeV$ reaction (the solid line gives the initial asymmetry). For both choices of the density stiffness of the symmetry term we clearly see the effects of the mass splitting, resulting in a reduced fast neutron emission when $m_n^* > m_p^*$.

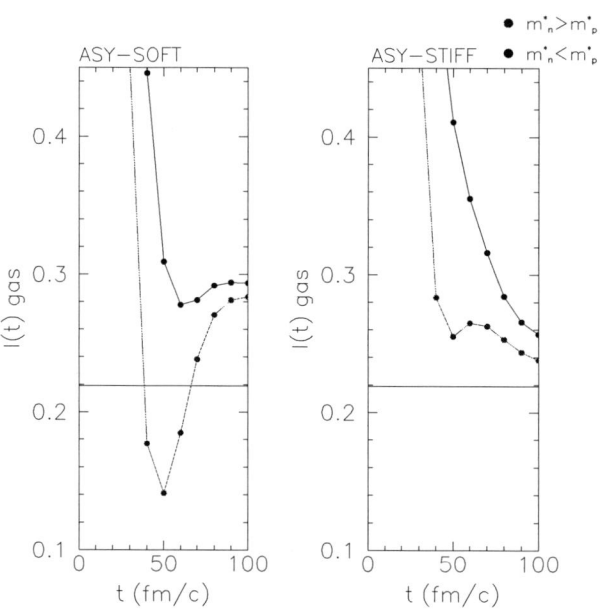

FIGURE 5. *Isospin gas content as a function of time in a central collision $^{132}Sn + ^{124}Sn$ at $50AMeV$ for two opposite choices of mass splitting. Left: asy-soft EOS; right: asy-stiff EOS.*

In order to better disentangle the mass splitting effect and to select the corresponding observables, in Figs. 6, 7 we report the N/Z of the "gas" at two different times, $t = 60 fm/c$, end of the pre-equilibrium emission, and $t = 100 fm/c$ roughly freeze out time. We have followed the transverse momentum dependence, for a fixed central rapidity $y^{(0)}$ (normalized to projectile rapidity). In Fig. 6 we show the results *without the mass splitting effect*, i.e. only taking into account the different repulsion of the symmetry term (the intial average asymmetry is $N/Z = 1.56$).

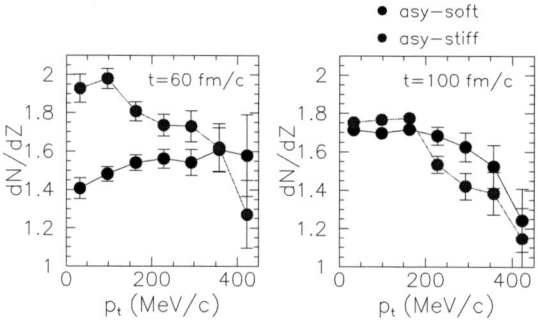

FIGURE 6. *Transverse momentum dependence of the neutron/proton ratio in the rapidity range $|y^{(0)}| \leq 0.3$ for a central reaction $^{132}Sn + ^{124}Sn$ at $50AMeV$. Comparison between two choices of symmetry energy stiffness at different times.*

At early times (left panel of Fig. 6), when the emission from high density regions is dominant, we see a difference due to the larger neutron repulsion of the the asy-stiff choice. The effect is reduced at higher p_ts due to the overall repulsion of the isoscalar momentum dependence. Finally at freeze out (right panel) the difference is almost disappearing even for the more efficient *Isospin Distillation* of the asy-soft choice during the expansion phase, [5] and refs. therein.

When we introduce the mass splitting the difference in the isotopic content of the gas, for the larger transverse momenta, is very evident at all times, in particular at the freeze-out of experimental interest, see Fig. 7. As expected the $m_n^* > m_p^*$ sharply reduces the neutron emission at high p_ts.

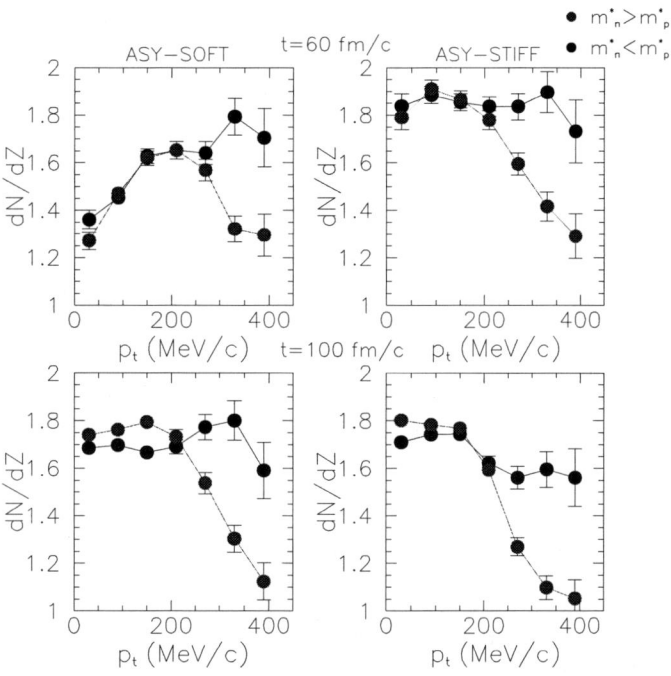

FIGURE 7. *Same as Fig. 6, for two opposite choices of mass splitting.*

We have repeated the analysis at higher energy, 100$AMeV$, and the effect appears nicely enhanced, see Fig. 8.

Similar results have been obtained in ref. [40] at 400$AMeV$ in a different $Iso-MD$ model, restricted to the $m_n^* > m_p^*$ choice. In conclusion it appears that the isospin content of pre-equilibrium emitted nucleons at high transverse momentum can solve the mass-splitting puzzle.

Differential collective flows and mass splittings

Collective flows are expected to be very sensitive to the momentum dependence of the mean field, see [5] and refs.therein. We have then tested the isovector part of the momentum dependence just evaluating the *Differential* transverse and elliptic flows

$$V_{1,2}^{(n-p)}(y,p_t) \equiv V_{1,2}^n(y,p_t) - V_{1,2}^p(y,p_t)$$

at various rapidities and transverse momenta in semicentral ($b/b_{max} = 0.5$) $^{197}Au + ^{197}Au$ collisons at 250$AMeV$, where some proton data are existing from the $FOPI$ collaboration at GSI [41, 42].

Transverse flows

For the differential transverse flows, see Fig. 9 the mass splitting effect is evident at all rapidities, and nicely increasing at larger rapidities and transverse momenta, with more neutron flow when $m_n^* < m_p^*$

Just to show that our simulations give realistic results we compare in lower right panel of Fig. 9 with the proton data of the $FOPI$ collaboration for similar selections of impact parameters rapidities and transverse momenta.

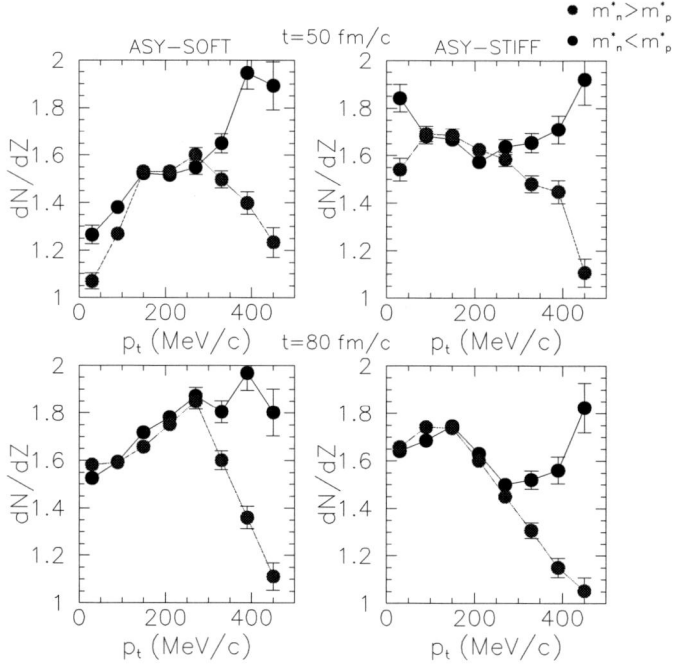

FIGURE 8. *Same as Fig. 7, but for an incident energy of* $100AMeV$.

The agreement is quite satisfactory. We see a slightly reduced proton flow at high transverse momenta in the $m_n^* < m_p^*$ choice, but the effect is too small to be seen fron the data. Our suggestion of the differential flows looks much more promising.

Elliptic flows

The same analysis has been performed for the differential elliptic flows, see Fig. 10. Again the mass splitting effects are more evident for higher rapidity and tranverse momentum selections. In particular the differential elliptic flow becomes sistematically negative when $m_n^* < m_p^*$, revealing a faster neutron emission and so more neutron squeeze out (more spectator shadowing).

In the lower right panel we also show a comparison with recent proton data from the *FOPI* collaboration. The agreement is still satisfactory. As expected the proton flow is more negative (more proton squeeeze out) when $m_n^* > m_p^*$. It is however difficult to draw definite conclusions only from proton data.

Again the measurement of differential flows appears essential. This could be in fact an experimental problem due to the difficulties in measuring neutrons. Our suggestion is to measure the difference between light isobar flows, like triton vs. ^3He and so on. We expect to clearly see the effective mass splitting effects, maybe even enhanced due to larger overall flows, see [5, 43].

REFERENCES

1. D.Vautherin and D.M.Brink, Phys.Rev. C3 (1972) 676.
2. E.Chabanat, P.Bonche, P.Haensel, J.Meyer, R.Schaeffer, Nucl.Phys. A627 (1997) 710.
3. H.Krivine, J.Treiner and O.Bohigas, Nucl.Phys. A336 (1980) 155.
4. E.Chabanat, P.Bonche, P.Haensel, J.Meyer, R.Schaeffer, Nucl.Phys. A635 (1998) 231.
5. V.Baran, M.Colonna, V.Greco, M.Di Toro, Physics Reports 410 (2005) 335-466.
6. F.Douchin, P.Haensel and J.Meyer, Nucl.Phys. A665 (2000) 419.
7. M.Colonna, M.Di Toro, A.Larionov, Phys.Lett. B428 (1998) 1.
8. M.Kutschera, W.Wojcik, Phys.Lett. B325 (1994) 217.

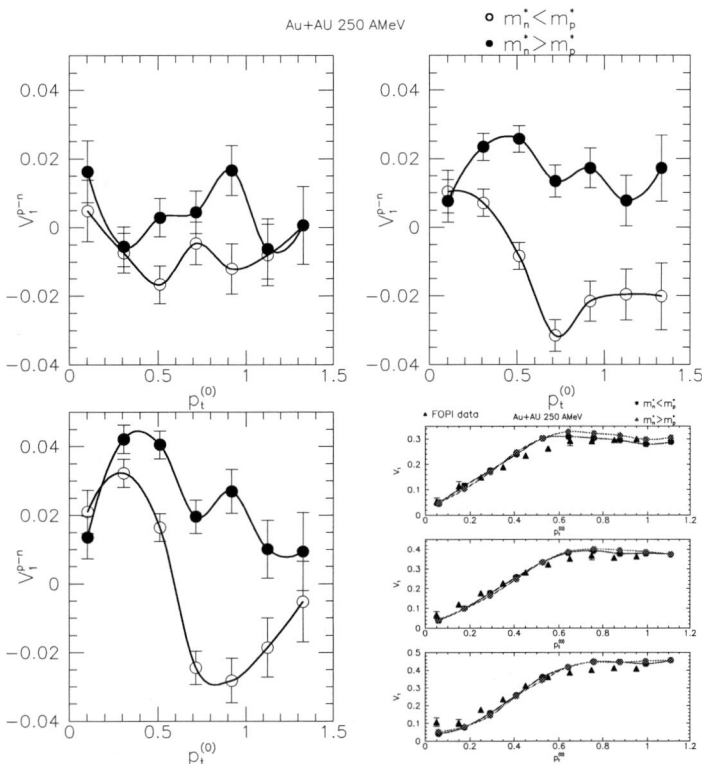

FIGURE 9. *Difference between proton and neutron V_1 flows in a semi-central reaction Au+Au at 250 AMeV for three rapidity ranges. Upper Left Panel: $|y^{(0)}| \leq 0.3$; Upper Right: $0.3 \leq |y^{(0)}| \leq 0.7$; Lower Left: $0.6 \leq |y^{(0)}| \leq 0.9$. Lower Right Panel: Comparison of the V_1 proton flow with FOPI data [41] for three rapidity ranges. Top: $0.5 \leq |y^{(0)}| \leq 0.7$; center: $0.7 \leq |y^{(0)}| \leq 0.9$; bottom: $0.9 \leq |y^{(0)}| \leq 1.1$.*

9. F.Hofmann, C.M.Keil, H.Lenske, Phys.Rev. C64 (2001) 034314 .
10. E.Schiller, H.Müther, Eur.Phys.J. A11 (2001) 15.
11. E.N.E.van Dalen, C.Fuchs, A.Faessler, Nucl.Phys. A744 (2004) 227.
12. B.Liu, V.Greco, V.Baran, M.Colonna, M.Di Toro, Phys.Rev. C65 (2002) 045201.
13. V.Greco, M.Colonna, M.Di Toro, F.Matera, Phys.Rev. C67 (2003) 015203.
14. I.Bombaci, "EOS for isospin-asymmetric nuclear matter for astrophysical applications", in [15] pp. 35-81 and refs. therein.
15. *Isospin Physics in Heavy-ion Collisions at Intermediate Energies*, Eds. Bao-An Li and W. Udo Schröder, Nova Science Publishers (2001, New York).
16. W.Zuo, I.Bombaci, U.Lombardo, Phys.Rev. C60 (1999) 24605.
17. A.M.Lane, Nucl.Phys. 35 (1962) 676.
18. F.D.Becchetti, G.W.Greenless, Phys.Rev. 182 (1969) 1190.
19. P.E.Hodgson, *The Nucleon Optical Model*, World Scientific, 1994
20. J.D.Walecka, Ann.Phys.(N.Y.) 83 (1974) 491.
21. B.D.Serot, J.D.Walecka in *Advances in Nuclear Physics* Vol. 16, Eds. J.M.Negele and E.Vogt, Plenum, New York, 1986.
22. B.D.Serot, J.D.Walecka, Int.J.Mod.Phys. E6 (1997) 515.
23. S.Kubis, M.Kutschera, Phys.Lett. B399 (1997) 191.
24. The $NL\rho$, $NL\rho\delta$ parametrizations used here have been derived from transport studies of relativistic heavy ion collisions, see Ch.8 of ref. [5].
25. A.Bouyssy, J.F.Mathiot, N.Van Giai and S.Marcos, Phys.Rev. C36 (1987) 380.
26. M.Jaminon, C.Mahaux, P.Rochus, Phys.Rev. C22 (1980) 2027.
27. M.Jaminon, C.Mahaux, Phys.Rev. C40 (1989) 354.
28. B.Blättel, V.Koch and U.Mosel, Rep.Prog.Phys. 56 (1993) 1.
29. We like to remind that an isovector scalar channel in the effective interaction "automatically" appear when correlations are accounted for, see discussion in [5].
30. E. van Dalen, C.Fuchs, A.Fässler, *Effective Nucleon Masses in Symmetric and Asymmetric Nuclear Matter*, arXiv:nucl-th/0502064.
31. Z-Y.Ma et al., Phys.Lett. B606 (2004) 170.
32. D.Alonso and F.Sammarruca, arXiv:nucl-th/0301032

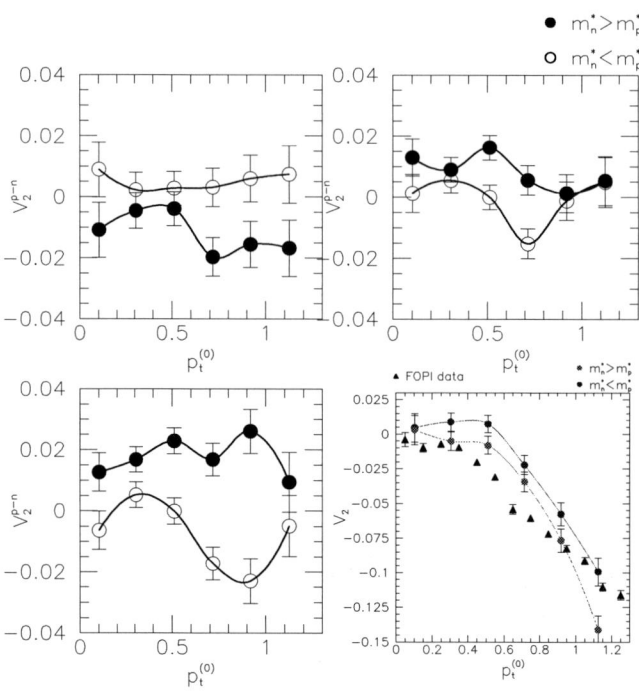

FIGURE 10. *Upper (left and right) and lower left panels: Difference between proton and neutron elliptic flows for the same reaction and rapidity ranges as in Fig. 9. Lower right panel: Comparison of the elliptic proton flow with FOPI data [42] (M3 centrality bin, $|y^{(0)}| \leq 0.1$).*

33. F.Sammarruca, W.Barredo, P.Krastev, arXiv:nucl-th/0411053v2.
34. C.Gale, G.F.Bertsch, S.Das Gupta, Phys.Rev. C41 (1990) 1545.
35. V.Greco, Diploma Thesis (1997);
    V.Greco, A.Guarnera, M.Colonna, M.Di Toro, Phys.Rev. C59 (1999) 810.
    Nuovo Cimento A111 (1998) 865.
36. J.Rizzo, M.Colonna, M.Di Toro, V.Greco, Nucl.Phys. A732 (2004) 202.
37. R.Kozack, D.G.Madland, Phys.Rev. C39 (1989) 1461.
38. R.Kozack, D.G.Madland, Nucl.Phys. A509 (1990) 664.
39. J.Klug et al., Phys.Rev. C67 (2003) 0316001R
    ibidem Phys.Rev. C68 (2003) 064605.
40. B.-A.Li, C.B.Das, S.Das Gupta, C.Gale, Nucl.Phys. A735 (2004) 563.
41. A.Andronic et al., FOPI Collab., Phys.Rev. C67 (2003) 034907.
42. A.Andronic et al., FOPI Collab., Phys.Lett. B612 (2005) 173.
43. L.Scalone, M.Colonna, M.Di Toro, Phys.Lett. B461 (1999) 9.

Isoscaling and symmetry energy in dynamical fragment formation

Akira Ono

Department of Physics, Tohoku University, Sendai 980-8578, Japan

Abstract. In medium energy heavy ion collisions, fragments are formed in an expanding system. The isospin composition of the produced fragments may reflect the symmetry energy of such low-density nuclear matter. In fact, the simulations by antisymmetrized molecular dynamics (AMD) show that the fragment isospin composition is basically consistent with the statistical expectations even in dynamically evolving system. Isoscaling is satisfied by the AMD results. The width of the fragment isotope distribution can be explained by the ratio of the symmetry energy to the temperature if the symmetry energy at a reduced density is assumed to be relevant. This assumption is justified by studying the dependence on the density-dependent symmetry energy. The symmetry energy extracted from the AMD results is almost independent of the fragment size, which suggests that the fragment isospin composition is governed by the symmetry energy of low-density uniform matter rather than the symmetry energies for isolated nuclei.

Keywords: Isoscaling, symmetry energy, multifragmentation, antisymmetrized molecular dynamics
PACS: 25.70.Pq

1. INTRODUCTION

In typical medium energy nuclear collisions, numerous fragments of intermediate size are produced in addition to light particles [1]. Dynamical simulations show that such fragments are produced in dynamically expanding nuclear matter when the density is lower than the normal nuclear matter density. Therefore it may be expected that fragments carry some information of the property of low density nuclear matter, especially the equation of state (EOS)

$$E(\rho,\delta)/A = E(\rho,\delta=0)/A + C_{\text{sym}}(\rho)\delta^2, \tag{1}$$

which shows the energy per nucleon at zero temperature as a function of $\rho = \rho_n + \rho_p$ and $\delta = (\rho_n - \rho_p)/\rho$, and $C_{\text{sym}}(\rho)$ is the symmetry energy which is a function of the density. By the experimental and theoretical studies of multifragmentation with various neutron-to-proton ratios, it may be possible to extract the information of the symmetry energy of low density nuclear matter. Multifragmentation may also be regarded as a mixed phase system of nuclei and nucleons, which is also found in supernova collapses [2] and in the inner crusts of neutron stars [3, 4].

One of the possible scenarios of multifragmentation is that, if the neutron-proton density difference $\rho_n(\mathbf{r}) - \rho_p(\mathbf{r})$ is governed by thermal equilibrium in the low density matter, the fluctuation of this quantity can be related to the ratio of the symmetry energy at this density to the temperature. And, if this density fluctuation is directly reflected in the formed fragments, the isospin composition of fragments can be related to the symmetry energy at low density. Another possibility is, as assumed in statistical multifragmentation models [5, 6, 7, 8], that the fragment isospin composition is governed by the equilibrium of fragments in a freeze-out volume, and then the symmetry energies of individual nuclei (rather than the symmetry energy of uniform matter) will be reflected in the observables. The surface of the fragments will play an important role in the symmetry energy, which may be somehow different from that of the ground state nuclei because of the finite temperature, in principle. Yet another possibility is that dynamics plays important role in the real collisions and that it is not possible to explain the fragment isospin composition by merely the thermal effect, which has been predicted by models based on mean field dynamics [9, 10]. Even in this case, the results reflect the density dependence of the symmetry energy, though the relation is not as simple as that under equilibrium.

In order to investigate the above issue and distinguish the possible scenarios, we perform the antisymmetrized molecular dynamics (AMD) simulations and analyze the fragment isospin composition paying attention to the dependence on the density-dependent symmetry energy. The AMD model does not assume any equilibrium and therefore we can answer the question whether equilibrium has any sense in the real collisions. It is also known that AMD can describe the equilibrium situations reasonably well if it is applied to thermalized ideal systems. It should be noted that some degrees of freedom (such as the fragment isospin composition) may be explained by the equilibrium idea even though other degrees of freedom (such as the nucleon kinetic energy) are not thermalized.

The isoscaling relation plays an important role in analyzing the fragment isospin composition and relating it to the symmetry energy. Isoscaling means the relation

$$Y_2(N,Z)/Y_1(N,Z) \propto e^{\alpha N + \beta Z}, \qquad (2)$$

where $Y_i(N,Z)$ are the fragment yields from two similar reaction systems $i = 1$ and 2 with different neutron-to-proton ratios. Isoscaling has been observed in the experimental data [11] and in the predictions by various statistical models [8] and dynamical models [12, 9]. The isoscaling parameter α (and β) can be related to the symmetry energy. Isoscaling is also useful to combine the results of different reaction systems to obtain universal quantities [13].

2. AMD MODEL

2.1. Framework

The description of dynamics of nuclear multifragmentation reactions is a very complicated quantum many-body problem. If the many-body time-dependent Schrödinger equation is solved for an initial state which may be roughly approximated by a single Slater determinant, the intermediate and final states will be a very complicated states containing a huge number of reaction channels corresponding to different fragmentation configurations. The AMD model respects the existence of channels, while neglecting some of the interference among them. Namely, the total many-body wave function $|\Psi(t)\rangle$ is approximated by

$$|\Psi(t)\rangle\langle\Psi(t)| \approx \int \frac{|\Phi(Z)\rangle\langle\Phi(Z)|}{\langle\Phi(Z)|\Phi(Z)\rangle} w(Z,t) dZ, \qquad (3)$$

where the channel wave function $|\Phi(Z)\rangle$ is parametrized by a set of parameters Z, and $w(Z,t)$ is the time-dependent weight of each channel.

In AMD, we choose the Slater determinant of Gaussian wave packets as the channel wave function

$$\langle \mathbf{r}_1 \ldots \mathbf{r}_A | \Phi(Z) \rangle \propto \det_{ij}\left[\exp\left\{ -\nu(\mathbf{r}_i - \mathbf{Z}_j/\sqrt{\nu})^2 \right\} \chi_{\alpha_j}(i) \right], \qquad (4)$$

where χ_{α_i} are the spin-isospin states with $\alpha_i = p\uparrow, p\downarrow, n\uparrow$, or $n\downarrow$. Thus the many-body state $|\Phi(Z)\rangle$ is parametrized by a set of complex variables $Z \equiv \{\mathbf{Z}_i\}_{i=1,\ldots,A}$, where A is the number of nucleons in the system. The width parameter $\nu = 0.16$ fm^{-2} is treated as a constant parameter common to all the wave packets. If we ignore the antisymmetrization effect, the real part of $\mathbf{Z}_i$ corresponds to the position centroid and the imaginary part corresponds to the momentum centroid. This choice of channel wave functions is very suitable for the study of fragmentation reactions, because we can easily avoid the spurious correlations among independent fragmentation channels.

Instead of directly considering the weight function $w(Z,t)$ in Eq. (3), we solve a stochastic equation of motion for the wave packet centroids Z, which may be symbolically written as

$$\frac{d}{dt}\mathbf{Z}_i = \{\mathbf{Z}_i, \mathcal{H}\}_{\text{PB}} + (\text{NN coll}) + \Delta \mathbf{Z}_i(t) + \mu(\mathbf{Z}_i, \mathcal{H}'). \qquad (5)$$

The first term $\{\mathbf{Z}_i, \mathcal{H}\}_{\text{PB}}$ is the deterministic term derived from the time-dependent variational principle with an assumed effective interaction. The second term represents the effect of stochastic two-nucleon collision process, where a parametrization of the energy-dependent in-medium cross section is adopted. The collisions are performed with the "physical nucleon coordinates" that take account of the antisymmetrization effects, and then the Pauli blocking in the final state is automatically introduced [14]. The third term $\Delta \mathbf{Z}_i(t)$ is a stochastic fluctuation term that has been introduced in order to respect the change of the shape of the single particle distribution even with the fixed width in each channel $|\Phi(Z)\rangle$ [15, 16, 17]. In other words, the combination $\{\mathbf{Z}_i, \mathcal{H}\}_{\text{PB}} + \Delta \mathbf{Z}_i(t)$ approximately reproduces the prediction by mean field theories (for a short time period) for the ensemble-averaged single-particle distribution, while each nucleon is localized in phase space for each channel. The term $\Delta \mathbf{Z}_i(t)$ is calculated practically by solving the Vlasov equation (for a short time period) with the same effective interaction as for the term $\{\mathbf{Z}_i, \mathcal{H}\}_{\text{PB}}$. The last term

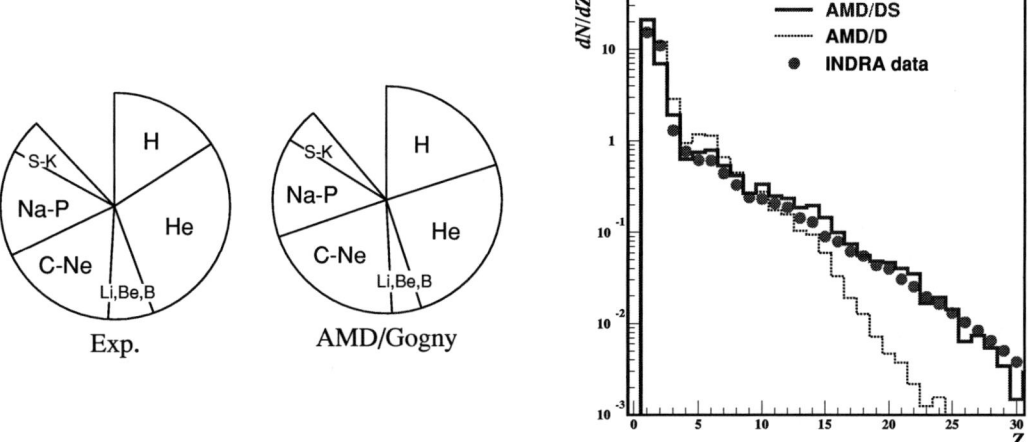

FIGURE 1. **Left:** Partition of the total charge of the system into the fragments for ^{40}Ca + ^{40}Ca collisions at 35 MeV/nucleon. The result of AMD with the Gogny force is compared with the experimental data [29] after applying the filter program to simulate the experimental conditions. **Right:** The charge distribution of the produced clusters in ^{129}Xe + Sn collisions at 50 MeV/nucleon with the impact parameter $0 < b < 4$ fm, after calculating the statistical decay of excited clusters and applying the experimental filter for the detector setup. Solid histogram shows the result of AMD/DS, while the dotted histogram shows the result of AMD/D [17]. The INDRA experimental data [30] are shown by solid points.

$\mu(\mathbf{Z}_i, \mathcal{H}')$ is a dissipation term related to the fluctuation term $\Delta \mathbf{Z}_i(t)$. The dissipation term is necessary in order to restore the conservation of energy that is violated by the fluctuation term. The coefficient μ is given by the condition of energy conservation. However, the form of this term is somehow arbitrary. We shift the variables Z to the direction of the gradient of the energy expectation value $\mathcal{H}$ under the constraints of conserved quantities (the center-of-mass variables and the total angular momentum) and global one-body quantities (monopole and quadrupole moments in coordinate and momentum spaces). The latter constraints should be imposed because the one-body time evolution is already considered by $\{\mathbf{Z}_i, \mathcal{H}\}_{\text{PB}} + \Delta \mathbf{Z}_i(t)$. A complete formulation of AMD can be found in Ref. [28].

2.2. Typical applications

AMD has been successfully applied to fragmentation reactions, such as ^{40}Ca + ^{40}Ca collisions at 35 MeV/nucleon [15, 18] and ^{129}Xe + Sn central collisions at 50 MeV/nucleon [17]. The copious fragment formation is very well explained by the AMD simulations as shown in Fig. 1 for these reaction systems. The fluctuation term $\Delta \mathbf{Z}_i(t)$ and the related dissipation term play an essential role to the proper description of fragment formation. The same version of AMD can also be applied to study ideally equilibrated systems [20]. For example, by solving the time evolution of a many-nucleon system confined in a box for a very long time, a microcanonical ensemble (with fixed energy, volume and nucleon number) is obtained. By generating ensembles with various energies and volumes, we can construct the microcanonical constant-pressure caloric curves, which shows the existence of negative heat capacity corresponding to the first-order liquid-gas phase transition in systems with a finite nucleon number.

2.3. Effective force

In the present work, we perform reaction simulations employing two different effective forces in order to study effects of the asymmetry term within the forces. One is the usual Gogny force [21], consistent with the saturation of symmetric nuclear matter at the incompressibility $K = 228$ MeV. The force is composed of finite-range two-body terms and of a density-dependent term of the form $t_3 \rho^{1/3}(1 + P_\sigma)\delta(\mathbf{r}_1 - \mathbf{r}_2)$, where P_σ is the spin exchange operator and t_3 is a coefficient. The second force (called Gogny-AS force) is obtained by modifying the Gogny force with

$$V_{\text{Gogny-AS}} = V_{\text{Gogny}} - (1-x)t_3 \left(\rho(\mathbf{r}_1)^{1/3} - \rho_0^{1/3} \right) P_\sigma \delta(\mathbf{r}_1 - \mathbf{r}_2), \tag{6}$$

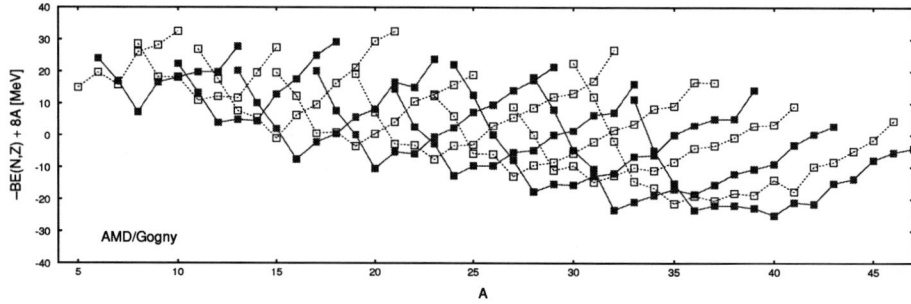

FIGURE 2. Binding energies (not per nucleon) of nuclei calculated by AMD with the Gogny force. The quantities $-\text{BE}(N,Z) + 8(N+Z)$ MeV are shown by filled and open squares for even-Z and odd-Z nuclei, respectively. Lines connect isotopes.

where $x = -\frac{1}{2}$ and $\rho_0 = 0.16$ fm^{-3}. The two forces coincide at $\rho = \rho_0$. Furthermore, they produce the same EOS of symmetric nuclear matter at all densities. However, the two forces produce different density dependences of the symmetry energy, as shown in the right panel of Fig. 8. The choice of $x = -\frac{1}{2}$ has been made to ensure that the part of the symmetry energy from the direct interaction term is proportional to the density [22]. At densities below ρ_0, the Gogny force has somewhat higher symmetry energy than the Gogny-AS force. At densities above ρ_0, the Gogny-AS symmetry energy continues to rise while the Gogny symmetry energy begins to fall, so that significant differences develop. However, the difference in the density dependence of symmetry energy for these forces is not as extreme as for those used in other works such as Ref. [9].

2.4. Ground state nuclei

Even though the AMD wave function of Eq. (4) may seem very simple, it reproduces ground state binding energies of nuclei to within about 0.5 MeV/nucleon. Figure 2 shows the binding energies of nuclei $\text{BE}(N,Z)$ with $A \lesssim 40$ which are obtained by minimizing the energy within AMD by adopting the Gogny force [13]. In order to remove the strong A-dependence, we plot $-\text{BE}(N,Z) + 8A$ MeV.

For the purpose of the present study, it is useful to extract the symmetry energy from these calculated binding energies by fitting them with a liquid-drop mass formula which contains a symmetry energy term such as

$$E_{\text{sym}}(N,Z) = c(A) \frac{(N-Z)^2}{A}. \tag{7}$$

In the simplest mass formula [23], the coefficient is independent of the nuclear size, $c(A) = a_{\text{sym}}$, which assumes the volume nature of the symmetry energy. On the other hand, advanced mass formulas [24, 25, 26] have introduced the A-dependence of $c(A)$ as the surface effect. The extraction of $c(A)$ from the nuclear binding energies is not quite straightforward even for ground state nuclei. For example, if one extracts the symmetry energy by using the energy difference of neighboring nuclei [27], the symmetry energy is largely a fluctuating function in the nuclear chart due to shell and paring effects. Nevertheless, the global fitting with the assumption of $c(A) = c_v + c_s A^{-1/3}$, together with the standard volume, surface, Coulomb and paring terms, usually results in a reasonable value of the coefficients. If we fit the AMD binding energies in Fig. 2 for $7 \leq A \leq 40$, we obtain $c_v = 30.9$ MeV and $c_s = -35.2$ MeV. The extracted value of c_v is very close to the symmetry energy in infinite nuclear matter at saturation density ρ_0, which is $C_{\text{sym}}(\rho_0) = 30.7$ MeV for the Gogny force (see the right panel of Fig. 8).

3. RESULTS AND DISCUSSIONS

In this section, we discuss how the density-dependent symmetry energy is reflected in the fragment isospin composition, by using the result of AMD simulations of Refs. [12, 13] for $^{40}\text{Ca} + ^{40}\text{Ca}$ ($i = 1$), $^{48}\text{Ca} + ^{48}\text{Ca}$ ($i = 2$), $^{60}\text{Ca} + ^{60}\text{Ca}$ ($i = 3$) and $^{46}\text{Fe} + ^{46}\text{Fe}$ ($i = 4$) collisions at zero impact parameter and an incident energy $E/A = 35$ MeV. We simulate collisions by boosting two nuclei whose centers are separated by 9 fm and calculating the dynamical evolution of each

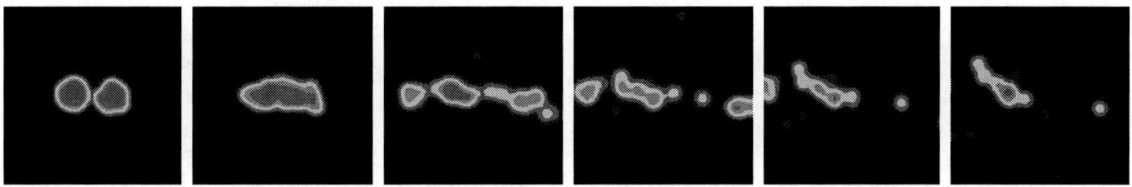

FIGURE 3. Time evolution of the density projected onto the reaction plain from $t = 0$ to 300 fm/c for an event of the central ^{40}Ca + ^{40}Ca collision at 35 MeV/nucleon. The size of the shown area is 40 fm × 40 fm.

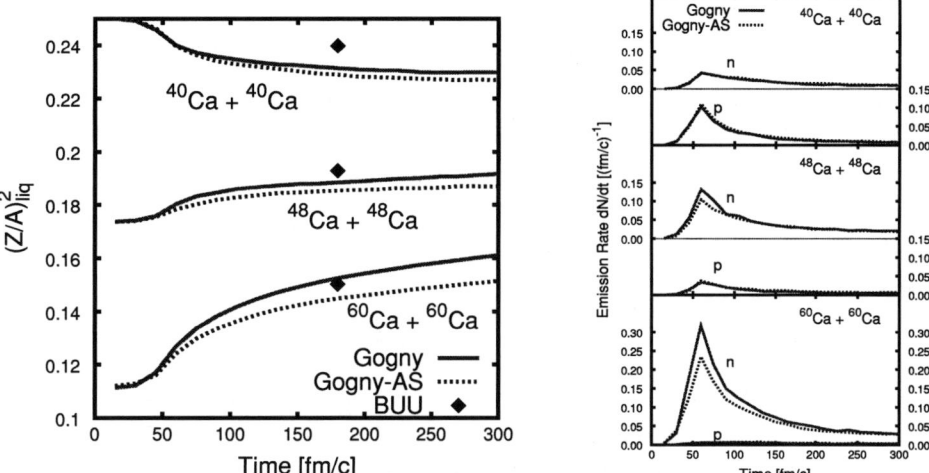

FIGURE 4. Left: $(Z/A)^2$ of the liquid part of the system as a function of time for the three reactions systems. The AMD results are represented by the solid and dashed lines, respectively, for the Gogny and Gogny-AS forces. Late-time BUU results are represented by filled diamonds. **Right:** Neutron and proton emission rates described by the left- and right-hand scales, respectively, for the three reaction systems as a function of time. The results of AMD simulations with the Gogny and Gogny-AS forces are, respectively, represented by the solid and dashed lines.

collision until $t = 300$ fm/c. The numbers of simulated events are 1040, 949, 978 and 1400, respectively, for the four systems. In central collisions, as shown in previous papers [15, 18], the projectile and target basically penetrate each other and many fragments are formed not only from the projectilelike and targetlike parts but also from a neck region between the two parts. The time evolution of the density profile is shown in Fig. 3 for a typical event. The nuclear matter seems to be strongly expanding one-dimensionally in the beam direction.

3.1. Isospin fractionation

Figure 4 shows the basic feature of isospin fractionation [19] where the neutron and proton densities are distributed inhomogeneously between liquid and gas phase, when we define the liquid part as the part of the system composed of the clusters with $A > 4$. The left panel shows the time evolution of the isospin asymmetry $(Z/A)^2$ of the liquid part for the three reaction systems. At the initial value ($t \sim 0$), $(Z/A)^2_{\text{liq}}$ is $(Z/A)^2$ of the initial nuclei. For the neutron-rich systems, $(Z/A)^2_{\text{liq}}$ increases rapidly before $t \sim 100$ fm/c, and then it continues to increase only gradually. This effect can be regarded as the isospin fractionation because the liquid part is getting less neutron-rich and the gas part is getting more neutron-rich. Such fractionation effect is much more dramatic in the gas part. The right panel shows the neutron and proton emission rates for the three systems. For the very neutron-rich system (^{60}Ca + ^{60}Ca) many more neutrons are emitted compared to nearly zero proton emission suggesting the existence of a very neutron-rich gas. This is contrasted with the ^{40}Ca + ^{40}Ca system where more protons are emitted than neutrons. These figures show a clear dependence on the asymmetry term of the effective force. The Gogny force (solid lines) always yields a larger

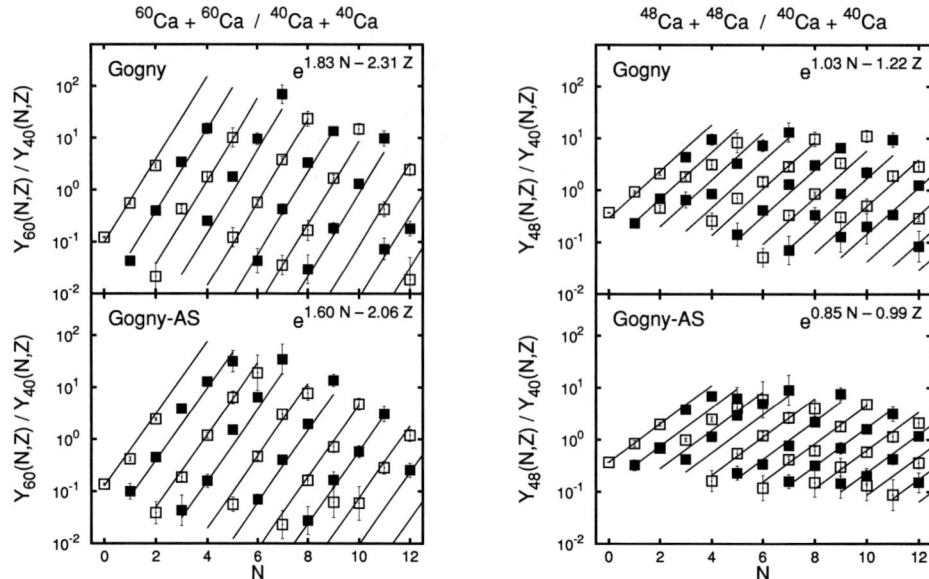

FIGURE 5. **Left:** The fragment yield ratio between the AMD simulations of central ^{60}Ca + ^{60}Ca and ^{40}Ca + ^{40}Ca collisions at 35 MeV/nucleon, at time $t = 300$ fm/c. The top and bottom panels show, respectively, the results obtained using the Gogny and Gogny-AS forces. **Right:** The same as Left but for the ^{48}Ca + ^{48}Ca and ^{40}Ca + ^{40}Ca collisions.

$(Z/A)^2_{\text{liq}}$ and a larger number of emitted neutrons (namely stronger fractionation) than the Gogny-AS force (dashed lines). Similar fractionation effects have been observed in other dynamical model simulations [22, 31].

3.2. Isoscaling

Figure 5 shows that the fragment yield ratios $Y_2(N,Z)/Y_1(N,Z)$ between different reaction systems satisfies isoscaling of Eq. (2). This is not a trivial consequence because AMD does not assume any statistical equilibrium, while isoscaling is usually understood in a statistical context. Therefore, Fig. 5 is the first implication that the fragment isospin composition is governed by a statistical law even in dynamically evolving reaction systems. However, further evidences are necessary to draw a decisive conclusion, which will be given later. The isoscaling relation of Fig. 5 is for the fragments present at $t = 300$ fm/c. The extracted scaling parameters α and β are provided in the individual panels. The isoscaling parameters clearly depend on the asymmetry term of the effective force. Their magnitude increases with increased differences in the asymmetry of the two systems.

3.3. Relation to the symmetry energy

Isoscaling is equivalent to the fact that the fragment yields of different reaction systems k are given by

$$Y_i(N,Z) = \exp[-K(N,Z) + \alpha_i N + \beta_i Z + \gamma_i], \quad (8)$$

where α_i, β_i and γ_i are constants that depend on the reaction system i, while $K(N,Z)$ is a function that is independent of the reaction system. By combining the fragment yields from the four reaction systems ($i = 1,2,3,4$), we can get the function $K(N,Z)$ for a wide region of (N,Z). The function $K(N,Z)$ extracted from the AMD simulation results is shown in Fig. 6 by solid and open squares for even-Z and odd-Z nuclei, respectively [13]. It can be seen that $K(N,Z)$ are obtained with good statistical precision for a large number of isotopes for each Z, even though the number of generated events is not very large.

So far, we have not assumed any specific form of $K(N,Z)$. Nevertheless, the results show a very smooth behavior of $K(N,Z)$ as a function of N and Z. The shell and paring effects are weak in $K(N,Z)$ compared to the ground state

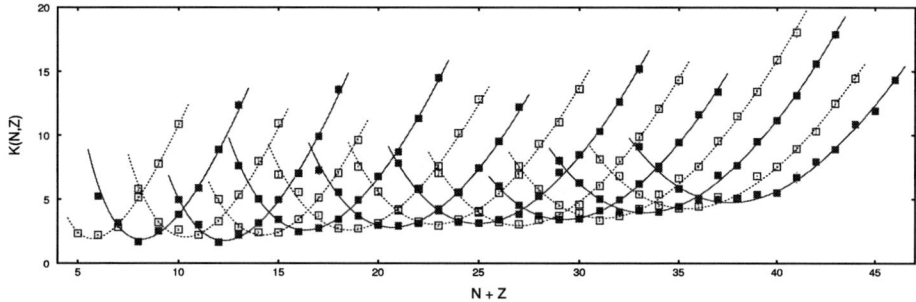

FIGURE 6. The values of $K(N,Z)$ for $3 \leq Z \leq 18$ are shown by symbols for the abscissa of $N+Z$. The values are obtained by combining the results of ^{40}Ca + ^{40}Ca, ^{48}Ca + ^{48}Ca, ^{60}Ca + ^{60}Ca and ^{46}Fe + ^{46}Fe simulations. The error bars show the statistical uncertainty due to the finite number of events. The curve for each Z was obtained by fitting $K(N,Z)$ using Eq. (9).

binding energies shown in Fig. 2. Each curve in Fig. 6 shows the fitting of $K(N,Z)$ for each Z by a function

$$K(N,Z) = \xi(Z)N + \eta(Z) + \zeta(Z)\frac{(N-Z)^2}{N+Z}, \quad (9)$$

where $\xi(Z)$, $\eta(Z)$ and $\zeta(Z)$ are the fitting parameters. The result of the simulations is fitted well by this functional form. We choose the quadratic term similar to Eq. (7) for convenience, so that the parameter $\zeta(Z)$ is directly related to the symmetry energy as shown below.

The meaning of $K(N,Z)$ is clear in a statistical context. For an equilibrated system, the fragment yields can be related with the nuclear free energy $G_{\text{nuc}}(N,Z)$ by

$$Y_i(N,Z) \propto \exp\left[-\frac{G_{\text{nuc}}(N,Z)}{T} + \frac{\mu_n^{(i)}}{T}N + \frac{\mu_p^{(i)}}{T}Z\right], \quad (10)$$

where $\mu_n^{(i)}$ and $\mu_p^{(i)}$ are the neutron and proton chemical potentials in the reaction system i. Therefore, $K(N,Z)$ can be identified with $G_{\text{nuc}}(N,Z)/T$, where T is the temperature. Especially, the fitting parameter $\zeta(Z)$ in Eq. (9) is related to the symmetry energy $C(Z)$ by

$$\zeta(Z) = C(Z)/T, \quad (11)$$

where the symmetry energy can depend on the size (or the charge Z) of the nuclei because of the presence of the surface effect.

We are now going to check whether the AMD result of $\zeta(Z)$, which is obtained without any assumption of equilibrium, behaves in a similar way to what one would expect for equilibrated systems.

The obtained values of $\zeta(Z)$ are shown in the left panel of Fig. 7 by solid points. Except for light fragments ($Z \lesssim 5$), $\zeta(Z)$ is a smooth function of Z which depends on Z very weakly. This situation may be contrasted with the fitting for the ground state masses, in which the fitting for each Z gives strongly fluctuating symmetry energy coefficient as a function of Z. In a statistical context, the very weak Z-dependence of $\zeta(Z)$ for $Z \gtrsim 5$ suggests a reduction of the surface contribution to the symmetry energy. The three curves in the left panel of Fig. 7 show functions $\zeta(Z) \propto 1 - k(2Z)^{-1/3}$ for different surface-to-volume ratios $k = 1.14$ (thick solid line), 0.5 (thin solid line) and 0 (horizontal dashed line), respectively. The curves are normalized at $Z = 10$. The Z-dependence of $\zeta(Z)$ cannot be explained by the surface-to-volume ratio $k = -c_s/c_v = 1.14$ for the symmetry energy of ground state nuclei. The result shows that the surface effect in $\zeta(Z)$ is reduced to between $k = 0$ and 0.5. Thus we can clearly exclude the possibility that the fragment isospin composition would be governed by the symmetry energy for the ground state nuclei in a statistical manner.

There can be several possible explanations for the weakening of the surface symmetry free energy. First of all, it is not very surprising that the coefficient in the free energy at a finite temperature is different from that at the zero temperature. As is well known, the surface tension reduces towards zero when the temperature is raised towards the critical temperature. A similar effect has been obtained for the symmetry free energy in Thomas-Fermi surface calculations [32, 33]. If we adopt the formula in Ref. [32], the reduction factor of the surface symmetry energy is approximately $[1-(T/T_c)^2]^2 = 0.91$ for $T = 3.4$ MeV (see bellow) and the critical temperature $T_c = 16$ MeV. This reduction is, however, not sufficient to explain the weak Z-dependence of $\zeta(Z)$. Another explanation may

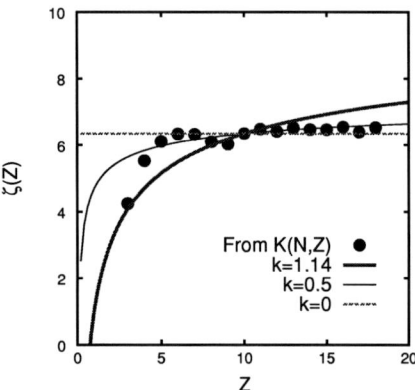 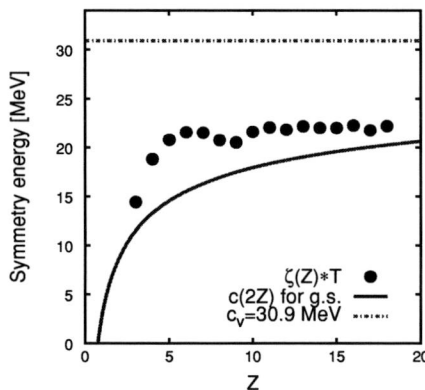

FIGURE 7. **Left:** The solid points are the extracted values of the coefficients $\zeta(Z)$ of Eq. (9) using the combined fragment yields of four systems shown in Fig. 6. The thick solid curve, the thin sold curve and the dashed line show functions $\zeta(Z) \propto 1 - k(2Z)^{-1/3}$ normalized at $Z = 10$ for $k = 1.14, 0.5$ and 0, respectively. **Right:** The solid points show $\zeta(Z)T$ when $T = 3.4$ MeV is assumed. The dot-dashed horizontal line shows the volume symmetry energy $c_v = 30.9$ MeV for the ground state nuclei calculated with AMD. The solid line shows the symmetry energy $c(A = 2Z) = c_v + c_s(2Z)^{-1/3}$ for the ground state nuclei ($c_v = 30.9$ MeV and $c_s = -35.2$ MeV).

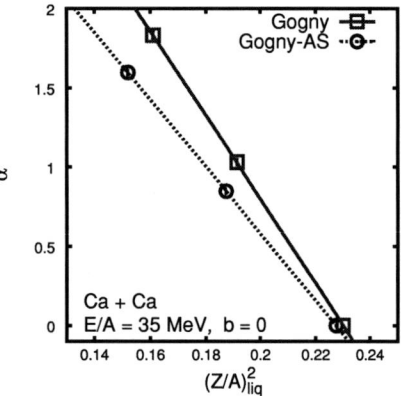 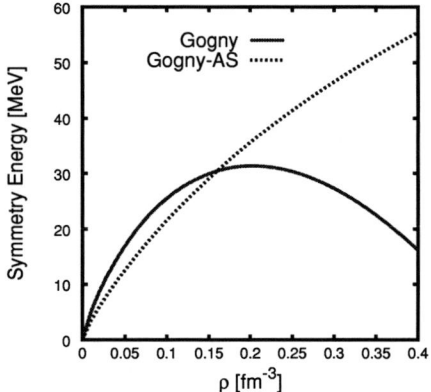

FIGURE 8. **Left:** Relation between $(Z/A)^2_{\text{liq}}$ and α for the three systems ^{60}Ca + ^{60}Ca, ^{48}Ca + ^{48}Ca and ^{60}Ca + ^{60}Ca (from left to right). Open squares and filled circles show the results of AMD with the Gogny force and Gogny-AS force, respectively. Both $(Z/A)^2_{\text{liq}}$ and α are calculated for fragments recognized at $t = 300$ fm/c. The straight lines are drawn so as to connect the points for ^{40}Ca + ^{40}Ca and ^{60}Ca + ^{60}Ca. The line slopes are -26.48 for the Gogny force and -21.16 for the Gogny-AS force, respectively. **Right:** Density dependence of the symmetry energy of nuclear matter for the Gogny force (solid line) and for the Gogny-AS force (dashed line).

be associated with the fact that fragments are not isolated when they are formed. When the density fluctuation is developing from a uniform low density matter, the fragments are still interacting with attractive force through their surfaces. Therefore, the surface free energies can be expected to be smaller for these fragments than for totally isolated fragments. Independent of the physical origin for the weakening of the surface symmetry free energy, it suggests that the volume quantity, which is the same as that in the infinite nuclear matter, can be directly obtained by the analysis of the fragmentation results even though the produced fragments are not very large.

Next, let us check how $\zeta(Z)$ is affected when we change the symmetry energy term in the effective interaction. For this purpose, it is noted that we can derive the equation [12]

$$\frac{\alpha_i - \alpha_j}{(Z/\bar{A}_j(Z))^2 - (Z/\bar{A}_i(Z))^2} = 4\zeta(Z), \qquad (12)$$

directly from Eqs. (8) and (9), where the numerator $\alpha_i - \alpha_j$ is the isoscaling parameter between the two reaction systems i and j, while the denominator $(Z/\bar{A}_j(Z))^2 - (Z/\bar{A}_i(Z))^2$ stands for the difference of the fragment isospin asymmetry for each Z. A similar equation has been pointed out [8] in the context of the expanding emitting source model [34], where the isospin asymmetry in the denominator is that of the emitting source rather than the function of Z of emitted fragments. In our case, since we have already seen that the Z-dependence of $\zeta(Z)$ is very weak, the Z-dependence of the denominator is also negligible. Therefore, in the left panel of Fig. 8, we show the relation between the isoscaling parameter α and the fragment isospin asymmetry $(Z/A)^2_{\text{liq}}$, where the latter is defined for the entity of the fragments with $A > 4$. The linear relation is clearly observed when the reaction system is changed. The slope corresponds to $\zeta(Z)$ because of Eq. (12). The results for the two different symmetry energy terms are shown by solid line (the Gogny force) and by dotted line (the Gogny-AS force). We obtain $4\zeta(Z) = 26.5 \pm 0.4$ for the Gogny force and $4\zeta(Z) = 21.2 \pm 0.3$ for the Gogny-AS force, where the shown uncertainties are due to statistics. In order that this dependence is consistent to the equilibrium assumption of Eq. (11), the ratio $C(\text{Gogny})/C(\text{Gogny-AS}) = 1.25 \pm 0.03$ is required. Considering the density dependence of the symmetry energy shown in the right panel of Fig. 8, this ratio of symmetry energy is possible if the fragments are formed when $\rho \sim 0.08$ fm^{-3}, which is consistent with the idea that fragmentation occurs at a reduced density. Furthermore, by using the absolute value of C at this density in the right panel of Fig. 8, we get the temperature $T \sim 3.4$ MeV, which is also in a reasonable range. Thus, by choosing a condition of $\rho \sim 0.08$ fm^{-3} and $T = 3.4$ MeV, the AMD result is well explained by the statistical assumption for uniform nuclear matter.

4. SUMMARY

We have analyzed AMD simulation results from nuclear collisions of various nuclei with different neutron-to-proton ratios. Isospin fractionation and isoscaling are observed in the dynamical AMD simulation. Fragment yields from different reaction systems are combined using the isoscaling relation. The availability of fragment yields over a wide range of N and Z allows us to extract the symmetry energy at low density when fragments are formed. The results, including the dependence on the density-dependent symmetry energy term, are consistent to the idea that the fragments are formed in uniform nuclear matter at a low density and at a finite temperature. The extracted symmetry energy shows almost no surface effect in it, which suggests that the property of infinite nuclear matter can be directly obtained from the information of fragmentation.

In this work, the fragments recognized at $t = 300$ fm/c were analyzed to get the information how the symmetry energy is reflected in dynamical collisions. In order to directly compare to the experimental data, the effect of the secondary decay of these excited fragments should be studied carefully in future works.

ACKNOWLEDGMENTS

The main part of this talk was based on the collaboration with M. B. Tsang, W. G. Lynch, P. Danielewicz and W. A. Friedman, which was supported by Japan Society for the Promotions of Science and the US National Science Foundation under the U.S.-Japan Cooperative Science Program (INT-0124186), by the High Energy Accelerator Research Organization (KEK) as the Supercomputer Project, and by grants from the US National Science Foundation, PHY-0245009, PHY-0070161 and PHY-01-10253, and a Grant-in-Aid for Scientific Research from the Japan Ministry of Education, Science and Culture. The work was also partially supported by RIKEN as a nuclear theory project.

REFERENCES

1. S. Das Gupta, A.Z. Mekjian and M.B. Tsang, Adv. Nucl. Phys. **26**, 91 (2001).
2. H.A. Bethe, Rev. Mod. Phys. 62, 801 (1990).

3. C.J. Pethick and D.G. Ravenhall, Ann. Rev. Nucl. Part. Sci. 45, 429 (1995).
4. J.M. Lattimer and M. Prakash, Ap. J., 550, 426 (2001).
5. D.H.E. Gross, Phys. Rep. 279, 119 (1997).
6. J.P. Bondorf, A.S. Botvina, A.S. Iljinov, l.N, Mishustin, and K. Sneppen, Phys. Rep. 257, 133 (1995).
7. W. P. Tan, S. R. Souza, R. J. Charity, R. Donangelo, W. G. Lynch, and M. B. Tsang, Phys. Rev. C 68, 034609 (2003).
8. M. B. Tsang, C. K. Gelbke, X. D. Liu, W. G. Lynch, W. P. Pan, G. Verde, H. S. Xu, W. A. Friedman, R. Donangelo, S. R. Souza, C. B. Das, S. Das Gupta, and D. Zhabinsky, Phys. Rev. C **64**, 054615 (2002).
9. T.X. Liu, X.D. Liu, M.J. van Goethem, W.G. Lynch, R. Shomin, W.P. Tan, M.B. Tsang, G. Verde, A. Wagner, H.F. Xi, H.S. Xu, M. Colonna, M. Di Toro, M. Zielinska-Pfabe, H.H. Wolter, L. Beaulieu, B. Davin, Y. Larochelle, T. Lefort, R.T. de Souza, R. Yanez, V.E. Viola, R.J. Charity, L.G. Sobotka, Phys. Rev. C **69**, 014603 (2004).
10. M. Colonna, F. Matera, nucl-th/0503018.
11. H. S. Xu, M. B. Tsang, T. X. Liu, X. D. Liu, W. G. Lynch, W. P. Tan, A. Vander Molen, G. Verde, A. Wagner, H. F. Xi, C. K. Gelbke, L. Beaulieu, B. Davin, Y. Larochelle, T. Lefort, R. T. de Souza, R. Yanez, V. E. Viola, R. J. Charity and L. G. Sobotka, Phys. Rev. Lett. **85**, 716 (2000).
12. A. Ono, P. Danielewicz, W.A. Friedman, W.G. Lynch and M.B. Tsang, Phys. Rev. C **68** 051601(R) (2003).
13. A. Ono, P. Danielewicz, W.A. Friedman, W.G. Lynch and M.B. Tsang, Phys. Rev. C **70** 041604(R) (2004).
14. A. Ono, H. Horiuchi, Toshiki Maruyama and A. Ohnishi, Phys. Rev. Lett. **68**, 2898 (1992); A. Ono, H. Horiuchi, Toshiki Maruyama and A. Ohnishi, Prog. Theor. Phys. **87**, 1185 (1992).
15. A. Ono and H. Horiuchi, Phys. Rev. C **53**, 2958 (1996).
16. A. Ono, Phys. Rev. C **59**, 853 (1999).
17. A. Ono, S. Hudan, A. Chbihi, J. D. Frankland, Phys. Rev. C **66**, 014603 (2002).
18. R. Wada, K. Hagel, J. Cibor, J. Li, N. Marie, W. Q. Shen, Y. Zhao, J. B. Natowitz and A. Ono, Phys. Lett. **B422**, 6 (1998).
19. H. Müller and B. D. Serot, Phys. Rev. C **52**, 2072 (1995).
20. T. Furuta, A. Ono, Prog. Theor. Phys. Suppl. **156**, 147 (2004).
21. J. Dechargé and D. Gogny, Phys. Rev. **C21**, 1568 (1980).
22. M. Colonna, M. Di Toro and A. B. Larionov, Phys. Lett. B **428**, 1 (1998); M. Colonna, M. Di Toro, G. Fabbri, and S. Maccarone, Phys. Rev. C **57**, 1410 (1998).
23. C. F. V. Weizsacker, Z. Physik 96, 431 (1935).
24. W. D. Myers and W. J. Swiatecki, Nucl. Phys. **A81**, 1 (1966).
25. P. Möller, J. R. Nix, W. D. Myers and W. J. Swiatecki, At. Data Nucl. Data Tables **59**, 185 (1995).
26. P. Danielewicz, Nucl. Phys. **A727**, 233 (2003).
27. J. Jänecke, T. W. O'Donnell, V. I. Goldanskii, Nucl. Phys. **A728**, 23 (2003).
28. A. Ono and H. Horiuchi, Porg. Part. Nucl. Phys. 53, 501 (2004).
29. K. Hagel, M. Gonin, R. Wada, J. B. Natowitz, F. Haddad, Y. Lou, M. Gui, D. Utley, B. Xiaó, J. Li, G. Nebbia, D. Fabris, G. Prete, J. Ruiz, D. Drain, B. Chambon, B. Cheynis, D. Guinet, X. C. Hu, A. Demeyer, C. Pastor, A. Giorni, A. Lleres, P. Stassi, J. B. Viano, and P. Gonthier, Phys. Rev. C **50**, 2017 (1994).
30. R. Nebauer, J. Aichelin and the INDRA Collaboration, Nucl. Phys. **A658**, 67 (1999).
31. Bao-An Li, Phys. Rev. Lett. 85, 4221 (2000).
32. J. M. Lattimer and F. D. Swesty, Nucl. Phys. **A535**, 331 (1991).
33. D. G. Ravenhall, C. J. Pethick and J. M. Lattimer, Nucl. Phys. A407, 571 (1983).
34. W. A. Friedman, Phys. Rev. Lett. **60**, 2125 (1988); Phys. Rev. C **42**, 667 (1990).

Determining Cross Sections for Reactions on Unstable Nuclei: A Consideration of Indirect Approaches

J. Escher[1] and F.S. Dietrich

Lawrence Livermore National Laboratory, P.O. Box 808, L-414, Livermore, CA 94551, USA

Abstract. An indirect method for determining cross sections for reactions proceeding through a compound nucleus is presented. The appropriate theoretical framework for applications of this method is reviewed and theoretical and experimental challenges that need to be addressed in applications of the method are outlined. Two approximations are considered and their advantages and limitations are discussed.

Keywords: statistical compound-nucleus reactions, indirect methods, unstable nuclei
PACS: 24.10.-i, 24.60.Dr, 25.40.Lw, 25.85.Ec

1. INTRODUCTION

A large number of nuclear reactions cannot be easily determined in the laboratory. Direct measurements encounter a variety of difficulties: The low-energy regime that is especially relevant for astrophysical reactions is often inaccessible and cross sections for charged-particle reactions become vanishingly small as the relative energy of the colliding nuclei decreases. Furthermore, many reactions involve unstable nuclear species which are too difficult to produce with currently available experimental techniques or too short-lived to serve as targets in present-day set-ups.

In order to overcome the experimental limitations, various indirect methods have been proposed in recent years. *Coulomb Dissociation* [1], e.g., has been used to extract cross sections for radiative-capture reactions, $A(a,\gamma)B$. In this approach, the Coulomb field of a highly charged target provides a virtual photon, which is absorbed by the projectile B. The cross section of the breakup $X(B,Aa)X$ is much larger than the capture cross section and can be related to the latter via the principle of detailed balance. The *ANC (Asymptotic Normalization Coefficient) Method* [2] has been explored for low-energy radiative-capture reactions $A(a,\gamma)B$ which are dominated by processes occurring far outside the nuclear radius. The cross section of such a reaction depends on the asymptotic behaviour of the overlap function I_{Aa}^{B} for $B \to A+a$. The radial shape of I_{Aa}^{B} is well known, and its normalization, the ANC, can be determined via a peripheral transfer reaction that involves the same asymptotic overlap, e.g. $d+A \to b+B$, where $d=a+b$ and $B=A+a$. The *Trojan-Horse method* [3] provides a mechanism for circumventing the Coulomb barrier which is responsible for the very small cross sections in low-energy two-body reactions with charged projectiles, $A(a,b)B$. It does so by selecting a reaction $d+A \to b+B+c$ with $d=a+c$ and kinematic conditions such that c can be considered a spectator in the reaction between a and A ('quasi-free scattering'). An approximate expression for the cross section of the three-body reaction then provides a link to the two-body reaction of interest and allows one to extract the energy-dependence of the latter. Thus it becomes possible to extrapolate absolute measurements carried out at higher energies to the relevant low-energy regime.

The focus of this contribution is an indirect method that complements the above approaches, the *Surrogate Nuclear Reactions* method. The Surrogate method combines experiment with reaction theory to obtain cross sections for compound-nucleus reactions involving difficult-to-produce targets. In this article, we discuss the Surrogate approach and two approximations to the method. In the next section the Surrogate idea is explained and the appropriate theoretical framework for describing a Surrogate reaction is outlined. In Section 3, we discuss briefly the Weisskopf-Ewing limit of the full Hauser-Feshbach theory and consider its implications for the Surrogate approach. In Section 4, we outline a new idea [4], which aims at extracting ratios of reaction cross sections from Surrogate experiments. An independent measurement of one of the reactions can then be used to obtain an unknown cross section. The "Ratio method" was recently employed to estimate the ^{237}U(n,f) cross section [5]. We discuss some of the advantages and

---

[1] Email: escher1@llnl.gov

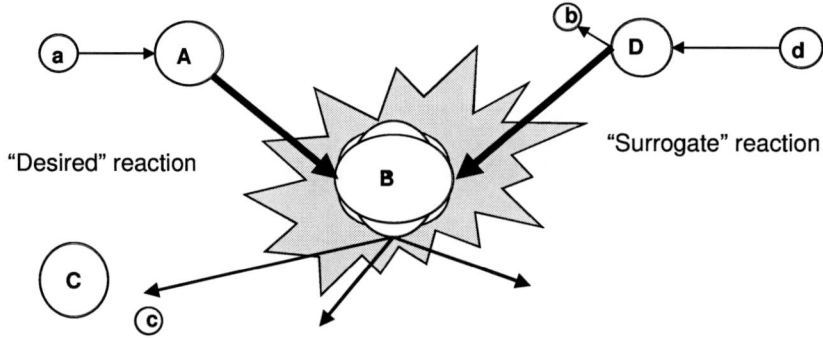

FIGURE 1. Schematic representation of the Surrogate reaction mechanism. The basic idea of the Surrogate approach is to replace the first step of the desired reaction by an alternative ("Surrogate") reaction that populates the same compound nucleus. The subsequent decay of the compound nucleus into the relevant channel can then be measured and used to extract the desired cross section.

limitations of this approach. Concluding remarks are given in Section 5.

2. THE SURROGATE METHOD

The Surrogate nuclear reaction technique is an indirect method for determining the cross section for a particular type of "desired" reaction, namely a two-step reaction, $a + A \to B^* \to c + C$, that proceeds through a compound nuclear state B^*, a highly excited state in statistical equilibrium (see Figure 1). In the Surrogate method, the compound nucleus B^* is produced by means of an alternative ("Surrogate") reaction, here $d + D \to b + B^*$, and the reaction cross section is obtained by combining the calculated cross section for the formation of B^* (from a and A) with the measured decay probabilities for this state. The Surrogate technique is particularly valuable when the target of interest, A, is short-lived and a suitable Surrogate reaction involving a stable target D can be identified. A simple version of the Surrogate idea was already used in the 1970s to estimate neutron-induced fission cross sections from transfer reactions [6]. More recently, this approach was refined [11] and applications to other reactions are now being considered, such as (n,γ) reactions on s-process branch points [7].

2.1. Hauser-Feshbach Formalism for Compound-Nucleus Reactions

The formalism appropriate for describing compound-nucleus reactions is the statistical Hauser-Feshbach theory (see, *e.g.*, chapter 10 of Ref. [8]). The average cross section per unit energy is given by:

$$\frac{d\sigma_{\alpha\chi}^{HF}(E_a)}{dE_\chi} = \pi \lambda_\alpha^2 \sum_{J\Pi} \omega_\alpha^J \sum_{lsl's'I'} \frac{T_{\alpha ls}^J T_{\chi l's'}^J \rho_{I'}(U')}{\sum'_{\chi''l''s''I''} T_{\chi''l''s''}^J + \sum_{\chi''l''s''I''} \int T_{\chi''l''s''}^J (E_{\chi''}) \rho_{I''}(U'') dE_{\chi''}}. \quad (1)$$

where it is assumed that the reaction proceeds to an energy region in the final nucleus described by a level density. Here α denotes the entrance channel $a + A$ and χ represents the relevant exit channel $c + C$, E_a is the kinetic energy of the projectile, and λ_α is the reduced wavelength in the incident channel (the inverse of the wave number). The spin of the incident particle is i, the target spin is I, the channel spin is $\vec{s} = \vec{i} + \vec{I}$, and the compound-nucleus angular momentum and parity are $J\pi$. The statistical-weight factor ω_α^J is $(2J+1)/[(2i+1)(2I+1)]$. Similarly, the spin of the outgoing particle is i', the spin of the residual nucleus is I', and the channel spin for χ' is $\vec{s}' = \vec{i}' + \vec{I}'$. The transmission coefficients are written as $T_{\alpha ls}^J$ and $\rho_{I'}(U')$ denotes the density of levels of spin I' at excitation energy U' in the residual nucleus. All energetically possible final channels χ'' have to be taken into account, thus the denominator includes contributions from decays to discrete levels in the residual nuclei (given by the first sum in the denominator, $\sum'$) as well as contributions from decays to regions of high level density in the residual nuclei (given by the second sum in the denominator which involves an energy integral of transmission coefficients and level densities in the residual nuclei). For simplicity, the parity quantum numbers have been suppressed in Equation 1. In realistic applications of the Hauser-Feshbach formalism, the level density depends on parity (even though this dependence tends to be weak), and all sums over quantum numbers respect parity conservation.

The above Hauser-Feshbach formula neglects correlations between the incident and outgoing reaction channels. These correlations can be taken into account formally by including width fluctuation corrections $W_{\alpha\chi}$ in the Hauser-Feshbach formula. The primary effect of the correlations is an enhancement of the elastic scattering cross section. Due to the requirement of flux conservation the inelastic and reaction cross sections are reduced, although this depletion rarely exceeds 10-20%, even at relatively low energies (below approximately 2 MeV). As the excitation energy of the compound nucleus increases and many reaction channels become available, the effect of the width fluctuations becomes quickly negligible for the non-elastic channels.

In the remainder of this contribution we will neglect these correlations and set $W_{\alpha\chi} = 1$. (The elastic channel will be considered separately wherever necessary.) This then allows us to rewrite the Hauser-Feshbach formula as:

$$\frac{d\sigma_{\alpha\chi}^{HF}(E_a)}{dE_\chi} = \sum_{J\pi} \sigma_\alpha^{CN}(E_{ex},J,\pi) G_\chi^{CN}(E_{ex},J,\pi), \qquad (2)$$

where $\sigma^{CN}(E_{ex},J,\pi) = \sigma(a+A \to B^*)$ denotes the cross section for forming the compound nucleus at excitation energy E_{ex} with angular-momentum and parity quantum numbers $J\pi$ and $G_\chi^{CN}(E_{ex},J,\pi)$ is the branching ratio for the decay of this compound state into the desired exit channel χ.

2.2. Hauser-Feshbach Formulation of the Surrogate Method

In the limit of negligible width fluctuation corrections considered here, the formation and decay of the compound nucleus are independent of each other, individually for each angular momentum and parity value. It is this independence that allows one to determine the desired cross section via a combination of theory and experiment in the Surrogate approach. In many cases the formation cross section σ_α^{CN} can be calculated to a reasonable accuracy by using optical potentials while the theoretical branching ratios G_χ^{CN} for the different channels χ are often quite uncertain. The objective of the Surrogate method is to determine or constrain these decay probabilities experimentally.

In a Surrogate experiment, the compound nucleus B^* is produced via an alternative (Surrogate), direct reaction $d+D \to b+B^*$ and the decay of B^* is observed in coincidence with the outgoing particle b. The direct-reaction particle is typically stopped in a detector which provides particle identification, as well as information on the kinetic energy and direction of b. The desired exit channel χ can be identified, e.g., by detecting fission fragments from B^* or γ rays from the desired residual nucleus C. The probability for forming B^* in the Surrogate reaction (with specific values for the excitation energy E_{ex}, angular momentum J, and parity π) is $F_\delta^{CN}(E_{ex},J,\pi)$, where δ refers to the entrance channel $d+D$. The quantity

$$P_{\delta\chi}(E_{ex}) = \sum_{J,\pi} F_\delta^{CN}(E_{ex},J,\pi) \, G_\chi^{CN}(E_{ex},J,\pi), \qquad (3)$$

which gives the probability that the compound nucleus B^* was formed with energy E_{ex} and decayed into channel χ, can be obtained experimentally. The direct-reaction probabilities $F_\delta^{CN}(E_{ex},J,\pi)$ have to be determined theoretically, so that the branching ratios $G_\chi^{CN}(E_{ex},J,\pi)$ can be constrained from the measurements. In practice, the decay of the compound nucleus is modeled and the $G_\chi^{CN}(E_{ex},J,\pi)$ are obtained by fitting the calculations to reproduce the measured decay probabilities $P_{\delta\chi}(E_{ex})$. Subsequently, the branching ratios obtained in this manner are inserted in Equation (2) to yield the desired reaction cross section. For simplicity, we have omitted the angular dependence of both the desired and the Surrogate reactions in the above discussion. The extension of the Hauser-Feshbach formulae is straight-forward [8].

2.3. Challenges for the Surrogate Method

In practice the procedure of determining the branching ratios is a difficult task due to several theoretical and experimental challenges: i) The experimental determination of the decay probability $P_{\delta\chi}(E_{ex}) = N_{\delta\chi}/N_\delta$ requires that both the number of b-χ coincidences, $N_{\delta\chi}$, and the number of reaction events, N_δ are accurately determined. If target contaminants are present, it becomes very difficult, if not impossible to determine a reliable value for N_δ. ii) The theoretical prediction of the direct-reaction probabilities $F_\delta^{CN}(E_{ex},J,\pi)$ requires a framework for calculating cross sections of direct reactions (stripping, pick-up, and inelastic scattering) to continuum states in B^*. iii) Extracting the branching ratios from measured decay probabilities $P_{\delta\chi}(E_{ex})$ requires modeling the decay of the compound nucleus produced in the Surrogate reaction and fitting the relevant parameters to reproduce the experimental results. iv) The possibility that the intermediate nucleus produced in the Surrogate reaction decays before statistical equilibrium is reached [9, 13, 14] has to be excluded or minimized.

3. THE WEISSKOPF-EWING LIMIT

The Hauser-Feshbach theory used in the previous section rigorously conserves total angular momentum J and parity π. Under certain conditions the branching ratios $G_\chi^{CN}(E_{ex}, J, \pi)$ can be treated as independent of J and π and the form of the cross section (for the desired reaction) simplifies to:

$$\frac{d\sigma_{\alpha\chi}^{WE}(E_a)}{dE_\chi} = \sigma_\alpha^{CN}(E_{ex}) \mathscr{G}_\chi^{CN}(E_{ex}) \quad (4)$$

where

$$\sigma_\alpha^{CN}(E_{ex}) = \sum_{J\Pi} \sigma_\alpha^{CN}(E_{ex}, J, \pi) = \pi \lambdabar_\alpha^2 \sum_{l=0}^{\infty} (2l+1) T_{\alpha l} \quad (5)$$

is the reaction cross section describing the formation of the compound nucleus at energy E_{ex} and $\mathscr{G}_\chi^{CN}(E_{ex})$ denotes the $J\pi$-independent branching ratio for the exit channel χ. This is the Weisskopf-Ewing limit of the Hauser-Feshbach theory. It is applicable when the following conditions are satisfied [9, 10]:

- The energy of the compound nucleus has to be sufficiently high, so that almost all channels into which the nucleus can decay are dominated by integrals over the level density (i.e. the first sum in the denominator of Equation 1, $\sum'$ has to be negligible).
- Width fluctuations have to be negligible. This will be the case if the previous condition is satisfied.
- The transmission coefficients $T_{\chi''l''s''}^J$ associated with the available exit channels have to be independent of the spin of the states reached in these channels. This condition is known to be satisfied since the dependence of transmission coefficients on target spin is generally weak.
- The level densities $\rho_{I''}(U'')$ in the available channels have to be independent of parity and their dependence on the spin I'' of the relevant nuclei has to be of the form $\rho_{I''}(U'') \propto (2I''+1)$. Using tools from statistical mechanics, it can be shown that for sufficiently high excitation energies U'', level densities are very weakly dependent on parity, so that the first of these conditions can be assumed to be satisfied. The second condition, which is a prerequisite for a rigorous derivation of the Weisskopf-Ewing limit from the full Hauser-Feshbach theory, is satisfied if the spin I'' is smaller than the spin cutoff parameter σ_{cut} in the relevant level density formula. In some cases, the spin cutoff parameter is not very large (e.g. $\sigma_{cut} \approx 6-7$ in the actinide region) but it is known empirically that the Weisskopf-Ewing limit is still roughly correct at higher spins.

The cross section in Equation 4 is expressed in differential form, with respect to the energy in the decay channel. For a comparison with a measurement the cross section needs to be integrated over all final-state energies, i.e. the quantity $\mathscr{G}_\chi^{CN}$ appearing on the right side has to be integrated over the energy E_χ. In the following only integrated quantities will be considered; the energy differential will be removed from the cross section expressions and the $\mathscr{G}_\chi^{CN}$ will represent integrated branching ratios.

3.1. The Surrogate Method in the Weisskopf-Ewing Limit

The Weiskopf-Ewing limit provides a simple and powerful approximate way of calculating cross sections for two-step reactions proceeding through a compound nucleus. In the context of Surrogate reactions, it greatly simplifies the application of the method: It becomes straightforward to obtain the $J\pi$-independent branching ratios $\mathscr{G}_\chi^{CN}(E_{ex})$ from measurements, since $P_{\delta\chi}(E_{ex}) = \mathscr{G}_\chi^{CN}(E_{ex}) \sum_{J\pi} F_\delta^{CN}(E_{ex}, J, \pi) = \mathscr{G}_\chi^{CN}(E_{ex})$, and to calculate the desired reaction cross section. Calculating the direct-reaction probabilities $F_\delta^{CN}(E_{ex},J,\pi)$ and modeling the decay of the compound nucleus are no longer required.

A Surrogate analysis in the Weisskopf-Ewing approximation was used in the 1970s to extract (n,f) cross sections for various actinides from transfer reactions with t and ^{3}He projectiles on neighboring nuclei, followed by fission [6]. Measured fission probabilities, P_f, were multiplied by an estimated cross section for the formation of the compound nucleus in the neutron-induced reaction of interest: $\sigma_{(n,f)} \approx \sigma_n^{CN} P_f$. The resulting (n,f) cross section estimates agreed with direct measurements (where available) to about 10-20% for incident neutron energies above about 1 MeV, but exhibited serious discrepancies below 1 MeV, which were attributed to i) large uncertainties in the low-energy optical-model calculations employed, and ii) the neglect of the difference in the angular-momentum populations of the compound nucleus in the Surrogate (direct) and "desired" (neutron-induced) reactions. A more recent analysis of the data used a simple direct-reaction model to account for the angular-momentum difference between the neutron-induced and direct reactions, i.e. it employed the full Surrogate framework, as well as improved optical-model calculations [11]. The results showed significant improvements over the earlier work in the Weisskopf-Ewing approximation.

The assumption that the Weisskopf-Ewing limit is valid implies a significant restriction for possible applications. At low excitation energies in the compound nucleus, the Weisskopf-Ewing description is not even approximately valid. For example, astrophysical (n,γ) reactions on s-process branch points cannot be estimated with this method since the relevant neutron energies are very low, $E_n \lesssim 100$ keV, i.e. the compound nucleus is excited to only slightly above the neutron separation threshold. At higher energies the assumption of $J\pi$-independent branching ratios, $G_\chi^{CN}(E_{ex},J,\pi) \approx \mathscr{G}_\chi^{CN}(E_{ex})$, breaks down for angular-momentum values larger than the relevant spin-cutoff parameter. This may affect the desired and Surrogate reactions differently, since the angular-momentum transferred is in general larger in the direct reaction than in the compound-nucleus reaction. For example, a Weisskopf-Ewing description might be applicable to an n-induced reaction on a target with small spin, while the Surrogate reaction that produces the same compound nucleus might populate states with spins much larger than the spin-cutoff parameter. In such situations a full Surrogate analysis, which takes into account conservation of spin and parity, is required.

Despite the simplifications that can be obtained in the Weisskopf-Ewing description, a number of issues remain to be resolved when applying this approximate version of the Surrogate approach. First, it is not a priori clear whether the Weisskopf-Ewing limit applies to a particular reaction in a given energy regime. This needs to be verified for each case of interest. In addition, even in the Weisskopf-Ewing limit it is necessary to consider the possibility that the intermediate nucleus which is produced in the Surrogate reaction can decay before it reaches equilibrium. Furthermore, Surrogate experiments in the Weisskopf-Ewing limit are still challenging since the requirement that both the number of $b - \chi$ coincidences and the number of reaction events be accurately determined remains.

4. RATIOS OF SURROGATE MEASUREMENTS

In a recent publication [5], a new approach was employed with the goal of determining the neutron-induced fission cross section for ^{237}U, for neutron energies up to approximately 14 MeV. This new approach, which we shall refer to as the "Ratio method", makes use of the Surrogate idea and requires the validity of the Weisskopf-Ewing limit. The primary motivation for using the Ratio method is the fact that it eliminates the need to accurately measure N_δ, the total number of reaction events, which has been the source of the largest uncertainty in Surrogate experiments performed recently.

The goal of the Ratio method is to experimentally determine the ratio

$$R(E) = \frac{\sigma_{\alpha_1 \chi_1}}{\sigma_{\alpha_2 \chi_2}} \qquad (6)$$

of the cross sections of two compound-nucleus reactions, $a_1 + A_1 \to B_1^* \to c_1 + C_1$ and $a_2 + A_2 \to B_2^* \to c_2 + C_2$, where the two reactions have to be "similar" in a sense that remains to be specified. An independent determination of one of these cross sections then allows one to infer the other by using the ratio R. In the Weisskopf-Ewing limit, the ratio R can be written as

$$R(E) = \frac{\sigma_{\alpha_1}^{CN}(E)\, \mathscr{G}_{\chi_1}^{CN}(E)}{\sigma_{\alpha_2}^{CN}(E)\, \mathscr{G}_{\chi_2}^{CN}(E)}, \qquad (7)$$

with branching ratios $\mathscr{G}_\chi^{CN}$ that are independent of the $J\pi$ population of the compound nuclei under consideration. For most cases of interest the compound-nucleus formation cross sections $\sigma_{\alpha_1}^{CN}$ and $\sigma_{\alpha_2}^{CN}$ can be calculated sufficiently reliably by using an optical model.

To determine $\mathscr{G}_{\chi_1}^{CN}/\mathscr{G}_{\chi_2}^{CN}$, two experiments are carried out. Both use the same direct-reaction mechanism, $D(d,b)B^*$, but different targets, D_1 and D_2, to create the relevant compound nuclei, B_1^* and B_2^*, respectively. For each experiment, the number of coincidence events, $N_{\delta_1\chi_1}^{(1)}$ and $N_{\delta_2\chi_2}^{(2)}$, is measured. The ratio of the branching ratios, for decay into the desired channel, for the compound nuclei created in the two reactions is given by:

$$\frac{\mathscr{G}_{\chi_1}^{CN}(E)}{\mathscr{G}_{\chi_2}^{CN}(E)} = \frac{N_{\delta_1\chi_1}^{(1)}(E)}{N_{\delta_2\chi_2}^{(2)}(E)} \times \frac{N_{\delta_2}^{(2)}(E)}{N_{\delta_1}^{(1)}(E)}. \tag{8}$$

In the Ratio approach the experimental conditions are adjusted such that both experiments give the same number of reaction events, $N_{\delta_1\chi_1}^{(1)} \approx N_{\delta_2\chi_2}^{(2)}$. This requires that the same setup be used in both experiments. Furthermore, the beam intensities and beam times have to be the same in both cases, and the number of atoms in each target must be equal or the differences have to be accounted for in the data analysis. Under those conditions, the ratio of the branching ratios simply equals the ratio of the coincidence events and the quantity R becomes:

$$R(E) = \frac{\sigma_{\alpha_1}^{CN}(E) N_{\delta_1\chi_1}^{(1)}(E)}{\sigma_{\alpha_2}^{CN}(E) N_{\delta_2\chi_2}^{(2)}(E)} \tag{9}$$

The definition of the energy E in the above equations remains to be specified. Typically, the energy-dependence of a compound-nucleus formation cross section, $\sigma_\alpha^{CN} = \sigma(a+A \rightarrow B^*)$ is characterized by the kinetic energy of the projectile, E_a, while a branching ratio is normally given as a function of the excitation energy of the compound nucleus, $\mathscr{G}_\chi^{CN}(E_{ex})$. In a compound-nucleus reaction, those two values are related via the separation energy S_a of the particle a in B^*: $E_{ex} = S_a + E_a$. While either E_{ex} or E_a can be used to uniquely specify the energy-dependence of such a reaction, it is important for the Ratio method that the comparison of the relevant reactions, $a_1 + A_1 \rightarrow B_1^* \rightarrow c_1 + C_1$ and $a_2 + A_2 \rightarrow B_2^* \rightarrow c_2 + C_2$, be made *at the same projectile energy E_a*. For a given projectile energy, $E_{a_1} = E_{a_2}$, small differences in the separation energies, S_{a_1} and S_{a_2}, will lead to different excitation energies in the compound nuclei, B_1^* and B_2^*, respectively. In typical applications of interest the branching ratios are less sensitive to such variation in excitation energy than the formation cross sections are to an energy variation of this magnitude. Thus, in the context of the Ratio approach, we take E to denote the kinetic energy of the projectile.

An illustration of the Ratio method. In Ref. [5], the Ratio method was used to obtain an estimate of the ^{237}U(n,f) cross section up to approximately 14 MeV. To this end, the ratio

$$R = \frac{\sigma(^{237}U(n,f))}{\sigma(^{235}U(n,f))} \approx \frac{\sigma_{n+^{237}U}^{CN}}{\sigma_{n+^{235}U}^{CN}} \frac{\mathscr{G}_{^{238}U \rightarrow f}^{CN}}{\mathscr{G}_{^{236}U \rightarrow f}^{CN}}, \tag{10}$$

was considered, where the cross section $\sigma(^{235}U(n,f))$ for neutron-induced fission of ^{235}U is known. The formation cross sections for the compound nuclei ^{238}U and ^{236}U were assumed to be very similar, $\sigma_{n+^{237}U}^{CN} \approx \sigma_{n+^{235}U}^{CN}$. To obtain information on the branching ratios $\mathscr{G}_{^{238}U \rightarrow f}^{CN}$ and $\mathscr{G}_{^{236}U \rightarrow f}^{CN}$, inelastic deuteron scattering experiments on ^{238}U and ^{236}U were carried out. Fission fragments from $^{238}U(d,d'f)$ and $^{236}U(d,d'f)$ were detected in coincidence with the outgoing deuterons and

$$\frac{\mathscr{G}_{^{238}U \rightarrow f}^{CN}}{\mathscr{G}_{^{236}U \rightarrow f}^{CN}} \approx \frac{N_{^{238}U(d,d'f)}}{N_{^{236}U(d,d'f)}} \tag{11}$$

was determined. Corrections were applied to account for differences in target thickness and beam intensity. The resulting cross section ratio was found to be in agreement with a theoretical estimate by Younes *et al.* [12].

4.1. Limitations of the Ratio Approach

The experiments required for a Ratio analysis are simpler than those that need to be carried out if a full Surrogate analysis (or a Surrogate analysis in the Weisskopf-Ewing limit) is planned. The primary advantage of considering relative branching ratios and relative cross sections lies in the fact that the number of direct-reaction events, N_δ, does not need to be determined for a Ratio analysis. Furthermore, unlike in the full Surrogate treatment, it is not necessary to calculate the direct-reaction probabilities, $F_\delta^{CN}(E,J,\pi)$, or to model the decay of the compound nucleus.

The Ratio method is based on the assumption that the Weisskopf-Ewing approximation is valid. It is therefore subject to the same restrictions that apply to the use of a Surrogate analysis in the Weisskopf-Ewing approximation, although small deviations from this assumption might affect the Ratio analysis to a lesser extent.

The Ratio method is also limited by the requirement that for obtaining an absolute result for an unknown cross section $\sigma(a_1 + A_1 \rightarrow B_1^* \rightarrow c_1 + C_1)$ a reliable independent cross-section measurement for a similar reaction, $a_2 + A_2 \rightarrow B_2^* \rightarrow c_2 + C_2$, at the same equivalent projectile energies, must be available. Furthermore, it is required that a direct-reaction mechanism, $D(d,b)B^*$, and target-projectile combinations can be identified that make it possible to produce the compound nuclei, B_1^* and B_2^*, respectively.

One can expect reliable cross section estimates from the Ratio approach only when the two reactions that are analyzed, $D_1(d,b)B_1^*$ and $D_2(d,b)B_2^*$, are sufficiently similar. When small systematic errors or small violations of the prerequisite assumptions, such as the validity of the Weisskopf-Ewing approximation or the absence of non-equilibrium decays, affect both reactions in the same manner, it is likely that the effects cancel in part in the Ratio analysis. Uncorrelated errors and deviations, on the other hand, will increase the overall uncertainty in the final result. Similarity in the present context implies that i) the same projectile initiates the compound-nucleus reactions that are compared, i.e. $a_1 = a_2$, and the same kind of decay (gamma emission, charged-particle emission, or fission) is considered in both cases; ii) the decays of the compound nuclei B_1^* and B_2^* have similar properties (number and kind of open channels, separation energies for the various channels, level densities in the residual nuclei, etc.); iii) the direct (Surrogate) reactions which produce the compound nuclei employ the same mechanism, $D(d,b)B^*$, and projectile-ejectile combination, $d-b$, in both cases.

5. CONCLUDING REMARKS

Indirect methods play an important role for obtaining many reaction cross sections of interest. A method which aims at extracting cross sections for reactions proceeding through a compound nucleus has been presented. While the Surrogate method is very general and can in principle be employed to determine cross sections of all types of compound-nucleus reactions on a large variety of nuclei, significant challenges remain to be addressed to establish the validity of this approach. For applications to (n,f) cross sections on actinide nuclei, Younes and Britt have studied some of these issues [11] and for applications to (n,γ) cross sections on lighter nuclei, work is currently underway [7]. In the interim, it is useful to investigate whether certain simplifications or approximations to the method can be utilized to determine the relevant cross sections. Two possible approximations to the Surrogate method have been considered here, the Weisskopf-Ewing limit of the Surrogate method and the Ratio approach. Both provide simple and potentially powerful ways of estimating cross sections that cannot be measured directly. Both are much simpler than a full Surrogate treatment and both are much more limited in their applicability.

ACKNOWLEDGMENTS

The authors gratefully acknowledge receiving a copy of Ref. [5] prior to publication. This work was performed under the auspices of the U.S. Department of Energy by the University of California, Lawrence Livermore National Laboratory (LLNL) under contract No. W-7405-Eng-48. Partial funding was provided by the Laboratory Directed Research and Development (LDRD) Program at LLNL under project 04-ERD-057.

REFERENCES

1. G. Baur, K. Hencken, and D. Trautmann, *Prog. Part. Nucl. Phys.* **51** 487 (2003);
   G. Baur and H. Rebel, *Annu. Rev. Nucl. Part. Sci.* **46** 321 (1996);

2. N.K. Timofeyuk, R.C. Johnson, and A.M. Mukhamedzhanov *Phys. Rev. Lett.* **91** 232501 (2003);
   L. Trache *et al.*, *Phys. Rev. C* **69** 032802(R) (2004);
   L. Trache *et al.*, *Phys. Rev. Lett.* **87** 271102 (2001);
   A. Azhari *et al.*, *Phys. Rev. Lett.* **82** 3960 (1999);
   A.M. Mukhamedzhanov and R.E. Tribble, *Phys. Rev. C* **59** 3418 (1999);
   H. M. Xu *et al.*, *Phys. Rev. Lett.* **73** 2027 (1994).
3. S. Typel and G. Baur, *Ann. Phys.* **305** 228 (2003);
   G. Baur and S. Typel, preprint nucl-th/0401054.
4. L. Bernstein, private communication.
5. C. Plettner *et al.*, Phys. Rev. C in press (2005).
6. J.D. Cramer and H.C. Britt, *Nucl. Sci. and Eng.* **41** 177 (1970);
   J.D. Cramer and H.C. Britt, *Phys. Rev. C* **2** 2350 (1970);
   B.B. Back *et al.*, *Phys. Rev. C* **9** 1924 (1974);
   B.B. Back *et al.*, *Phys. Rev. C* **10** 1948 (1974);
   H.C. Britt and J.B. Wilhelmy, *Nucl. Sci. and Eng.* **72** 222 (1979).
7. J. Escher *et al.*, Nucl. Phys. A, in press;
   C. Forssén *et al.*, Nucl. Phys. A, in press
8. P. Fröbrich and R. Lipperheide, *Theory of Nuclear Reactions*, Oxford University Press, Oxford, U.K., 1996.
9. E. Gadioli and P. E. Hodgson, *Pre-Equilibrium Nuclear Reactions* (Clarendon Press, Oxford, 1992).
10. F. S. Dietrich, *Simple Derivation of the Hauser-Feshbach and Weisskopf-Ewing Formulae, with Application to Surrogate Reactions*, Technical Report No. UCRL-TR-201718, Lawrence Livermore National Laboratory, Livermore, CA, 2004 (unpublished).
11. W. Younes W and H.C. Britt, *Phys. Rev. C* **67** 024610 (2003);
    W. Younes W and H.C. Britt, *Phys. Rev. C* **68** 034610 (2003).
12. W. Younes, H. C. Britt, J. A. Becker, and J. B. Wilhelmy, *Initial estimate for the $^{237}U(n,f)$ cross section for $0.1 \leq E_n$*, Technical Report No. UCRL-ID-154194, Lawrence Livermore National Laboratory, Livermore, CA, 2003 (unpublished).
13. A. K. Kerman and K. W. McVoy, Ann. Phys. (N.Y.) **122** 197 (1979).
14. W. Parker *et al.*, Phys. Rev. C **52** 252 (1995).

Recent progress on understanding "pasta" phases in dense stars

Gentaro Watanabe[*,†] and Hidetaka Sonoda[**]

[*]*NORDITA, Blegdamsvej 17, DK-2100 Copenhagen Ø, Denmark*
[†]*The Institute of Chemical and Physical Research (RIKEN), Saitama 351-0198, Japan*
[**]*Department of Physics, University of Tokyo, Tokyo 113-0033, Japan*

Abstract. In cores of supernovae and crusts of neutron stars, nuclei can adopt interesting shapes, such as rods or slabs, etc., which are referred to as nuclear "pasta." Recently, we have been studying the pasta phases focusing on their dynamical aspects with quantum molecular dynamic (QMD) approach. We review our findings on the following topics: dynamical formation of the pasta phases by cooling down the hot uniform nuclear matter; a phase diagram on the density versus temperature plane; structural transitions between the pasta phases induced by compression and their mechanism. Properties of the nuclear interaction used in our works are also discussed.

I. INTRODUCTION

In ordinary matter, atomic nuclei are roughly spherical. This may be understood in the liquid drop picture of the nucleus as being a result of the forces due to the surface tension of nuclear matter, which favors a spherical nucleus, being greater than those due to the electrical repulsion between protons, which tends to make the nucleus deform. When the density of matter approaches that of atomic nuclei, i.e., the normal nuclear density ρ_0, nuclei are closely packed and the effect of the electrostatic energy becomes comparable to that of the surface energy. Consequently, at subnuclear densities around $\rho \simeq \rho_0/2$, the energetically favorable configuration is expected to have remarkable structures: the nuclear matter region (i.e., the liquid phase) is divided into periodically arranged parts of rodlike or slablike shape, embedded in the gas phase and in a roughly uniform electron gas. Besides, there can be phases in which nuclei are turned inside out, with cylindrical or spherical bubbles of the gas phase in the liquid phase. These phases with nonspherical nuclei are often referred to as nuclear "pasta" phases because nuclear slabs and rods look like "lasagna" and "spaghetti." Likewise, spherical nuclei and bubbles are called "meatballs" and "cheese," respectively.

In equilibrium dense matter in supernova cores and neutron stars, existence of the pasta phases has been predicted by Ravenhall *et al.* [27] and Hashimoto *et al.* [11]. Since these seminal works, properties of the pasta phases in equilibrium states have been investigated with various nuclear models. They include studies on phase diagrams at zero temperature [16, 23, 31, 34, 39] and at finite temperatures [15]. These earlier works have confirmed that, for various nuclear models, the nuclear shape changes as: sphere → cylinder → slab → cylindrical hole → spherical hole → uniform, with increasing density.

In these earlier works, however, a liquid drop model or the Thomas-Fermi approximation is used with an assumption on the nuclear shape (except for Ref. [39]). Thus the phase diagram at subnuclear densities and the existence of the pasta phases should be examined without assuming the nuclear shape. It is also noted that at temperatures of several MeV, which are relevant to the collapsing cores, effects of thermal fluctuations on the nucleon distribution are significant. However, these thermal fluctuations cannot be described properly by mean-field theories such as the Thomas-Fermi approximation used in the previous work [15].

In contrast to the equilibrium properties, dynamical or non-equilibrium aspects of the pasta phases had not been studied until recently except for some limited cases [13, 24]. Thus it had been unclear even whether or not the pasta phases can be formed and the transitions between them can be realized during the collapse of stars and the cooling of neutron stars, which have finite time scales.

To solve the above problems, molecular dynamic approaches for nucleon many-body systems are suitable. They treat the motion of the nucleonic degrees of freedom and can describe thermal fluctuations and many-body correlations beyond the mean-field level.

Using the framework of QMD [1], which is one of the molecular dynamic methods, we have solved the following two major questions [35, 36, 37].

- *Question* 1: Whether or not the pasta phases are formed by cooling down hot uniform nuclear matter in a finite time scale much smaller than that of the neutron star cooling?
- *Question* 2: Whether or not transitions between the pasta phases can occur by the compression during the collapse of a star?

The pasta phases have recently begun to attract the attention of many researchers (see, e.g., Refs. [5, 17] and references therein). The mechanism of the collapse-driven supernova explosion has been a central mystery in astrophysics for almost half a century (e.g., Ref. [4]). Previous studies suggest that the revival of the shock wave by neutrino heating is a crucial process. As has been pointed out in Refs. [34, 36] and elaborated in Refs. [12, 32], the existence of the pasta phases instead of uniform nuclear matter increases the neutrino opacity of matter in the inner core significantly due to the neutrino coherent scattering by nuclei [9, 28]; this affects the total energy transferred to the shocked matter. Thus the pasta phases could play an important role in the future study of supernova explosions. Our recent work [35] strongly suggests the possibility of dynamical formation of the pasta phases from a crystalline lattice of spherical nuclei; effects of the pasta phases on the supernova explosions should be seriously discussed in the near future.

II. METHOD: QUANTUM MOLECULAR DYNAMICS

Among various versions of the molecular dynamic models, Quantum Molecular Dynamics (QMD) [1] is the most practical one for investigating the pasta phases. Rodlike and slablike nuclei are mesoscopic entities of nuclei themselves and they contain a large number of nucleons. QMD, which is a less elaborate in the treatment of the exchange effect, allows us to study such large systems with several nonspherical nuclei. The typical length scale r_c of half of the inter-structure is $r_c \sim 10$ fm and the density region of interest is around half of the normal nuclear density ρ_0. The total nucleon number N necessary to reproduce n structures in the simulation box is $N \sim \rho_0 (2r_c n)^3$ (for slabs). It is thus desirable to use $\sim 10^4$ nucleons in order to reduce boundary effects. Such large systems are difficult to be handled by other molecular dynamic models such as FMD [8] and AMD [22], whose calculation costs increase as $\sim N^4$, but are tractable for QMD, whose calculation costs increase as $\sim N^2$.

It is also noted that the exchange effect is less important for the nuclear pasta structures, which are in the macroscopic scale for nucleons. This can be seen by comparing the typical values of the exchange energy and of the energy difference between pasta phases. Suppose there are two identical nucleons, $i = 1$ and 2, bound in different nuclei. The exchange energy between these particles is calculated as an exchange integral: $I = \int U(\mathbf{r}_1 - \mathbf{r}_2) \, \varphi_1(\mathbf{r}_1) \varphi_1^*(\mathbf{r}_2) \varphi_2(\mathbf{r}_2) \varphi_2^*(\mathbf{r}_1) \, d\mathbf{r}_1 d\mathbf{r}_2$, where U is the potential energy. An asymptotic form of the wave function is given by $\varphi_i \sim \exp(-k_i r)$ with $k_i = \sqrt{2mE_i}/\hbar$, $(i = 1, 2)$, where E_i is the binding energy and m is the nucleon mass. The exchange integral reads $I \sim \exp[-(k_1 + k_2)R] \sim 5 \times 10^{-6}$ MeV for the internuclear distance $R \simeq 10$ fm and $E_i \simeq 8$ MeV, which is extremely smaller than the typical energy difference per nucleon between different pasta phases of order 0.1 keV (for neutron star matter) - 10 keV (for supernova matter). Therefore, it is expected that QMD is not a bad approximation for investigating the pasta phases.

1. Model Hamiltonian and its Properties

In our studies on the pasta phases, we have used a nuclear force given by a QMD model Hamiltonian with the medium-equation-of-state parameter set in Ref. [18]. This model Hamiltonian consists of six parts:

$$\mathcal{H} = K + V_{\text{Pauli}} + V_{\text{Skyrme}} + V_{\text{sym}} + V_{\text{MD}} + V_{\text{Coulomb}}, \qquad (1)$$

where K is the kinetic energy; V_{Pauli} is the momentum-dependent "Pauli potential," which reproduces the effects of the Pauli principle phenomenologically; V_{Skyrme} is the Skyrme potential which consists of an attractive two-body term and a repulsive three-body term; V_{sym} is the symmetry potential; V_{MD} is the momentum-dependent potential introduced as two Fock terms of the Yukawa interaction; V_{Coulomb} is the Coulomb energy between protons.

The parameters in the Pauli potential are determined to fit the kinetic energy of the free Fermi gas at zero temperature. The above model Hamiltonian reproduces the binding energy of symmetric nuclear matter, 16 MeV per nucleon, at the normal nuclear density $\rho_0 = 0.165$ fm^{-3} and other saturation properties: the incompressibility is set to be 280 MeV and the symmetry energy is 34.6 MeV. This model Hamiltonian also well reproduce the properties of stable nuclei relevant to our interest: the binding energy except for light nuclei from ^{12}C to ^{20}Ne [18], and the rms radius of the ground state of heavy ones with $A \gtrsim 100$ [14]. It is also confirmed that another QMD Hamiltonian close to this model provides a good description of nuclear reactions including the low energy region (several MeV per nucleon) [21].

Let us then examine other properties of the nuclear interaction (at zero temperature), which have not been determined accurately yet but have important effects on inhomogeneous structure of matter at subnuclear densities. Such quantities are the nuclear surface tension E_{surf}, the energy per nucleon ε_n of the pure neutron matter, and the proton chemical potential $\mu_p^{(0)}$ in the pure neutron matter. The surface tension E_{surf}, which is the most important among these three quantities, controls the size of the nuclei and bubbles, and hence the sum of the Coulomb and surface energies. With increasing E_{surf} and so this energy sum, the density ρ_m at which matter becomes uniform is lowered. There is a tendency, especially in a case of neutron star matter, that the higher ε_n, ρ_m is lowered. This is because larger ε_n tends to favor uniform nuclear matter without dripped neutron gas regions than mixed phases. In neutron star matter, there is also a tendency that the lower $\mu_p^{(0)}$, the smaller ρ_m. This is because $-\mu_p^{(0)}$ represents the degree to which the neutron gas outside the nuclei favors the presence of protons in itself.

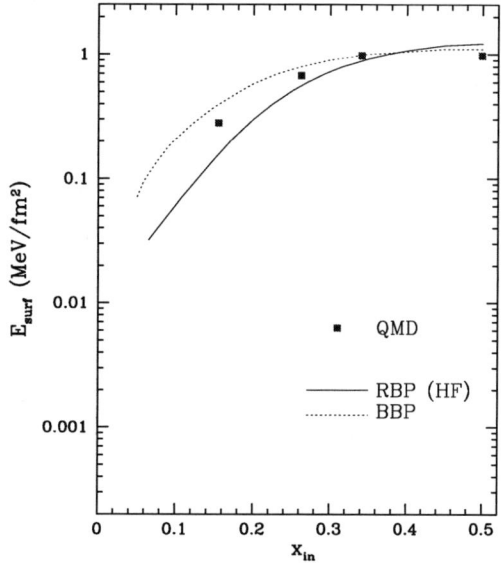

FIGURE 1. (Color) The nuclear surface energy per unit area (the surface tension) E_{surf} versus the proton fraction x_{in} in the nuclear matter region. The red solid squares are the values of the present QMD Hamiltonian [18]; the solid curve is the result of the Skyrme-Hartree-Fock calculation with a modified version of 1' parameter set done by Ravenhall, Bennett and Pethick (RBP) [26]; the dotted curve is from Baym, Bethe and Pethick (BBP) [3]. This figure is adapted from Ref. [36].

In Figs. 1 and 2, we have plotted the results of these quantities for the present model Hamiltonian. We can say that, on the whole, they give reasonable values within uncertainties of each quantities. It is noted that E_{surf} of the present model shows moderate values between the RBP and BBP results. The behavior of $\mu_p^{(0)}$ is also in a reasonable agreement with various Skyrme-Hartree-Fock calculations except for higher densities of $\rho_n \gtrsim 0.1$ fm^{-3} relevant only for neutron star matter just below ρ_m. The quantity ε_n, however, shows some deviation from the major trend of Skyrme-Hartree-Fock results and of the values of the microscopic calculations. The behavior of ε_n of the present model is similar to that of the SkM interaction and of the interaction by Myers *et al.* [20].

In the following, for each case of different value of the proton fraction x of matter, we summarize the consequences of the features of E_{surf}, ε_n and $\mu_p^{(0)}$ of the present QMD interaction.

1. For symmetric nuclear matter ($x = x_{\text{in}} = 0.5$)
   According to E_{surf} at $x_{\text{in}} = 0.5$, the present model is consistent with the other results, and is an appropriate effective interaction for studying the pasta phases at $x = 0.5$.
2. For neutron star matter ($x \lesssim 0.1$)

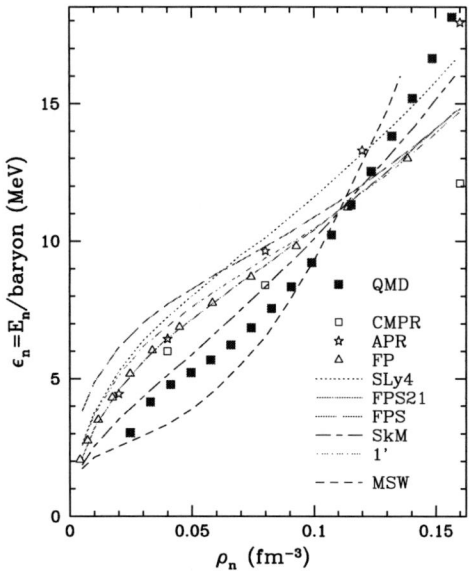 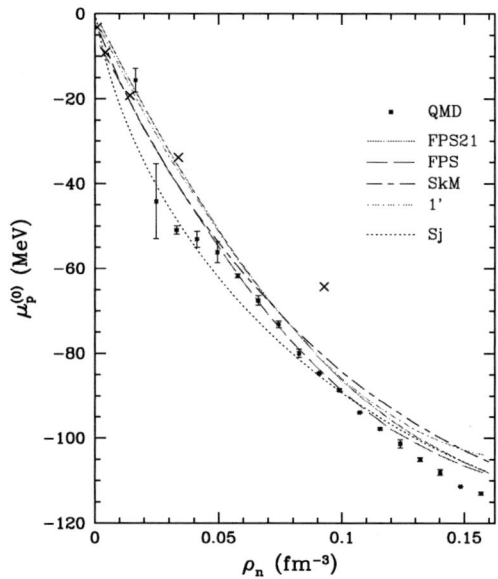

FIGURE 2. (Color) The neutron density ρ_n dependence of the energy per nucleon ε_n (left panel) and the proton chemical potential $\mu_p^{(0)}$ (right panel) of pure neutron matter. The red solid squares show the result of the present QMD model Hamiltonian [18]. The red dotted line denoted by SLy4 is the result from Ref. [7], and the colored broken lines as marked by the other Skyrme interactions (FPS21, 1', FPS and SkM) are the results summarized by Pethick, Ravenhall and Lorenz [25]. The black dotted line is the result of Sjöberg [30] and the black dashed line is that of Myers, Swiatecki and Wang [20]. The open squares shows the result of the GFMC calculation by Carlson et al. [6]; the open stars denote the values obtained by Akmal, Pandharipande and Ravenhall [2]; the triangles are those from Friedman and Pandharipande [10]; the crosses from Siemens and Pandharipande [29]. The large error bars and the scatter of $\mu_p^{(0)}$ of QMD in the low density region $\rho \lesssim 0.3\rho_0$ are due to the local roughness of the density in the neutron matter, which is caused by the fixed width of the wave packet. This figure is adapted from Ref. [36].

The melting density ρ_m is lowered by larger E_{surf}, steep rise of ε_n at $\rho \gtrsim 0.12$ fm^{-3} and larger negative values of $\mu_p^{(0)}$ at $\rho \gtrsim 0.1$ fm^{-3} compared to various Skyrme-Hartree-Fock calculations.

3. For supernova matter ($x \sim 0.3$)

Relatively low ε_n at low neutron densities $\lesssim 0.1$ fm^{-3} acts to favor mixed phases rather than the uniform phase and E_{surf} acts in the opposite way in comparison with the Skyrme-Hartree-Fock result by RBP [26].

III. SIMULATIONS AND RESULTS

Using the framework of QMD, we have solved the two major questions posed in the beginning of this article [35, 36, 37]. In the present section, we will review these works [1]. Hereafter, we set the Boltzmann constant $k_B = 1$.

In our simulations, we treated the system which consists of neutrons, protons, and electrons in a cubic box with periodic boundary condition. The system is not magnetically polarized, i.e., it contains equal numbers of protons (and neutrons) with spin up and spin down. The relativistic degenerate electrons which ensure charge neutrality are regarded as a uniform background [24, 33]. The Coulomb interaction is calculated by the Ewald method taking account of the Gaussian charge distribution of the proton wave packets.

---

[1] This section is based on our recent review article [38].

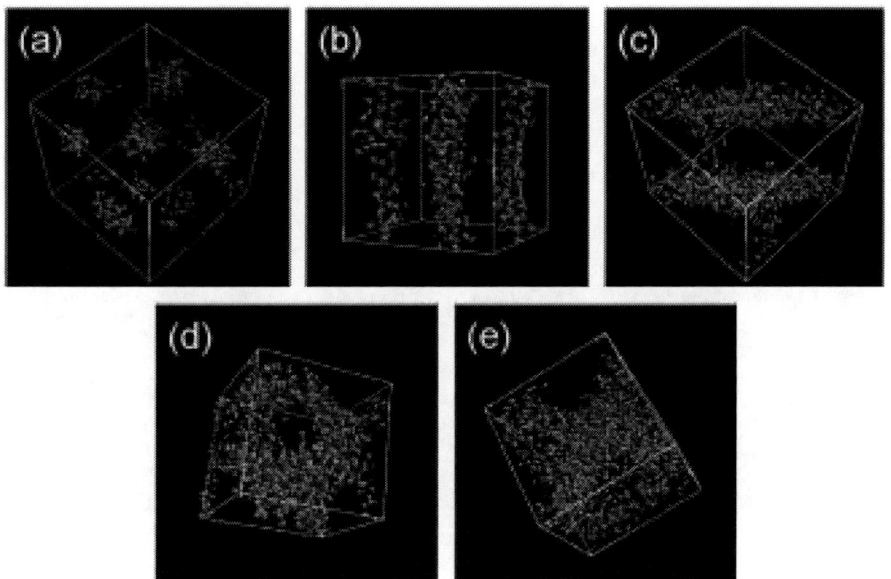

FIGURE 3. (Color) Nucleon distributions of the pasta phases in cold matter at $x = 0.5$; (a) sphere phase, $0.1\rho_0$ ($L_{\text{box}} = 43.65$ fm, $N = 1372$); (b) cylinder phase, $0.225\rho_0$ ($L_{\text{box}} = 38.07$ fm, $N = 2048$); (c) slab phase, $0.4\rho_0$ ($L_{\text{box}} = 31.42$ fm, $N = 2048$); (d) cylindrical hole phase, $0.5\rho_0$ ($L_{\text{box}} = 29.17$ fm, $N = 2048$) and (e) spherical hole phase, $0.6\rho_0$ ($L_{\text{box}} = 27.45$ fm, $N = 2048$), where L_{box} is the box size and N is the total number of nucleons. The whole simulation box is shown in this figure. The red particles represent protons and the green ones neutrons. Taken from Ref. [37].

1. Realization of the Pasta Phases and Equilibrium Phase Diagrams

In Refs. [36, 37], we have reproduced the pasta phases from hot uniform nuclear matter and discussed phase diagrams at zero and finite temperatures. In these works, we first prepared a uniform hot nucleon gas at the temperature $T \sim 20$ MeV as an initial condition, which is equilibrated for $\sim 500 - 2000$ fm/c in advance. To realize the ground state of matter, we then cooled it down slowly until the temperature got ~ 0.1 MeV or less for $O(10^3 - 10^4)$ fm/c, keeping the nucleon number density constant. In the cooling process, we mainly used the frictional relaxation method (equivalent to the steepest descent method), which is given by the QMD equations of motion plus small friction terms. In the case of finite temperatures, we also used thermostat to reproduce the equilibrium states.

The resultant typical nucleon distributions of cold matter at subnuclear densities are shown in Fig. 3 for proton fraction of matter $x = 0.5$. We can see from these figures that the phases with rodlike and slablike nuclei, cylindrical and spherical bubbles, in addition to the phase with spherical nuclei are reproduced. The above simulations have shown that the pasta phases can be formed dynamically from hot uniform matter within a time scale of $\sim O(10^3 - 10^4)$ fm/c.

We show snapshots of nucleon distributions at $T = 1, 2$ and 3 MeV for a density $\rho = 0.225\rho_0$ in Fig. 4. This density corresponds to the phase with rodlike nuclei at $T = 0$. From these figures, we can see the following qualitative features: at $T \simeq 1.5 - 2$ MeV (but snapshots for $T \simeq 1.5$ MeV are not shown), the number of the evaporated nucleons becomes significant; at $T \gtrsim 3$ MeV, nuclei almost melt and the spatial distribution of the nucleons are rather smoothed out.

When we try to classify the nuclear structure systematically, the integral mean curvature and the Euler characteristic (see, e.g., Ref. [19] and references therein) are useful. Suppose there is a set of regions R, where the density is higher than a threshold density ρ_{th}. The integral mean curvature and the Euler characteristic for the surface of this region ∂R are defined as surface integrals of the mean curvature $H = (\kappa_1 + \kappa_2)/2$ and the Gaussian curvature $G = \kappa_1 \kappa_2$, respectively; i.e., $\int_{\partial R} H dA$ and $\chi \equiv \frac{1}{2\pi} \int_{\partial R} G dA$, where κ_1 and κ_2 are the principal curvatures and dA is the area element of the surface of R. The Euler characteristic χ depends only on the topology of R and is expressed as $\chi =$ (number of isolated regions) $-$ (number of tunnels) $+$ (number of cavities). Using a combination of these two quantities

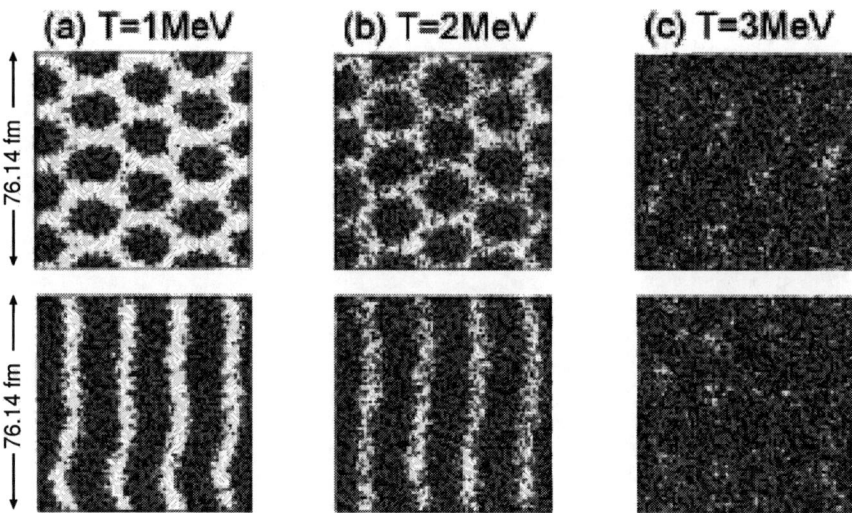

FIGURE 4. (Color) Nucleon distributions for $x = 0.5$, $\rho = 0.225\rho_0$ at the temperatures of 1, 2 and 3MeV. The total number of nucleons $N = 16384$ and the box size $L_{\text{box}} = 76.14$ fm. The upper panels show the top views along the axis of the cylindrical nuclei at $T = 0$, the lower ones the side views. Protons are represented by the red particles, and neutrons by the green ones. Taken from Ref. [37].

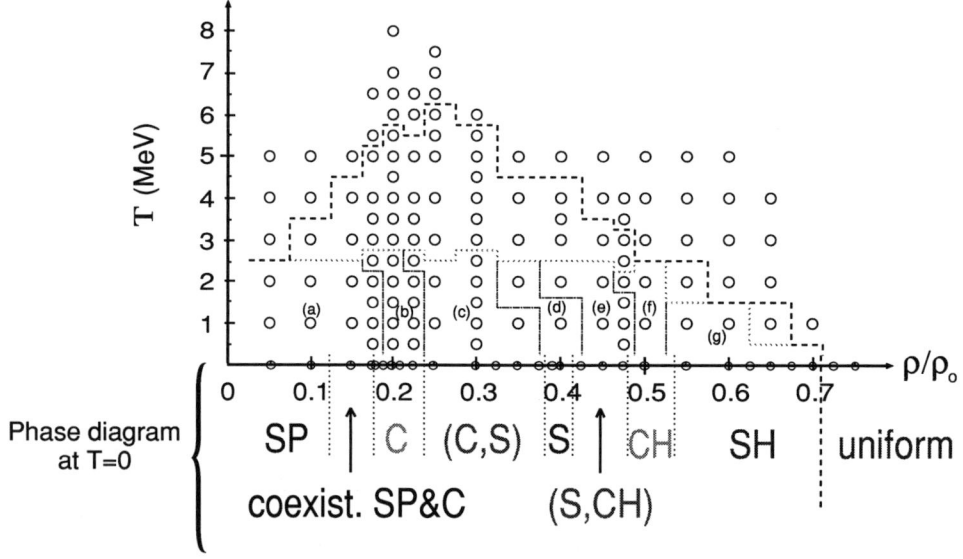

FIGURE 5. (Color) Phase diagram of matter at $x = 0.5$ plotted in the ρ - T plane. The dashed and the dotted lines on the diagram show the phase separation line and the limit below which the nuclear surface can be identified, respectively. The dash-dotted lines are the phase boundaries between the different nuclear shapes. The symbols SP, C, S, CH, SH, U stand for nuclear shapes, i.e., sphere, cylinder, slab, cylindrical hole, spherical hole and uniform, respectively. The parentheses (A,B) show intermediate phases between A and B-phases with negative χ, which are different from coexistence phases of them. The regions (a)-(g) correspond to the nuclear shapes characterized by $\int_{\partial R} H dA$ and χ as follows: (a) $\int_{\partial R} H dA > 0$, $\chi > 0$; (b) $\int_{\partial R} H dA > 0$, $\chi = 0$; (c) $\int_{\partial R} H dA > 0$, $\chi < 0$; (d) $\int_{\partial R} H dA = 0$, $\chi = 0$; (e) $\int_{\partial R} H dA < 0$, $\chi < 0$; (f) $\int_{\partial R} H dA < 0$, $\chi = 0$; (g) $\int_{\partial R} H dA < 0$, $\chi > 0$. Simulations have been carried out at points denoted by circles. Adapted from Ref. [36]

calculated for nuclear surface [2], each pasta phase can be represented uniquely, i.e., for the phase with spherical nuclei: $\int_{\partial R} H dA > 0$, $\chi > 0$, cylindrical nuclei: $\int_{\partial R} H dA > 0$, $\chi = 0$, slablike nuclei: $\int_{\partial R} H dA = 0$, $\chi = 0$, cylindrical bubbles: $\int_{\partial R} H dA < 0$, $\chi = 0$, and spherical bubbles: $\int_{\partial R} H dA < 0$, $\chi > 0$. We note that the value of χ for the ideal pasta phases is zero except for the phase with spherical nuclei or spherical bubbles with positive χ; negative χ is not obtained for the pasta phases.

The phase diagram obtained for $x = 0.5$ is plotted in Fig. 5. As shown above, nuclear surface can be identified typically at $T \lesssim 3$ MeV (see the dotted lines) in the density range of interest. Thus the regions between the dotted line and the dashed line correspond to some non-uniform phase, which is however difficult to be classified into specific phases because the nuclear surface cannot be identified well.

In the region below the dotted lines, where we can identify the nuclear surface, we have obtained the pasta phases with spherical nuclei [region (a)], rodlike nuclei [region (b)], slablike nuclei [region (d)], cylindrical holes [region (f)] and spherical holes [region (g)]. It is noted that in addition to these pasta phases, structures with negative χ have been also obtained in the regions of (c) and (e); matter consists of multiply connected nuclear and bubble regions (i.e., spongelike structure) with branching rodlike nuclei, perforated slabs and branching bubbles, etc. A detailed discussion on the phase diagrams is given in Ref. [37].

2. Structural Transitions between the Pasta Phases

In Ref. [35], we have approached the second question asked at the beginning of this article. We have performed QMD simulations of the compression of dense matter and have succeeded in simulating the transitions between rodlike and slablike nuclei and between slablike nuclei and cylindrical bubbles.

The initial conditions of the simulations are samples of the columnar phase ($\rho = 0.225\rho_0$) and of the laminar phase ($\rho = 0.4\rho_0$) of 16384-nucleon system at $x = 0.5$ and $T \simeq 1$ MeV. These are obtained in Ref. [37], which are presented in the last section. We then adiabatically compressed the above samples by increasing the density at the average rate of $\simeq 1.3 \times 10^{-5} \rho_0/(\text{fm}/c)$ for the initial condition of the columnar phase and $\simeq 7.1 \times 10^{-6} \rho_0/(\text{fm}/c)$ for that of the laminar one. According to the typical value of the density difference between each pasta phase, $\sim 0.1\rho_0$ (see Fig. 5), we increased the density to the value corresponding to the next pasta phase taking the order of 10^4 fm/c, which was much longer than the typical time scale of the nuclear fission, ~ 1000 fm/c. Thus the above rates ensured the adiabaticity of the simulated compression process with respect to the change of nuclear structure. Finally, we relaxed the compressed sample at $\rho = 0.405\rho_0$ for the former case and at $0.490\rho_0$ for the latter one. These final densities are those of the phase with slablike nuclei and cylindrical bubbles, respectively, in the equilibrium phase diagram at $T \simeq 1$ MeV (see Fig. 5).

The resulting time evolution of the nucleon distribution is shown in Figs. 6 and 7. As can be seen from Fig. 6, the phase with slablike nuclei is finally formed [Fig. 6-(6)] from the phase with rodlike nuclei [Fig. 6-(1)]. The temperature in the final state is $\simeq 1.35$ MeV. It is noted that the transition is triggered by thermal fluctuation, not by the fission instability: when the internuclear spacing becomes small enough and once some pair of neighboring rodlike nuclei touch due to thermal fluctuations, they fuse [Figs. 6-(2) and 6-(3)]. Then such connected pairs of rodlike nuclei further touch and fuse with neighboring nuclei in the same lattice plane like a chain reaction [Fig. 6-(4)]; the time scale of the each fusion process is of order 10^2 fm/c, which is much smaller than that of the density change.

The transition from the phase with slablike nuclei to the phase with cylindrical holes is shown in Fig. 7. When the internuclear spacing decreases enough, neighboring slablike nuclei touch due to the thermal fluctuation as in the above case. Once nuclei begin to touch [Fig. 7-(2)] bridges between the slabs are formed at many places on a time scale (of order 10^2 fm/c) much shorter than that of the compression. After that the bridges cross the slabs nearly orthogonally for a while [Fig. 7-(3)]. Nucleons in the slabs continuously flow into the bridges, which become wider and merge together to form cylindrical holes. Afterwards, the connecting regions consisting of the merged bridges move gradually, and the cylindrical holes relax to form a triangular lattice [Fig. 7-(6)]. The final temperature in this case is $\simeq 1.3$ MeV.

Trajectories of the above processes on the plane of the integral mean curvature $\int_{\partial R} H dA$ and the Euler characteristic χ are plotted in Fig. 8. This figure shows that the above transitions proceed through a transient state with "spongelike" structure, which gives negative χ. As can be seen from Fig. 8-(a) [Fig. 8-(b)], the value of the Euler characteristic

---

[2] Nuclear surface generally corresponds to an isodensity surface for the threshold density $\rho_{\text{th}} \simeq 0.5\rho_0$ in our simulations.

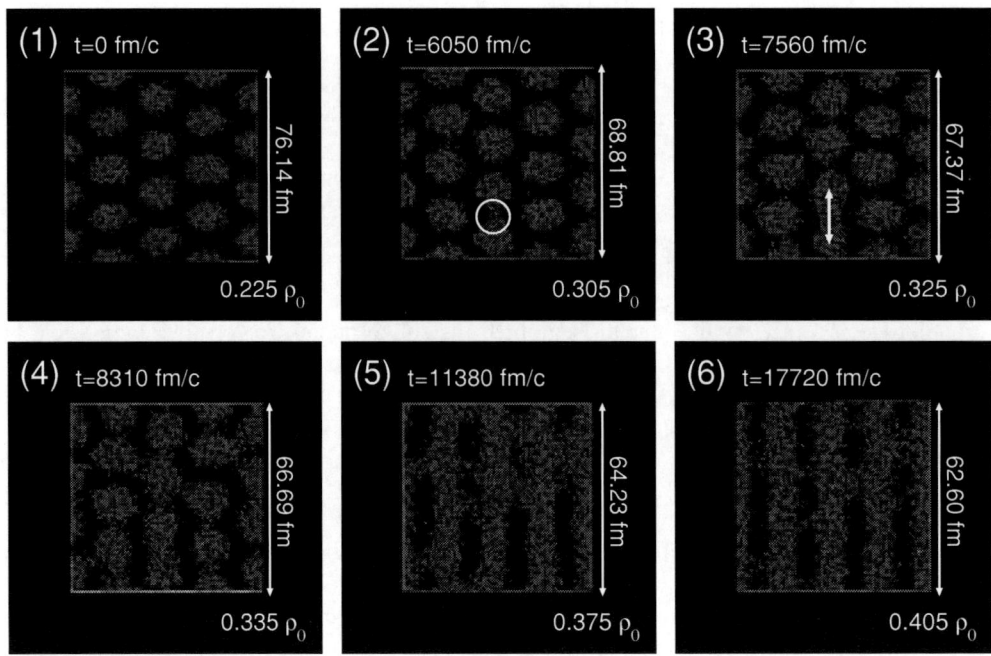

FIGURE 6. (Color) Snapshots of the transition process from the phase with rodlike nuclei to the phase with slablike nuclei (the whole simulation box is shown). The red particles show protons and the green ones neutrons. After neighboring nuclei touch as shown by the circle in Fig. 6-(2), the "compound nucleus" elongates along the arrow in Fig. 6-(3). The box size is rescaled to be equal in this figure. Adapted from Ref. [35].

begins to decrease from zero when the rodlike [slablike] nuclei touch. It continues to decrease until all of the rodlike [slablike] nuclei are connected to others by small bridges at $t \simeq 9840$ fm/c [$\simeq 12000$ fm/c]. Then the bridges merge to form slablike nuclei [cylindrical holes] and the value of the Euler characteristic increases towards zero. Finally, the system relaxes into a layered lattice of the slablike nuclei [a triangular lattice of the cylindrical holes]. Thus the whole transition process can be divided into the "connecting stage" and the "relaxation stage" before and after the moment at which the Euler characteristic is minimum; the former starts when the nuclei begin to touch and it takes $\simeq 3000 - 4000$ fm/c and the latter takes more than 8000 fm/c.

3. Formation of the Pasta Phases

In closing the present article, let us briefly show our recent results of a study on the formation process of pasta nuclei from spherical ones; i.e., a transition from the phase with spherical nuclei to that with rodlike nuclei. Time evolution of the nucleon distribution in the transition process is shown in Fig. 9. The initial condition of this simulation is a nearly perfect bcc unit cell with 409 nucleons (202 protons and 207 neutrons) at $T \simeq 1$ MeV. We compressed the system in a similar way to that of the simulations explained in the previous section. The average rate of the density change in the present case is $\simeq 4.4 \times 10^{-6} \rho_0/(\text{fm}/c)$. Since the two nuclei start to touch [see the circles in Fig. 9-(2)], the transition process completes within $\simeq 2500$ fm/c and the rodlike nucleus is formed. The final state [Fig. 9-(4)] is a triangular lattice of the rodlike nuclei.

The present simulation has been performed using a rather small system; effects of the finite system size in this simulation should be examined. Detailed investigation of the transition using a larger system will be presented in a future publication [32].

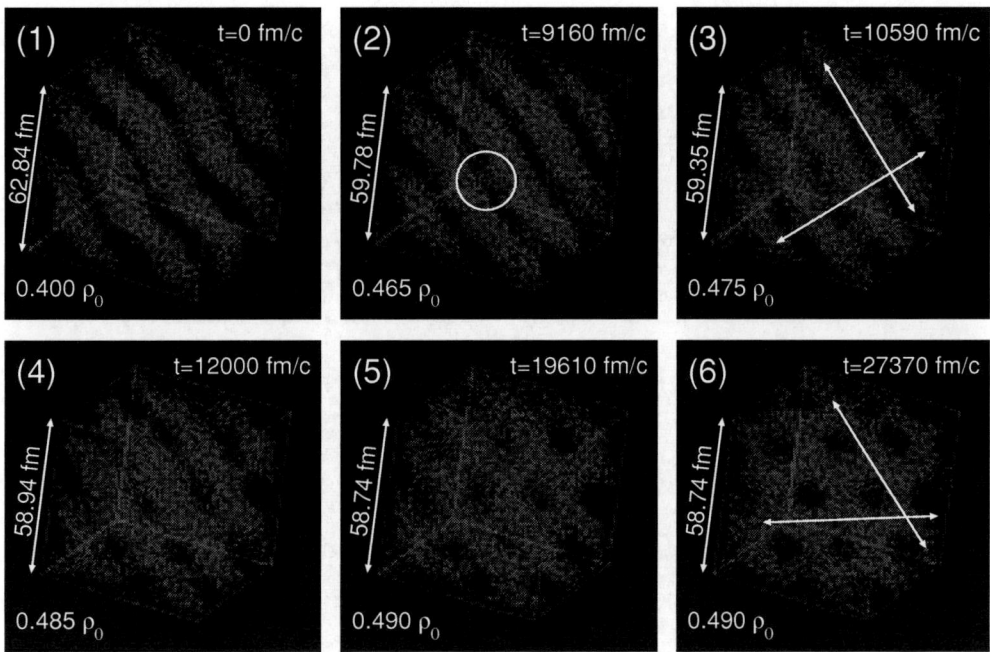

FIGURE 7. (Color) The same as Fig. 1 for the transition from the phase with slablike nuclei to the phase with cylindrical holes (the box size is not rescaled in this figure). After the slablike nuclei begin to touch [see the circle in Fig. 7-(2)], the bridges first crosses them almost orthogonally as shown by the arrows in Fig. 7-(3). Then the cylindrical holes are formed and they relax into a triangular lattice, as shown by the arrows in Fig. 7-(6). Adapted from Ref. [35].

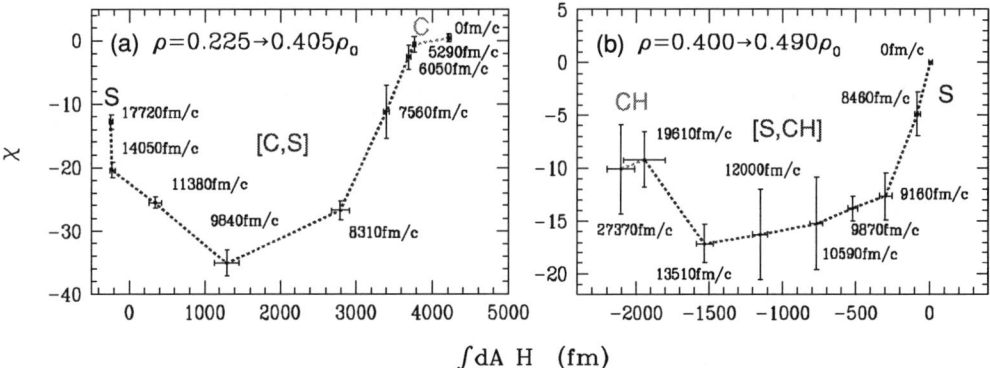

FIGURE 8. (Color) Time evolution of $\int_{\partial R} H dA$ and χ during the simulations. The data points and the error bars show, respectively, the mean values and the standard deviations in the range of the threshold density $\rho_{th} = 0.3 - 0.5\rho_0$, which includes typical values for the nuclear surface. The panel (a) is for the transition from cylindrical (C) to slablike nuclei (S) and panel (b) for the transition from slablike nuclei to cylindrical holes (CH). Transient states are shown as [C,S] and [S,CH] for each transition. Adapted from Ref. [35].

IV. CONCLUSION

We approached the two questions posed in Section I using the framework of QMD. According to the results of our simulations, our answer is strongly affirmative for both questions. The nuclear interaction used in our simulations shows generally reasonable properties at subnuclear densities not only for symmetric nuclear matter but also for neutron matter. This result also supports our conclusion.

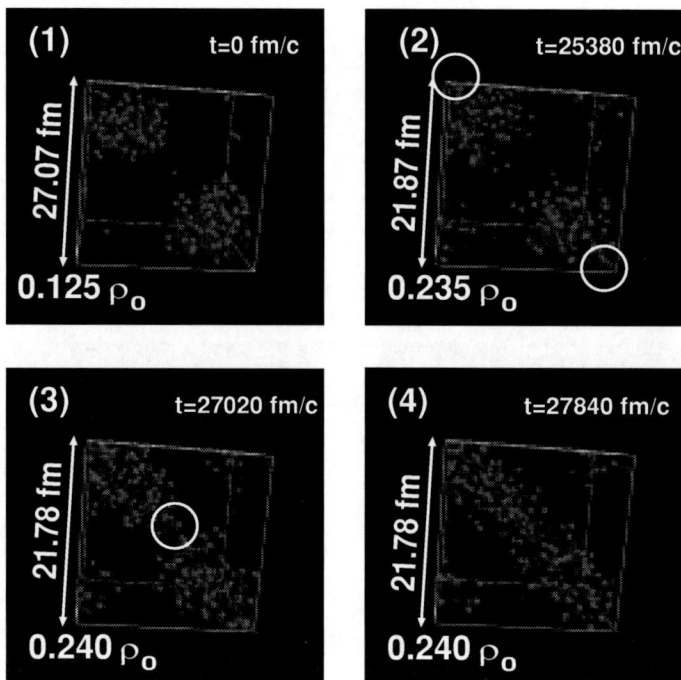

FIGURE 9. (Color) Snapshots of the transition process from the bcc lattice of spherical nuclei to the triangular lattice of rodlike nuclei (the whole simulation box is shown). The red particles show protons and the green ones neutrons. The box size is rescaled to be equal in this figure.

ACKNOWLEDGMENTS

The research reported in this article grew out of collaborations with Kei Iida, Toshiki Maruyama, Katsuhiko Sato, Kenji Yasuoka and Toshikazu Ebisuzaki. Further research currently in progress is performed using RIKEN Super Combined Cluster System with MDGRAPE-2. This work was supported in part by the Nishina Memorial Foundation, by the Japan Society for the Promotion of Science, by the Ministry of Education, Culture, Sports, Science and Technology through Research Grant No. 14-7939, and by RIKEN through Research Grant No. J130026.

REFERENCES

1. J. Aichelin and H. Stöcker, Phys. Lett. **B176**, 14 (1986); J. Aichelin, Phys. Rep. **202**, 233 (1991).
2. A. Akmal, V. R. Pandharipande and D. G. Ravenhall, Phys. Rev. C **58**, 1804 (1998).
3. G. Baym, H. A. Bethe and C. J. Pethick, Nucl. Phys. **A175**, 225 (1971).
4. H. A. Bethe, Rev. Mod. Phys. **62**, 801 (1990).
5. A. Burrows, S. Reddy and T. A. Thompson, Nucl. Phys. **A**, in press (astro-ph/0404432).
6. J. Carlson, J. Morales, Jr., V. R. Pandharipande and D. G. Ravenhall Phys. Rev. C **68**, 025802 (2003).
7. F. Douchin and P. Haensel, Phys. Lett. **B485**, 107 (2000).
8. H. Feldmeier, Nucl. Phys. **A515**, 147 (1990); H. Feldmeier and J. Schnack, Prog. Part. Nucl. Phys. **39**, 393 (1997).
9. D. Z. Freedman, Phys. Rev. D **9**, 1389 (1974).
10. B. Friedman and V. R. Pandharipande, Nucl. Phys. **A361**, 502 (1981).
11. M. Hashimoto, H. Seki and M. Yamada, Prog. Theor. Phys. **71**, 320 (1984).
12. C. J. Horowitz, M. A. Pérez-García, and J. Piekarewicz, Phys. Rev. C **69**, 045804 (2004).
13. K. Iida, G. Watanabe and K. Sato, Prog. Theor. Phys. **106**, 551 (2001); Erratum, *ibid.* **110**, 847 (2003).
14. T. Kido, T. Maruyama, K. Niita and S. Chiba, Nucl. Phys. **A663 & 664**, 877c (2000).
15. M. Lassaut, H. Flocard, P. Bonche, P.H. Heenen and E. Suraud, Astron. Astrophys. **183**, L3 (1987).
16. C. P. Lorenz, D. G. Ravenhall and C. J. Pethick, Phys. Rev. Lett. **70**, 379 (1993).
17. G. Martinez-Pinedo, M. Liebendoerfer, D. Frekers, astro-ph/0412091.
18. T. Maruyama, K. Niita, K. Oyamatsu, T. Maruyama, S. Chiba and A. Iwamoto, Phys. Rev. C **57**, 655 (1998).
19. K. Michielsen and H. De Raedt, Phys. Rep. **347**, 461 (2001).

20. W. D. Myers, W. J. Swiatecki and C. S. Wang, Nucl. Phys. **A436**, 185 (1985).
21. K. Niita, in the Proceedings of the Third Simposium on *"Simulation of Hadronic Many-body System"*, A. Iwamoto *et al.*, Eds., JAERI-conf. **96-009**, 22 (1996) (in Japanese).
22. A. Ono, H. Horiuchi, T. Maruyama and A. Ohnishi, Prog. Theor. Phys. **87**, 1185 (1992); Phys. Rev. Lett. **68**, 2898 (1992).
23. K. Oyamatsu, Nucl. Phys. **A561**, 431 (1993).
24. C. J. Pethick and D. G. Ravenhall, Annu. Rev. Nucl. Part. Sci. **45**, 429 (1995).
25. C. J. Pethick, D. G. Ravenhall and C. P. Lorenz, Nucl. Phys. **A584**, 675 (1995).
26. D. G. Ravenhall, C. D. Bennett and C. J. Pethick, Phys. Rev. Lett. **28**, 978 (1972).
27. D. G. Ravenhall, C. J. Pethick and J. R. Wilson, Phys. Rev. Lett. **50**, 2066 (1983).
28. K. Sato, Prog. Theor. Phys. **53**, 595 (1975); *ibid.* **54**, 1325 (1975).
29. P. J. Siemens and V. R. Pandharipande, Nucl. Phys. **A173**, 561 (1971).
30. O. Sjöberg, Nucl. Phys. **A222**, 161 (1974).
31. K. Sumiyoshi, K. Oyamatsu and H. Toki, Nucl. Phys. **A595**, 327 (1995).
32. G. Watanabe *et al.*, to be published.
33. G. Watanabe and K. Iida, Phys. Rev. C **68**, 045801 (2003).
34. G. Watanabe, K. Iida and K. Sato, Nucl. Phys. **A676**, 455 (2000); *ibid.* **A687**, 512 (2001); Erratum, *ibid.* **A726**, 357 (2003).
35. G. Watanabe, T. Maruyama, K. Sato, K. Yasuoka and T. Ebisuzaki, Phys. Rev. Lett. **94**, 031101 (2005).
36. G. Watanabe, K. Sato, K. Yasuoka and T. Ebisuzaki, Phys. Rev. C **66**, 012801(R) (2002); *ibid.* **68**, 035806 (2003).
37. G. Watanabe, K. Sato, K. Yasuoka and T. Ebisuzaki, Phys. Rev. C **69**, 055805 (2004).
38. G. Watanabe and H. Sonoda, to appear in "Soft Condnsed Matter: New Research", ed. F. Columbus (cond-mat/0502515).
39. R. D. Williams and S. E. Koonin, Nucl. Phys. **A435**, 844 (1985).

Production of complex particles in low energy spallation and in fragmentation reactions by in-medium random clusterization

Denis Lacroix* and Dominique Durand*

Laboratoire de Physique Corpusculaire, ENSICAEN and Université de Caen, IN2P3-CNRS, Blvd du Maréchal Juin 14050 Caen, France

Abstract.
Rules for in-medium complex particle production in nuclear reactions are proposed. These rules have been implemented in two models to simulate nucleon-nucleus and nucleus-nucleus reactions around the Fermi energy [1, 2]. Our work emphasizes the effect of randomness in cluster formation, the importance of the nucleonic Fermi motion as well as the role of conservation laws. The concepts of total available phase-space and explored phase-space under constraint imposed by the reaction are clarified. The compatibility of experimental observations with a random clusterization is illustrated in a schematic scenario of a proton-nucleus collision. The role of randomness under constraint is also illustrated in the nucleus-nucleus case.

Keywords: nuclear reaction, models, cluster production
PACS: 25.40.h,24.10.-i,28.20.-v

INTRODUCTION

Nuclear reactions around the Fermi energy have revealed that nuclei can break into several pieces of various sizes: the so-called multifragmentation process[3]. A striking feature of experimental observation is the large number of charge and energy partitions that can be accessed. In order to understand the statistical aspects of the explored phase-space, several physical origins have been proposed. Among them, the nuclear liquid-gas phase transition appears as one of the best candidate. However, due to the complexity of nuclear reactions including impact parameter mixing, pre-equilibrium emission and thermal decay, it is hard to trace-back the process of cluster formation. Nowadays, the complexity of experimental analyses increases constantly [4, 5]. Conjointly, more and more elaborated models have been proposed to simulate reactions [6, 7, 8, 9, 10, 11, 12, 13, 14, 15]. However, the issue concerning cluster formation remains a highly debated question. In this work, we would like to contribute to the discussion on particle emission during the pre-equilibrium stage. We have tested a large number of hypothesis for the formation of clusters in the nuclear medium in order to provide event generators for the study of nuclear reactions. Guided by the experimental observation, surprising conclusions concerning the way cluster are formed may be assessed. Simple rules have been found for the formation and the emission of complex particles. The hypothesis retained are not only fully compatible with experiments on multifragmentation, but seems also to be adequate for nucleon-nucleus reactions in the same energy range.

The paper is organized as follows: first the rules for cluster formation and emission are introduced and illustrated in a schematic scenario for experiments. In a second part, additional effects that should be accounted to compare quantitatively with measurements are discussed. Finally, the compatibility of the rules with data are illustrated. Conclusions and perspectives are drawn at the end of the paper.

RULES FOR THE FORMATION AND THE EMISSION OF CLUSTERS

In this section, rules for the cluster formation and the production of fragmentation partitions are defined. In order to illustrate these rules we consider a proton colliding a heavy target with an energy close to the Fermi energy [1]. Then

---

[1] The energy is chosen small enough to avoid strong influence of direct two-body nucleon-nucleon collisions

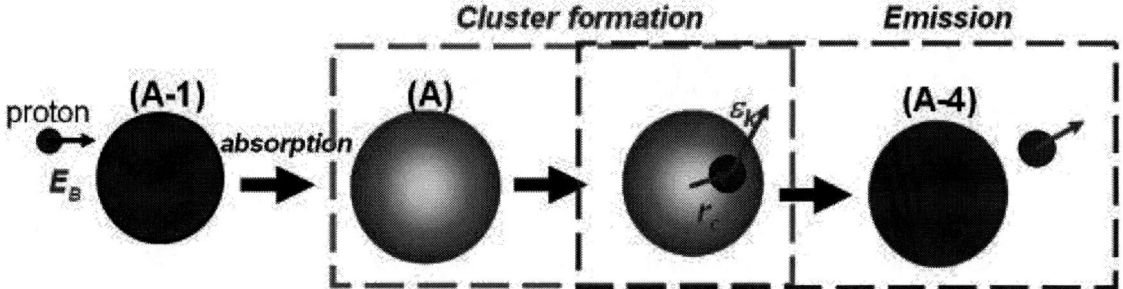

FIGURE 1. *Schematic representation of a three step nucleon-induced reaction. A nucleon with beam energy close to the Fermi energy is first absorbed by a nucleus. Then two steps are identified for pre-equilibrium emission: the formation of the cluster and its emission.*

a simplified three steps scenario is considered (see figure 1). First the incident nucleon is absorbed. The second step corresponds to the in-medium formation of the cluster while the last step is the emission in the continuum.

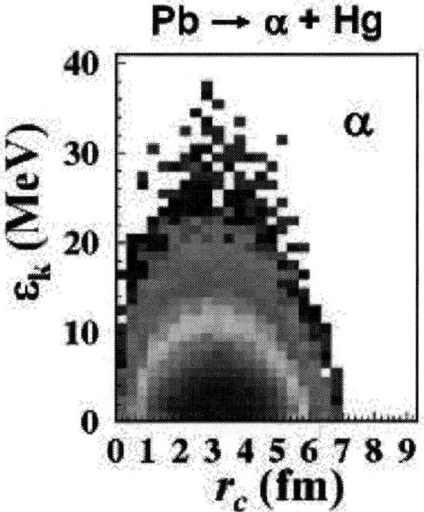

FIGURE 2. Correlation between the position and the kinetic energy per nucleon for the α particles using a random sampling assumption for the nucleons forming the α particle. This two-dimensional plot corresponds to the total "accessible" phase-space for the considered particle.

The hypothesis retained to describe the last two stages can be summarized as follow:

- **Cluster formation:** Considering a cluster of mass A_c and charge Z_c formed in the medium, we assume that the cluster is composed of nucleons chosen randomly in the target[2]. Thus, the kinematics of the cluster is directly linked to the kinematics of the nucleons. This defines the *"total accessible phase-space"* for the cluster (A_c, Z_c) in the medium. Figure 2 displays the correlation between the position r_c and the kinetic energy ε_k of an α particle produced in a Pb target obtained with the random sampling assumption. We would like to stress that the random assumption is a generalization of the pioneering work of Goldhaber [16].

- **Cluster emission:** we dissociate here the total accessible phase-space from the explored phase-space because the latter must take into account the constraints of the reaction. Indeed, while the first rule described above leads to a large set of configurations, all configurations will not necessarily lead to the emission of a cluster. Two constraints can be identified. The first one, which is independent of the entrance channel, is due to the mutual interaction

---

[2] In the energy range considered here, a Thomas-Fermi distribution corresponding to the ground state of the target is assumed. This means that in-medium nucleon-nucleon collisions are neglected in this first approach. Such a sudden approximation is partially relaxed in a more realistic situation.

between the cluster and the heavy emitter [3]. Figure 3 shows an example of such an interaction in the case of an α particle and a Hg nucleus. In a classical picture, the cluster cannot escape from the heavy nucleus if its energy is below the emission barrier. Let $V_{A+A_c}(r_c)$ be the interaction potential and V_B the associated barrier. We have the "local" condition[4]

$$\varepsilon_k(r_c) \geq V_B - V_{A+A_c}(r_c) \qquad (1)$$

leading to a lower limit on the cluster kinetic energy.

The second constraint is directly dependent on the reaction type and is due to the energy balance. Indeed, the accessible configuration is further reduced due to the total energy available in the reaction. In the simplified scenario presented here (accounting for the fact that the initial nucleon is absorbed), we have the following inequality

$$E_B - Q - V_{A+A_c}(r_c) \geq \varepsilon_k(r_c) \qquad (2)$$

which gives an upper limit. Here E_B denotes the incident energy while Q is the Q-value of the reaction. It is worth to notice that the second condition depends not only on the beam energy but also on the configuration itself. Therefore, only a fraction of the total phase-space accessible for the cluster will indeed lead to emission in the continuum. This fraction corresponds to the *"explored phase-space"* which takes into account the energy constraints induced by the reaction.

These two constraints are shown in Fig. 3 (top left) for a proton-induced reaction at $E_B = 39$ MeV. There, an α particle can only be emitted in a small interval of kinetic energy (called "escape window" in the following) leading to a significant restriction in phase-space. The fraction of the phase-space available for the cluster emission is shown in Figure 3 (top-right). According to the energy constraint, all configurations between the two lines lead to the emission of an α particle.

Direct application of the rules

The prescription described above are compatible with experimental data as shown in the case of a proton-induced reaction at $E_B = 39$ MeV. Assuming that the proton is absorbed by the target, a Monte-Carlo sampling (using the cluster creation rules) of the α particle is obtained in order to obtain the initial configurations in the medium. Then, using the emission rules, only those configurations allowed by the energy constraint are conserved. Last, each conserved configuration is propagated in the target potential. Thus, the α kinetic energy distribution is obtained and successfully compared with the experimental data (from [18]). This is shown in Figure 3 (bottom part) where the calculated spectrum (open square) is compared to experimental data (filled circles). A similar agreement is found for the emission of protons, deuterons and tritons (see Fig. 4). However in this case, direct reactions are also present in the experimental data leading to an additional contribution at high energy.

TOWARDS NUCLEAR REACTIONS

Direct application of the rules to the simplified three steps scenario described above allows only qualitative comparisons with data. In order to provide quantitative comparisons, additional effects must be considered. Two phenomenological models (called n-IPSE[5] [2] and HIPSE[6] [1]) based on the very same assumptions have been developed and confronted with the experimental data. We only give here the physical effects that have been added on top of the rules:

- **In medium nucleon-nucleon collisions:** At energies below the Fermi energy, the effect of in medium two-body collisions is small. However, as the beam energy increases, such collisions must be taken into account.

---

[3] Since, we are considering here rather low beam energy leading to small available energy, we do not expect that two clusters are emitted at the same time, thus the outgoing channels are essentially binary. In addition, the use of a heavy target is very helpful since in that case, due to the small available energy in entrance channel, no particle can be emitted in the secondary decay stage. Therefore, in experimental data, detected clusters are issued from the pre-equilibrium stage only.

[4] Due to the large mass assymetry, it is assumed for simplicity that the heavy target is at rest in the laboratory frame.

[5] n-IPSE: nucleon-Ion Phase-Space Exploration

[6] HIPSE: Heavy-Ion Phase-Space Exploration

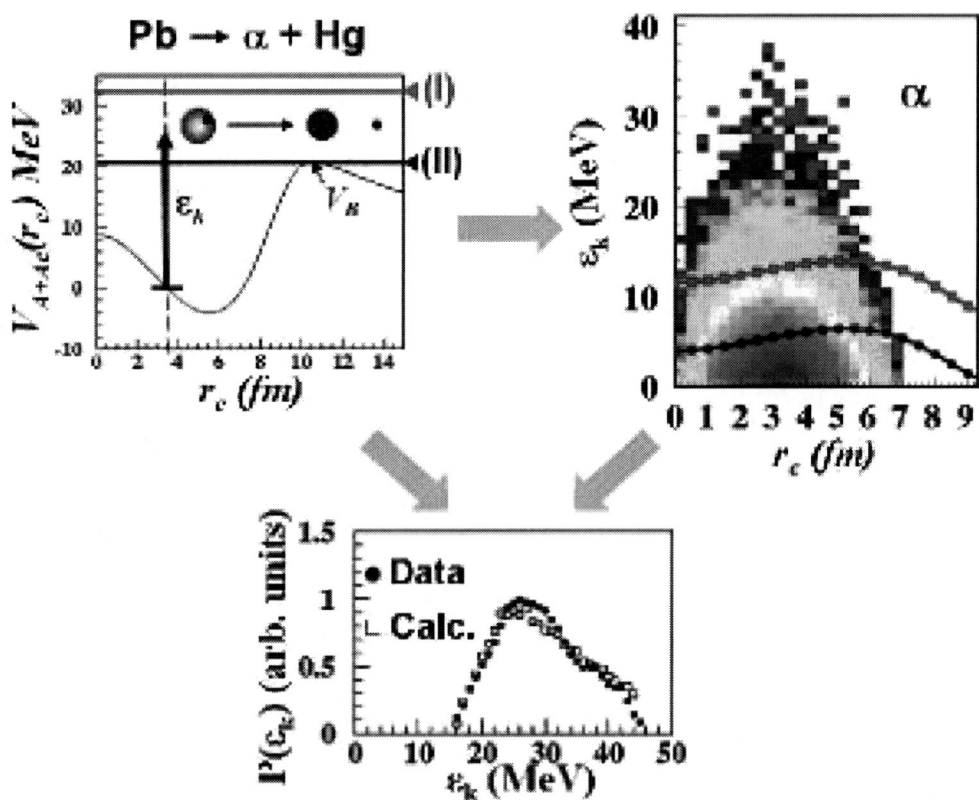

FIGURE 3. Top-left: Two-body Potential between the α and the emitter. Above the line (I), the cluster cannot be emitted. This upper limit is directly given by the energy balance of the reaction. Below the line (II), the cluster cannot overcome the barrier (since here, quantum tunnelling is not considered). In between the two lines, there is a small "escape window" for the emission of the cluster. Top-right: Total available phase-space of the cluster. This latter is significantly reduced due to the energy constraint. The two curves correspond respectively to the lower and upper limit in the kinetic energy. Bottom: Calculated kinetic energy distribution (open squares) of the α particle obtained by propagating each configuration in the "escape window" up to infinity. The calculated spectrum is compared with the experimental data (black circles).

Accordingly, the initial Thomas-Fermi distributions are distorted by two-body effects.

- **Influence of the impact parameter:** In nucleus-nucleus collisions, geometrical aspects associated with the impact parameter are accounted for by using a participant-spectator picture. In nucleon-induced reactions, this picture is replaced by the "influence area" (equivalent to the participant region) notion which defines the number of nucleons of the target affected by the projectile.
- **One-body dissipation and nucleon absorption:** Depending on the incident energy, particles are exchanged by the two partners of the reaction: This is treated by means of a phenomenological parameter (see [1]). In nucleon induced collisions, this process is replaced by a probability that the incoming nucleon be absorbed by the target.
- **Application of** *cluster formation rules* **and "nucleosynthesis" in the medium:** We have described previously a rule to obtain cluster properties when a single cluster is formed in the medium. However, the number of clusters is not a priori fixed. In order to solve this difficulty in both models, a coalescence algorithm is used to form clusters starting from the nucleons in the participant region. In the nucleon-induced reaction case, it was possible to show explicitly that this coalescence is equivalent to a random sampling assumption.
- **Application of** *cluster emission rules* **and Final-State interaction (FSI):** After the coalescence stage, many configurations are accessible. However due to the energy-balance generalized to the many cluster case, only part of the accessible phase-space is really explored. In addition, if the relative energy between two clusters is lower than the barrier associated with their mutual interaction they will not separate during the expansion. In a realistic model, the recombination of fragments is allowed. In HIPSE, possible "re-fusion" of fragments is accounted for before the freeze-out configuration is reached. This process can lead to important FSI's and may relax completely

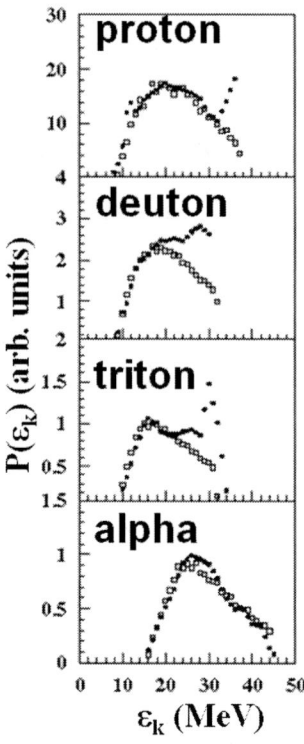

FIGURE 4. From top to bottom, calculated kinetic energy distributions (open squares) obtained for proton, deuteron, triton and alpha particles. Distributions are compared to experimental data (open circles).

the participant-spectator picture. For instance, the quasi-target and the quasi-projectile can fuse.
- **Freeze-out and the after-burner stage:** When the available energy is large, fragments are excited and once the chemical and thermal freeze-out are reached, the possible in-flight de-excitation of each cluster must be taken into account. This induces a complex mixing of pre- and post-equilibrium emission.

More details on technical aspects can be found in ref. [1, 2]. The important point we would like to stress is that rather different experimental data can be described using the same hypothesis on the production and the emission of clusters.

CONTACT WITH EXPERIMENTAL DATA: TWO ILLUSTRATIVE EXAMPLES

A detailed comparison of the models with the experimental data can be found in [1, 2]. Here, we first concentrate on nucleon-induced reactions. Figure 5 shows a comparison between the kinetic energy differential cross-section of light clusters calculated with n-IPSE and data [19]. Note that there is no normalization between the data and the calculation. As a reference, we also show (right part of the figure) the calculated spectra obtained with GNASH[20]. A good agreement between the results of n-IPSE and the experimental data is obtained. This is true for a wide range of beam energy from 37 MeV to 135 MeV in both proton and neutron induced reactions on medium and heavy nuclei.

Concerning nucleus nucleus reactions, a systematic comparison with the INDRA data[17] have demonstrated that the HIPSE model is able to reproduce not only the average properties [2] but also the fluctuations of the experimental observables[21]. It appears that besides mean properties and fluctuations, "internal" correlations inside each event are also correctly reproduced as shown in Figure 6 [22] where the distribution of the relative velocity (top) and the relative angle (bottom) between the three largest fragments taken two-by-two are presented for the reaction Xe+Sn at three different beam energies ($E_B = 25, 50, 80$ MeV/A). In each case, the calculated spectra are compared with the INDRA data [17]. The very good agreement between HIPSE and the INDRA data gives additional proof of the compatibility between the rules for the production and the emission of complex particles and the experimental observables. Note

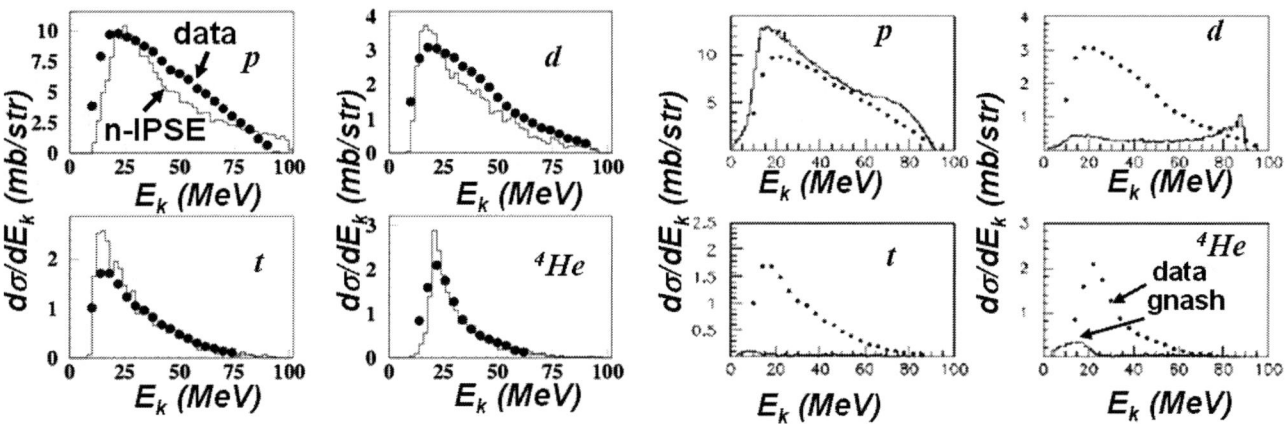

FIGURE 5. Left: Kinetic energy differential cross-section of proton, deuteron, triton and alpha particles, obtained with the n-IPSE model calculation (solid line) for neutron induced reaction on ^{208}Pb at beam $E_B = 96$ MeV (from [19]). Right: distributions obtained using the GNASH model[20].

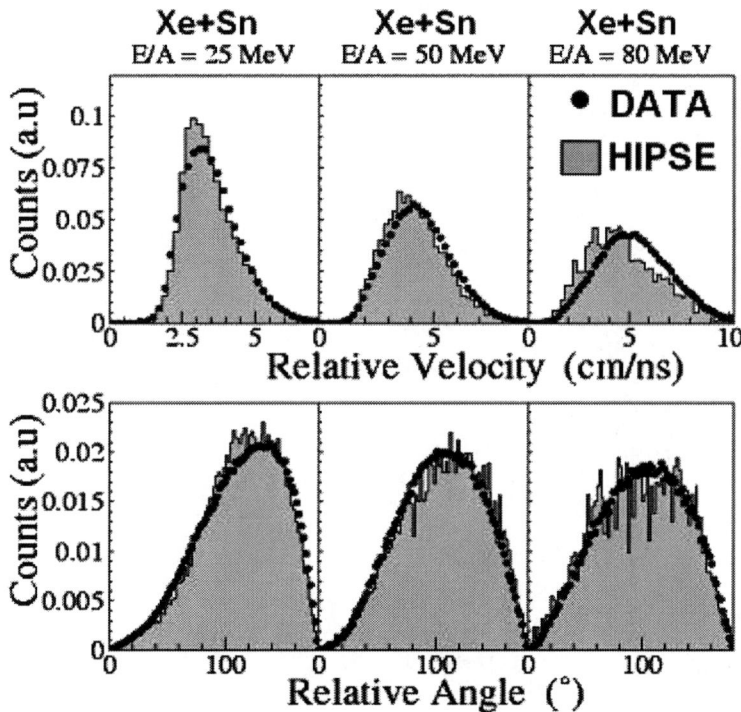

FIGURE 6. Distributions of the relative velocity (top) and the relative angle (bottom) between the three largest fragments taken two-by-two for the $Xe + Sn$ system at three different beam energies. From left to right, the beam energies $E_B = 25$, 50 and 80 MeV/A are considered. In each case, the calculated spectra are compared with the INDRA data (filled circle). Events considered here correspond to well detected events (80 % of the total charge and impulsion). The same selection is used for the "filtered" HIPSE events.

that a similar agreement has been is found in different symmetric systems[7].

---

[7] Please note that, in both models, a few free parameters are used: they are related to the description of the participant region, the nucleon-nucleon collision rate , the exchange and absorption nucleon processes. Such parameters depend only on the beam energy. This means, that a single set of parameters is used to reproduce simultaneously nucleon reactions on Fe, Pb and U. Similarly, the parameters adjusted on Xe+Sn reactions have

CONCLUSION

We have described simple rules that may be used to describe the pre-equilibrium emission of clusters in the course of nuclear reactions. These rules are based on a random sampling of the nucleons taking into account the Fermi motion and a proper account of nuclear effects as well as the conservation laws. Using these rules leads to a good agreement with data obtained from nucleus-nucleus reactions around the Fermi energy and surprisingly also for nucleon-nucleus reactions.

We would like to mention, that even if the complete randomness hypothesis appears compatible with the experimental data, this do not give any indication on the physical origin of randomness and several effects can be invoked: phase-transition, turbulence, self-organized criticality, quantum decoherence ...

ACKNOWLEDGMENTS

We thank warmly the INDRA collaboration for permission to use its data. We would like to thank V. Blideanu, O. Lopez, A. Van Lauwe and E. Vient for their collaboration in this work.

REFERENCES

1. D. Lacroix, A. Van Lauwe and D. Durand, *Phys. Rev. C* **69**, 054604 (2004).
2. D. Lacroix, V. Blideanu, and D. Durand, *Phys. Rev. C* **71**, 024601 (2005).
3. D. Durand, E. Suraud, B. Tamain, "Nuclear Dynamics in the Nucleonic Regime", IOP Publishing, (2001)
4. Proceedings of the IWM 2003 conference "International Workshop on Multifragmentation and related Topics", Caen, France, 2003.
5. Proceedings of the 18th Nuclear Physics Division Conference of the EPS, "Phase transitions in strongly interacting matter" edited by D. Adamova, A. Kugler and P. Tlusty, Prague, (2004) Czech Republic, Nucl. Phys. A749, Pages 3-371 (2004).
6. S. Ayik and C. Gregoire, *Phys. Lett.* **B212**, 269(1988) ; ibid, *Nucl Phys.* **A513** , 187 (1990).
7. J. Randrup and B. Remaud, *Nucl. Phys.* **A514**, 339 (1990).
8. A. Guarnera, M. Colonna, Ph. Chomaz, *Phys. Lett.* **B373**, 267 (1996).
9. J. P. Bondorf, A. S. Botvina, A. S. Ilinov, I. N. Mishustin and K. Sneppen, *Phys. Rept.* **257**, 133 (1995).
10. D. H. E. Gross, *Rept. Prog. Phys.* **53**, 605 (1990).
11. P. Danielewicz and G. F. Bertsch, *Nucl. Phys.* A **533**, 712 (1991).
12. H. Stocker and W. Greiner, *Phys. Rept.* **137**, 277 (1986).
13. J. Aichelin, *Phys. Rept.* **202**, 233 (1991).
14. H. Feldmeier, *Nucl. Phys.* A **515**, 147 (1990).
15. A. Ono, H. Horiuchi, H. Takemoto, R.Wada, *Nucl.Phys.* **A630**, 148c (1998).
16. A.S. Goldhaber, *Phys. Lett.* B53, 306 (1974).
17. S. Hudan et al, *Phys. Rev.* **C67**, 064613 (2003).
    see also the INDRA Web site: http://infodan.in2p3.fr
18. F. E. Bertrand and R. W. Peelle, *Phys. Rev. C* 8, 1045 (1973).
19. V. Blideanu et al., *Phys. Rev. C* **70**, 014607 (2004).
20. P.G. Young, E.D. Arthur, and M. B. Chadwick, "Comprehensive Nuclear Model Calculations: Introduction to the theory and Use of GNASH Code", Report No LA-12343-MS, 1992.
21. A. Van Lauwe, D. Lacroix, and D. Durand, Proceedings of the IWM 2003 conference, Caen, France, 2003.
22. A. Van Lauwe, Thèse de l'Université de Caen, France (2003).

---

been used to the Ni+Ni and the Au+Au cases giving reasonable agreement with the data.

Isospin effects on fragmentation mechanisms

M.Colonna*, V.Baran*,†, M.Di Toro* and R.Lionti*

Laboratori Nazionali del Sud INFN, via S.Sofia 62, I-95123 Catania, Italy and Dipartimento di Fisica e Astronomia, Universita' di Catania
†*NIPNE-HH and Bucharest University, Romania*

Abstract. We discuss fragmentation mechanisms, in neutron-rich systems, occurring in central and mid-peripheral collisions at Fermi energies. The most relevant isospin effects are the distillation mechanism, that leads to to the formation of more symmetric fragments, surrounded by a neutron-rich gaseous phase, in central collisions, and isospin transport mechanims that are responsible for the neutron enrichment of the overlap (neck) region in mid-peripheral collisions, where fragments are formed. The origin of fluctuations in the isotopic composition of fragments and the link to the low density properties of the symmetry energy are discussed.

Keywords: instabilities, fragmentation, symmetry energy, isospin distillation, isospin transport mechanisms
PACS: 21.30.Fe, 25.70.-z, 25.70.Lm, 25.70.Pq.

INTRODUCTION

Heavy ion collisions at Fermi energies offer the possibility to learn about properties of the nuclear effective interaction in situations of density and temperature different from normal ones. In particular, considering charge asymmetric systems, one can access information on the behaviour of the symmetry energy, that is poorly known at low and high densities. In this way one can also put some constraints on the effective interactions used in the astrophysical context [1, 2], where such information is essential for the understanding of the life of supernovae and neutron stars [3, 4, 5, 6, 7, 8]. Moreover we like to remark that the same symmetry energy is of relevant importance for structure properties, being clearly linked to the thickness of the neutron skin in n-rich (stable and/or unstable) nuclei (see [9] and the discussion in refs. [10, 11]).

Fragmentation mechanisms at Fermi energies can be used to study the symmetry energy at densities below and around the normal value. In violent central collisions, where the full disassambly of the system into many fragments is observed, one can study new properties of liquid-gas phase transitions occurring in asymmetric matter. In neutron-rich matter, phase co-existence leads to a different asymmetry in the liquid and gaseous phase: fragments (liquid) appear more symmetric with respect to the initial matter, while light particles (gas) are more neutron-rich [12, 13, 14, 15]. This effect, that is simple related to the fact that the symmetry energy decreases when the density gets lower, depends on the isovector part of the nuclear EOS, namely on the derivative of the symmetry energy. Hence fragmentation studies in central collisions can allow to get information on this low-density properties of the symmetry energy.

In mid-peripheral collisions intermediate mass fragments (IMF) are formed in the overlap (neck) region between projectile and target. This region may reach density values larger than normal one, during the first stage of the reaction, and then expand, before the system re-separates. IMF's form during this expansion phase, in some analogy with what happens, for the full system, in central collisions. However, in this case, the low density region (neck) is in contact with regions of larger density (the spectators), hence one observes an "isospin migration" from projectile and target towards the participant zone, leading to a neutron enrichment of the neck. Also this effect is nicely connected to the derivative of the symmetry energy, but around density closer to normal one [16, 17].

Hence the two fragmentation mechanisms, from central to peripheral collisions, nicely allow to explore the behaviour of the symmetry energy from low to normal values of the density.

We will describe the reaction dynamics using a stochastic mean-field approach. We follow semi-classical transport theories, that describe the time evolution of the one-body distribution function according to the nuclear mean-field (plus the Coulomb repulsion for protons) and including the effects of two-body nucleon-nucleon collisions, with some degree of stochasticity. This description allows the treatment of physical situations where bifurcations and larger fluctuations occur. This is certainly the case of phase separation and multi-fragment formation. In our scenario, fragmentation is due to the development of volume (spinodal) instabilities during the expansion phase that follows the initial collisional shock. The thermal agitation of the system generates density fluctuations, that are amplified by the unstable mean-field. Fragment properties will be related to the features of the most important unstable iso-scalar oscillations [16, 18]. In particular, we will discuss how these properties evolve with the charge asymmetry of the considered matter and how they depend on the symmetry energy of the nuclear EOS.

DESCRIPTION OF FRAGMENTATION MECHANISMS

Stochastic mean-field approaches

Theoretically the evolution of complex systems under the influence of fluctuations can be described by a transport equation with a stochastic fluctuating term, the so-called Boltzmann-Langevin equation (BLE):

$$\frac{df}{dt} = \frac{\partial f}{\partial t} + \{f, H\} = I_{coll}[f] + \delta I[f], \quad (1)$$

where $f(\mathbf{r}, \mathbf{p}, t)$ is the one-body distribution function, $H(\mathbf{r}, \mathbf{p}, t)$ is the one-body Halmitonian and $\delta I[f]$ represents the stochastic part of the two-body collision integral [19, 20]. Exact numerical solutions of the BLE are very difficult to be obtained and they have only been calculated for schematic models in two dimensions [21]. Therefore various approximate treatments of the BLE have been introduced [22, 23, 24].

Here we will follow the method presented in Ref.[24].

Within the assumption of local thermal equilibrium, the stochastic term of Eq.(1) essentially builds the kinetic equilibrium fluctuations typical of a Fermi gas: $\sigma_f^2 = f(1-f)$, where $f(\mathbf{r}, \mathbf{p}, t)$ can be approximated by a Fermi-Dirac distribution, $f(\mathbf{r}, \mathbf{p}, t) = 1/(1 + e^{(\varepsilon - \mu)/T})$ with local chemical potential and temperature, $\mu(\mathbf{r}, t), T(\mathbf{r}, t)$. We project on density fluctuations obtaining:

$$\sigma_{\rho,eq}^2(\mathbf{r}, t) = \frac{1}{V} \int \frac{d\mathbf{p}}{h^3/4} \sigma_f^2(\mathbf{r}, \mathbf{p}, t) = \frac{T}{V} \frac{3\rho}{2\varepsilon_F} (1 - \frac{\pi^2}{12}(\frac{T}{\varepsilon_F})^2 + ...), \quad (2)$$

where we have used the Sommerfeld expansion around $\varepsilon = \mu$ for small T/ε_F (ε_F denotes the Fermi energy).

Actually this kinetic equilibrium value is reached asymptotically for an ideal gas of fermions. After a small time step Δt, the value reached for the density fluctuations can be approximated by:

$$\sigma_\rho^2(\mathbf{r}, t, \Delta t) = \sigma_{\rho,eq}^2(\mathbf{r}, t) \frac{2\Delta t}{\tau_{coll}(\mathbf{r}, t)}, \quad (3)$$

where τ_{coll} is the damping time associated with the two-body collision process. At temperatures around 4 MeV, that are typically reached in fragmentation processes, τ_{coll} is of the order of 50 fm/c [25]. In our calculations fluctuations are injected each $\Delta t = 5\ fm/c$ and their amplitude is scaled accordingly (see Eq.(3)). In the cell of $\mathbf{r}$ space being considered, the density fluctuation $\delta \rho$ is selected randomly according to the gaussian distribution $exp(-\delta \rho^2/2\sigma_\rho^2)$. This determines the variation of the number of particles contained in the cell. A few left-over particles are randomly distributed again to ensure the conservation of mass. Momenta of all particles are finally slightly shifted to ensure momentum and energy conservation.

The fluctuations introduced are then amplified by the unstable mean-field. It is important to notice that the characteristic growth time for the unstable modes is smaller ($\tau \approx 30\ fm/c$) [26] than the collisional damping time, so two-body collisions act just as a seed to perturbe the density profile of the system, but the dynamics is essentialy driven by the propagation of the mean-field instabilities.

Reaction mechanisms

Calculations have been performed using the TWINGO code, where the test particle method is adopted to solve Eq.(1) [27]. A typical example of the dynamical evolution of a central collision is represented in Fig.1 (left), where we consider the reaction $^{124}Sn + ^{124}Sn$ at 50 Mev/A. After the initial collisional shock, the system expands and reaches a low-density unstable phase, where fragments start to form. Hence, according to stochastic-mean field calculations fragmentation happens through the development of volume (spinodal) instabilities. The same scenario applies, but in

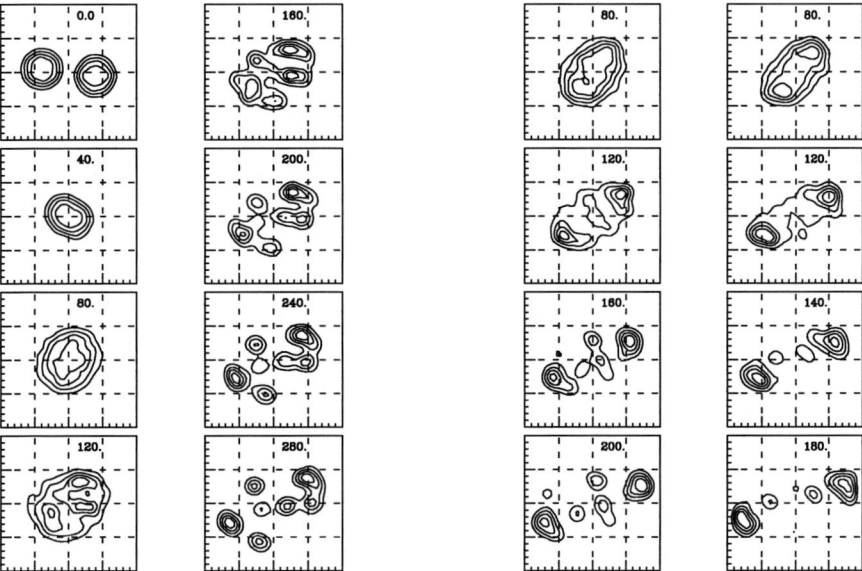

FIGURE 1. *Left: $^{124}Sn +^{124}Sn$ collision at 50AMeV : time evolution of the nucleon density projected on the reaction plane. Semi-central $b = 2fm$ collision: approaching, compression, separation, and fragmentation phases. The times in fm/c are written in each panel. The iso-density lines are plotted every $0.02 fm^{-3}$ starting from $0.02 fm^{-3}$. Right: First column: $b = 4fm$, second column: $b = 6fm$, separation phase up to the freeze-out.*

a reduced region, the overlap (neck) zone, in mid-peripheral collisions, as shown in Fig.1 (right) for the same system at b= 4,6 fm. The fact that also in neck fragmentation spinodal instabilities are encountered is confirmed by the role of the nuclear matter compressibility K in the observed fragmentation path. Changing K from 200 MeV to 380 MeV, in fact, nuclear matter becomes stiffer, density oscillations are reduced and the the neck fragmentation mechanism is quenched [17].

KINEMATICAL PROPERTIES OF NECK FRAGMENTS

Before discussing isospin effects in fragmentation mechanisms it is worth to mention some important properties of these fragments originated through instabilities. Properties of neck fragments have been recently investigated in the context of the reactions $^{124}Sn + ^{64}Ni$ and $^{112}Sn + ^{58}Ni$ at 35 MeV/A, for which nice experimental data exist [28]. It can be shown unambiguously that neck fragments are not related to the statistical emission of Projectile-like (PLF) or Target-like (TLF) residues, but they appear produced by a low density source that develops in between PLF and TLF. This can be seen by performing an analysis in terms of the deviation of the fragment velocity with respect to the Viola systematics. A scatter plot in the $r - r_1$ plane is shown in Fig.2. r (r_1) represents the ratio between the IMF-PLF(TLF) relative velocity and the velocity given by the Viola systematics. One can see that the points are mostly located far from the axes $r = 1, r_1 = 1$.

In Fig.3 we represent the velocity of fragments (including PLF and TLF) in the plane determined by the beam parallel and perpendicular velocities, in the center of mass frame. It is possible to easily recognize the points associated with PLF and TLF. Neck framents (points with velocity closer to the center of mass velocity) have mostly a negative velocity along the direction perpendicular to the beam. In fact, due to the mass asymmetry of the reaction considered, the projectile contributes by a larger proportion to the matter that forms in the overlap neck zone. This appears as a nice memory effect of the entrance channel. Hence this neck source will present a given degree of alignment with respect to the PLF-TLF direction.

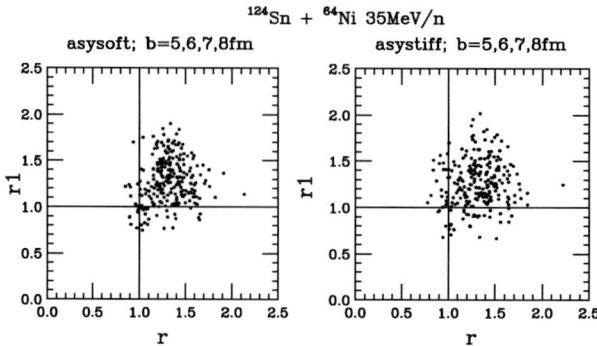

FIGURE 2. *Wilczynski-2 Plot: correlation between deviations from Viola systematics (see the text for definitions and ref.[28]). Results are shown for two asy – EOS (see text).*

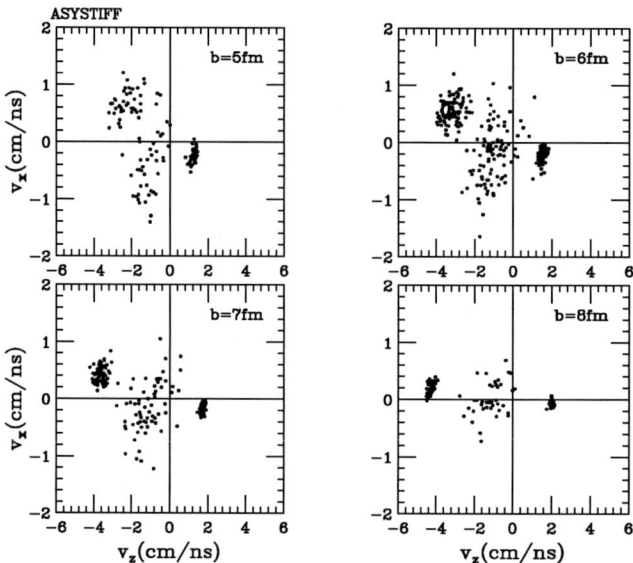

FIGURE 3. *Center of mass z,x-velocity component distributions.*

ISOSPIN PROPERTIES OF FRAGMENTS

Calculations have been performed adopting two different prescriptions for the behaviour of the symmetry energy, namely an "asy-soft" dependence, where the symmetry energy tends to saturate and eventually decreases at high density, and an "asy-stiff" dependence, where the symmetry energy increases linearly with density [16].

Central collisions

The properties of fragments obtained in a central collision, ^{124}Sn + ^{124}Sn at 50 MeV/A, are summarized in Fig.4. Let us look in particular at the dynamical evolution of the isospin properties of the considered system (panels (b)). By looking at the asymmetry of the liquid (fragment) phase, one cas see a first decrease, to to pre-equilibrium particle emission, that is neutron-rich in neutron-rich systems. Then, when the system expands and enters the low-density region, fragments are formed and the asymmetry continues to decrease. This second phase can be associated with the isospin-distillation mechanism: while fragments are formed, the neutron excess is transferred to the low density gaseous phase, while the liquid (fragment) phase results more symmetric. This effect depends on the EOS used, and in particular on the derivative of the symmetry energy, resulting more important in the asy-soft case, where the derivative is larger at low density. In fact the fragment asymmetry results smaller in the asy-soft case (panels (e)). We notice that properties of fragments are given at the time when they do not interact anymore, apart from the Coulomb repulsion (freeze-out time).

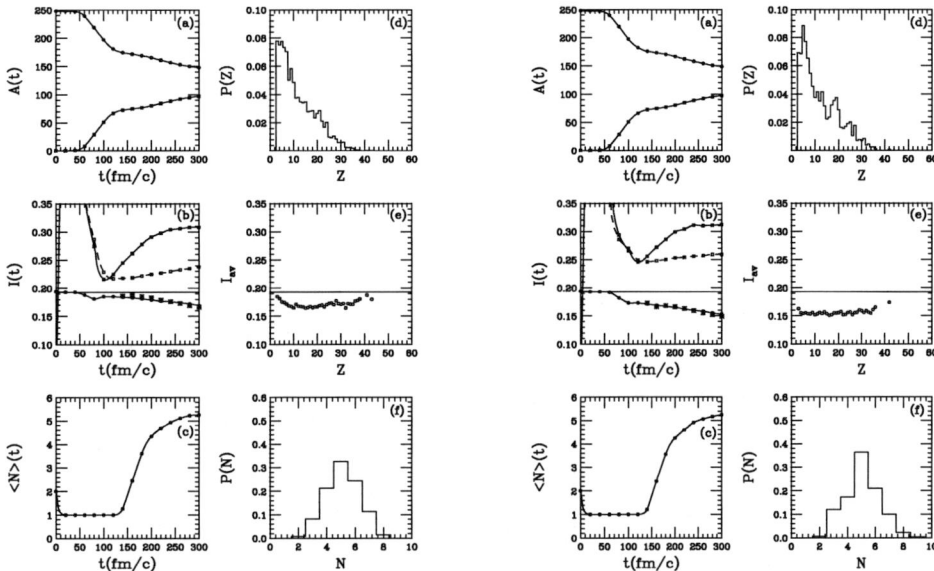

FIGURE 4. Left: $^{124}Sn + ^{124}Sn$ $b = 2 fm$ collision: Time evolution of total mass of the liquid (top curve) and gas (bottom curve) phases (panel (a)), asymmetry of liquid (bottom curve) and gas (top curves) phases (panel (b)), average IMF multiplicity (panel(c)). IMF charge distribution (panel (d)), average asymmetry (panel (e)) and multiplicity distribution (panel (f)), calculated at the freeze-out time. Results correspond to the asy-stiff EOS. Rigth: the same as in the left panels, but for the asy-soft EOS.

Mid-peripheral collisions

In neck fragmentation the isospin dynamics appears quite different, with respect to central collisions. Indeed now the formation of a low-density region in betwen projectile and target induces isospin migration towards the neck. So finally neck IMF's appear more neutron rich with respect to fragments emitted by the projectile or the target. So, after the pre-equilibrium emission, the asymmetry of the central liquid phase rises again, due to the formation of neutron-rich IMF's (see Fig.5). Of course also this effect depends on the adopted EOS and is more pronounced in the case of the asy-stiff interaction, since now we are testing the derivative of the symmetry energy close to normal density. So the asymmetry of neck fragments increases with the stiffness of the symmetry energy.

Effects in symmetric and charge-symmetric reactions

We will discuss the reactions $^{58}Fe + ^{58}Fe$ (charge asymmetric $N/Z = 1.23$) and $^{58}Ni + ^{58}Ni$ (charge symmetric $N/Z = 1.07$) at 47 MeV/A [29]. The interest of these reactions is that, due to the uniform N/Z distribution, we do not have isospin gradients initially. So the study of the full dynamics and of the possible occurrence of density gradients is essential also to understand the isospin dynamics. Fig. 6 reports the N/Z ratio of each fragment vs. the charge Z at the freeze-out time, obtained at the impact parameter $b = 4 fm$, using the asy-stiff EOS.

Residual nuclei, i.e PLF and TLF (large Z range), show a different behaviour in the two reactions: we note, in fact, that the points of Fe system are along the dashed line, that represents the initial system asymmetry, while for Ni reaction points lie above that line. The *IMF* behaviour is similar for the two reactions: the points, for both reactions, lie above the dashed line, although for the reaction $^{58}Ni + ^{58}Ni$ the difference between the N/Z of IMF's and large nuclei seems to be less pronounced.

We can conclude that: i) In the neutron rich reaction, neutron evaporation due to pre-equilibrium goes in the same direction of the neck neutron enrichment, caused by the isospin-migration. So finally we observe slightly more

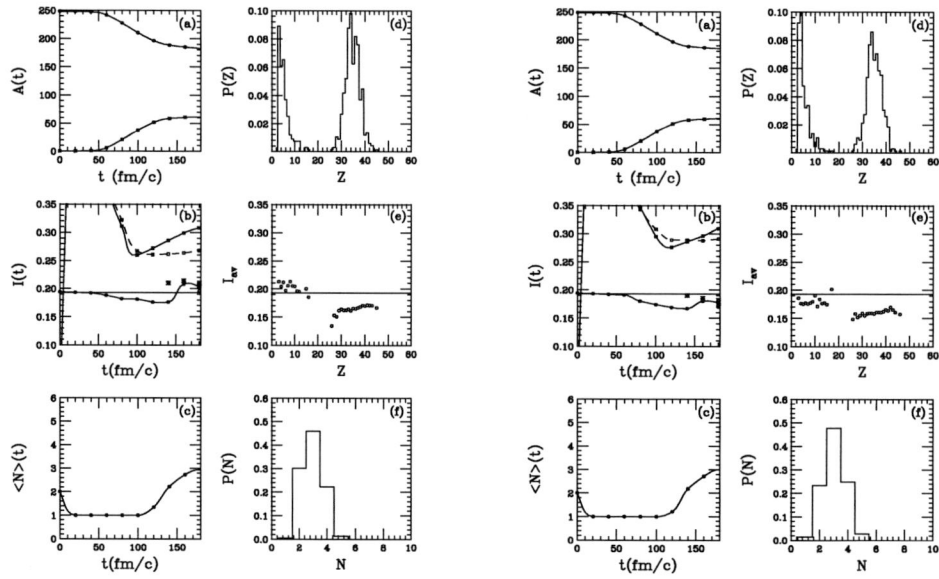

FIGURE 5. *The same as in Fig.4, but for b= 6 fm*

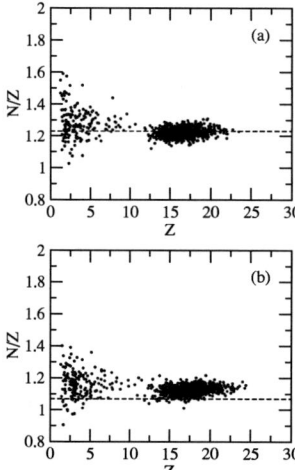

FIGURE 6. *Asymmetry vs. charge of each nuclei arising from the simulation of the reaction $^{58}Fe + {}^{58}Fe$ (a) and $^{58}Ni + {}^{58}Ni$ (b) with an asy−stiff EOS. Horizontal dashed lines are the initial asymmetries of the colliding systems.*

symmetric residues accompanied by neutron-richer IMF's; ii) In the neutron-poor collision, pre-equilibrium proton emission enhances the N/Z of large fragments. The acquired asymmetry is then transferred to the neck region.

To better disantangle between the two mechanisms, pre-equilibrium and isospin migration effects, one can study the charge composition of residues distinguishing between binary and ternary events.

For both reactions, we note that the N/Z ratio of residual nuclei in ternary events is lower than the value for binary events, since in the second case the isospin-migration effect does not apply. The isospin dynamics effect is rather evident from the comparison with the asymmetry values at the time corresponding to the end of the pre-equilibrium phase ($t \approx 100 fm/c$). For the Fe + Fe system the N/Z of residues changes from 1.22 to 1.19, in ternary events; for the Ni + Ni reaction this difference is not so evident (from 1.12 to 1.125) because the isospin-migration competes with proton evaporation. On the other hand, in binary events, we observe a neutron enrichment of residues (N/Z = 1.17) in the *Ni* reaction, due to a favorite proton pre-equilibrium emission.

It is interesting to look also for some correlations between asymmetry, mass, velocity and direction of outgoing

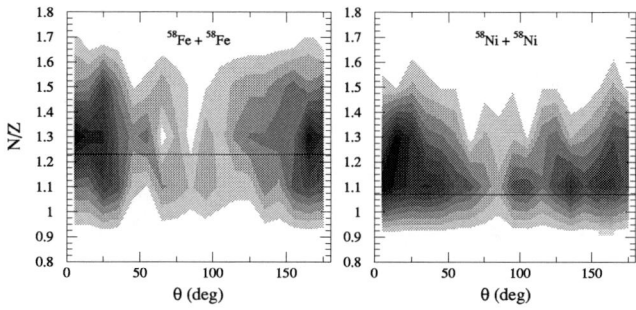

FIGURE 7. *Contour plots of the N/Z distribution versus the emission angle*

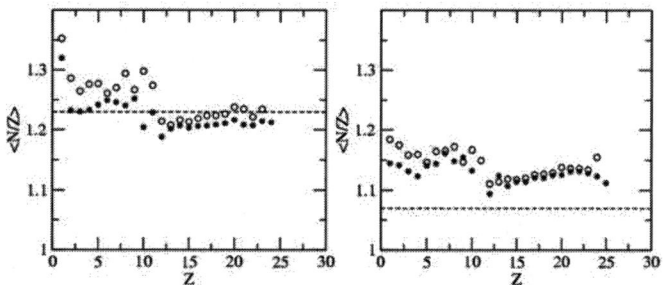

FIGURE 8. *Average asymmetry vs. Z for nuclei from the reactions $^{58}Fe + ^{58}Fe$ (left panel) and $^{58}Ni + ^{58}Ni$ (right panel) for an asy − stiff (circles) and asy − soft (stars) EOS. Horizontal dashed lines represent initial asymmetry of colliding systems [16].*

fragments. Small masses are emitted at all angles, while large fragments appear closer to PLF and TLF emission angles. The analysis of the N/Z distribution versus the emission angle reveals that larger fluctuations are present close to forward and backward angles, suggesting that IMF's that are more correlated to the spectator matter may become more neutron-rich, since they interact for a longer time with the system and the distillation mechanism becomes more effective (see Fig.7).

We stress again that the *neck − IMF*'s always present a neutron enrichment, *even in the case of a n-poor system*. The latter paradox is due to the pre-equilibrium isospin dynamics. In fact, as explained above, due to pre-equilibrium, the system will loose some protons and acquire a N/Z larger than the initial one. Then, before the di-nuclear system reseparates, the neutron excess is transferred to the neck region that is at low density. Some evidence has been found in recent data on ^{58}Ni induced fragmentation [30, 31, 32].

In Fig. 8, the average asymmetry of products arising from the two reactions ($b = 4fm$) is shown for the two choises, *asy − stiff* and *asy − soft EOS*.

We note that both fragments and residual nuclei are more symmetric in the asy-soft case. The fact that the N/Z of large fragments is smaller in the asy-soft case is due to pre-equilibrium effects, i.e. with the soft EOS more neutrons are emitted. On the other hand, the more pronounced difference between IMF and large fragment asymmetries with the stiff parameterization is a consequence of a more effective isospin migration in this case.

FLUCTUATIONS OF FRAGMENT ISOTOPIC CONTENT

Until now we have described mechanisms that are mainly responsible of the average fragment asymmetry. What is the origin of the fluctuations observed in the isotopic composition of fragments ? Fragment formation is driven by the development of instabilities, and so by unstable isoscalar-like mode, however isovector oscillations also contribute to the width of the isotopic distributions.

In asymmetric matter isoscalar and isovector modes are coupled, and some times isovector-like oscillations are even suppressed. The isospin distillation mechanism, that is a feature of isoscalar-like oscillations, creates a given

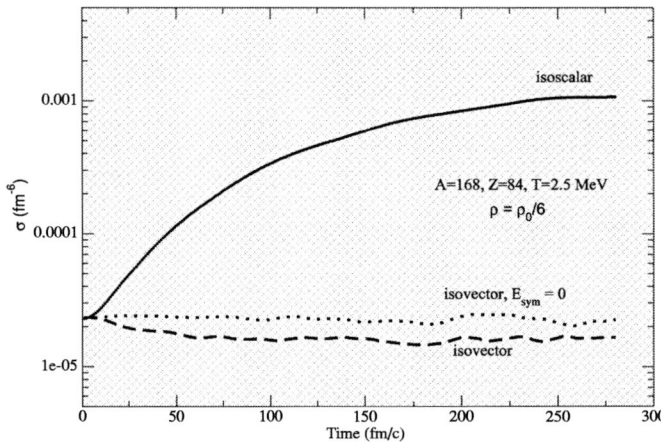

FIGURE 9. *Time evolution of isoscalar (full line) and isovector (dashed line) fluctuations, in a given cell of volume V, for unstable matter inside a box with periodic boundary conditions. The dotted line refers to calculations whith zero symmetry energy.*

dispersion in the N/Z of fragments. In fact the amplitide of the effect depends on the considered unstable modes and, since fragments come up from the beating of several modes, this leads to a given width in the isotopic distributions. The more effective the distillation is (as in the asy-soft case), the smaller the width, since all unstable modes will behave in a similar way. However, also the contribution of the isovector-like vibrations has to be considered [33].

To illustrate their role in spinodal decomposition we have performed very schematic calculations for unstable low-density matter inside a box with periodic boundary conditions, with $\rho_{in} = \rho_0/6$ and temperature $T = 2.5 MeV$. For the sake of simplicity, we consider here symmetric matter, for which isoscalar and isovector vibrations are de-coupled. In this case N/Z fluctuations are only due to iso-vector vibrations. We study the temporal evolution of fluctuations, inside a given cell, of volume V, of the densities $\rho = \rho_n + \rho_p$ and $\rho_i = \rho_n - \rho_p$, where ρ_n and ρ_p are neutron and proton densities. Some fluctuations are present at the initial time, due to the fact that particles are randomly distrinuted inside the box. As it can be seen from Fig.9, isoscalar fluctuations grow since fragments are formed, while iso-vector fluctuations rapidly converge to an "equilibrium" value, that depends on the initial density, temperature and on the value of the symmetry energy. If the symmetry potential is removed, iso-vector fluctuations keep their initial value. While the density grows inside a fragment, isovector fluctuations do not grow anymore. This is because the fragment is constructed according to isoscalar vibrations, where neutrons and protons move in phase, with the same amplitude, while isovector fluctuations maintain their equilibrium value, that is associated with the initial density ρ_{in}.

The variance associated with equilibrium isovector fluctuations in a given volume V can be written as:
$\sigma_{\rho_i} = <\delta\rho_i\delta\rho_i> = \frac{\mathscr{F}}{2}\frac{\rho_{in}}{V}$,
where the function $\mathscr{F}$ depends on the temperature and on the symmetry energy at the considered density.

Now the probability to observe a given fluctuation $\delta\rho_i$ can be expressed as:

$$P \approx exp(-\delta\rho_i^2/2\sigma_{\rho_i}) \qquad (4)$$

and then, for a fragment of given mass A, the distribution $P(N-Z)$ can be written as:

$$P(N-Z) \approx exp(-[N-Z-(\bar{N}-\bar{Z})]^2/(\mathscr{F}\rho_{in}V)), \qquad (5)$$

where $\bar{N}$ and $\bar{Z}$ are the average neutron and proton numbers. Indicating by $\rho_{fin} = A/V$ the actual density of the fragment and introducing the parameter $\alpha = \rho_{in}/\rho_{fin}$, the formula can be recast as follows

$$P(N-Z) \approx exp(-A[I-\bar{I}]^2/(\mathscr{F}\alpha)), \qquad (6)$$

where $I = (N-Z)/A$. Hence one can see that, in the spinodal decomposition process, the variance of the fragment isotopic distribution is reduced by a factor α with respect to the equilibrium value $\mathscr{F}$.

It is easy to realize that, in log scale, the ratio of the yields of a given isotope with charge Z, obtained in two different systems having asymmetries $\bar{I}_1$ and $\bar{I}_2$ depends linearly on the neutron number N. Hence this property, that is named isoscaling [34], and has been observed in several experimental data, is present also in spinodal decomposition processes.

CONCLUSIONS

We have discussed some properties of fragmentation mechanisms in charge-asymmetric systems. The isospin effects can be related to the behaviour of the symmetry energy below and around normal density, thus allowing to extract important information on fundamental quantities of the nuclear interaction. It appears that the isospin dynamics is strictly linked to the presence of density gradients. Indeed, even in symmetric reactions, one can observe isospin transport phenomena, due to pre-equilibrium emission and to the fact that the neutron excess is transferred to the low density regions, since this situations is energetically favourable for the system. This effect appears connected to the derivative of the symmetry energy at density lower than normal one, in the case of violent multifragmentation phenomena, up to around normal density, for more peripheral reactions. The results observed are compatible with the occurrence of spinodal instabilities and phase co-existence in charge-asymmetric nuclei. Comparison of the fragment isotopic distributions with experimental data can allow to extract information on the value of the symmetry energy at the moment when fragmentation occurs.

REFERENCES

1. I. Bombaci, T.T.S. Kuo and U. Lombardo, Phys. Reports. 242 (1994) 165 .
   I. Bombaci, Phys. Rev. C55 (1997) 1
2. M.Prakash et al., Phys.Rep.280 (1997) 1
3. I.M.Irvine, "Neutron Stars" Oxford Univ. Press 1978
4. J.M.Lattimer et al., Phys.Rev.Lett. 66 (1991) 2701
5. D.Pines, R.Tamagaki and S.Tsuruta Ed.s, "Neutron Stars", Addison-Wesley N.Y.1992
6. K.Sumiyashi and H.Toki, Astro.Phys.Jour. 422 (1994) 700
7. C.J.Pethick and D.G.Ravenhall in "The Lives of Neutron Stars" Ed.s M.A.Alpar et al., NATO ASI Series C Vol.450 (1995) pp.59-70
8. C.H.Lee, Phys.Rep. 275 (1996) 255
9. "It is important to put the properties of *neutron* matter at *subnuclear* densities on as firm a footing as possible, not only for astrophysical applications, but also for interpreting terrestrial experiments with coming radioactive beam facilities" C.J. Pethick and D.G. Ravenhall, in ref.[7].
10. B.A.Brown, Phys.Rev.Lett. 85 (2000) 5296
11. C.J.Horowitz and J.Piekarevicz, Phys.Rev.Lett. 86 (2001) 5647.
12. H.Mueller and B.D.Serot, Phys.Rev. C52 (1995) 2072
13. Bao-An Li and C.M.Ko, Nucl.Phys. A618 (1997) 498
14. M.Colonna, M.Di Toro and A.Larionov, Phys.Lett. B428 (1998) 1
15. V.Baran, M. Colonna, M. DiToro and V.Greco, Phys.Rev.Lett. 86 (2001) 4492
16. V. Baran *et al., Nucl. Phys.* **A 703**, 603-632 (2002).
17. V. Baran, M.Colonna, M.Di Toro, *Nucl. Phys.* **A 730**, 329-354 (2004).
18. M.Colonna, Ph.Chomaz, S.Ayik, Phys.Rev.Lett. **88** (2002) 122701; M.Colonna et al., Nucl. Phys.A742 (2004) 337-347.
19. S.Ayik, C.Gregoire, Phys. Lett. **B212** (1998) 269
20. J.Randrup and B.Remaud, Nucl. Phys. A514 (1990) 339
21. G.F.Burgio, Ph.Chomaz, J.Randrup, Phys. Rev. Lett.
22. S.Ayik, E.Suraud, M.Belkacem, D.Boilley, Nucl. Phys. A545 (1992) 35c
23. Ph.Chomaz et al., Phys. Rev. Lett. **73** (1994) 3512; A.Guarnera et al., Phys. Lett. **B403** (1997) 191
24. M.Colonna et al., Nucl. Phys. **A642** (1998) 449
25. J.Randrup and S.Ayik, Nucl. Phys. A572 (1994) 489
26. Ph. Chomaz, M. Colonna and J.Randrup, Phys. Rep. **389** (2004) 263-440.
27. A.Guarnera, M.Colonna, Ph.Chomaz, Phys. Lett. **B373** (1996) 297
28. A. Pagano et al.: Nucl. Phys. **A 681**, 331c (2001); E. De Filippo et al., Phys. Rev. C, in press.
29. D. V. Shetty *et al., Phys. Rev.* **C68**, 021602 (R)(2003), D. V. Shetty et al., arXiv:nucl-ex/0406008.
30. P. Milazzo *et al., Phys. Lett.* **B509**, 204 (2001).
31. L.Gingros *et al., Phys. Rev.* **C65**, 061604 (2002).
32. R.Moustbchir et al., *Nucl. Phys.* **A 739**, 15 (2004).
33. M.Colonna and F.Matera, arXiv:nucl-th/0503018, to appear in Phys. Rev. C
34. H.Xu et al., Phys.Rev.Lett. 85 (2000) 1908

Adiabatic approximation for nucleus-nucleus scattering

R. C. Johnson

Department of Physics, School of Electronics and Physical Sciences, University of Surrey, Guildford, Surrey, GU2 7XH, UK

Abstract. Adiabatic approximations to few-body models of nuclear scattering are described with emphasis on reactions with deuterons and halo nuclei (frozen halo approximation) as projectiles. The different ways the approximation should be implemented in a consistent theory of elastic scattering, stripping and break-up are explained and the conditions for the theory's validity are briefly discussed. A formalism which links few-body models and the underlying many-body system is outlined and the connection between the adiabatic and CDCC methods is reviewed.

Keywords: Adiabatic, nuclear, scattering, deuteron, nucleon
PACS: 21.60, 24.10.Eq, 24.10.Ht, 24.50.+g, 25.45.-z, 25.60.-t

1. ADIABATIC APROXIMATIONS

The use of adiabatic approximations in nuclear reaction theories has a long history. The key idea is to separate the relevant degrees of freedom into 'slow' and 'fast' categories. The 'slow' variables are treated as fixed during the collision and the associated scattering problem for the 'fast' variables is treated quantum mechanically or semiclassically depending on the masses and energies involved in the reaction. This is the analogue of the Bohr-Oppenheimer approximation for bound states of molecules, where the electronic motion for fixed nuclear positions is calculated quantum mechanically.

An early example is Barrett's treatment of neutron scattering by a nucleus regarded as a rigid rotor[1]. He calculated the scattering amplitude for the scattering of a neutron by a deformed potential as a function of the orientation of the nuclear body-fixed axes. Scattering amplitudes were then calculated by taking matrix elements of this amplitude between the nuclear states of interest. Here the 'slow' motion is obviously the nuclear collective rotation and the incident neutron motion is regarded as 'fast'. Other applications of the adiabatic approximation to theories of elastic and inelastic scattering in nuclear physics are reviewed in ref.[2], pages 83-84 and elsewhere in that book.

The approximation in the energy domain which complements this time picture, is that the energy associated with relevant rotational excitations is assumed to be small compared to the translational energy of the neutron. In the adiabatic limit all excited rotational states are assumed to be effectively degenerate on a scale determined by the incident energy. We will see how this comes about formally below. This way of thinking tends to give too conservative an idea of the usefulness of the adiabatic approximation because it fails to take into account the crucial role played by absorption in nuclear reactions. In addition, we now have a better understanding of the spatial regions in which adiabatic solutions of the few-body Schrodinger equation are expected to be most valid and how to exploit this knowledge in applications to particular reaction channels.

Historically an important feature of the adiabatic approximation was that it is a cheap way of doing a complicated coupled channels calculation. For example, in Barrett's calculation, channels with a fast neutron and the rotor in any one of its excited states are taken into account coherently and non-perturbatively. With modern computing power this may not be a big advantage, but when the projectile is loosely bound and the relevant excitation spectrum is in the continuum the adiabatic approximation can be a powerful tool as well as frequently providing important insight and checks of more complete calculations. CDCC calculations, which were pioneered and developed in [3],[4],[5],[6], discretise the continuum and are in principle an improvement over the adiabatic approximation, but CDCC codes are generally available only for 2-body projectiles. See [7] and [8] for recent reviews. The first CDCC calculations for a 3-body projectile have only just been published[9]. An adiabatic code which treats 3-body projectiles has been available for some time[10].

In this historical context we note that the adiabatic approximation in the sense we use here is the basis of Glauber's theory[11] of high energy composite particle scattering which has been widely used in the analysis of reaction experiments with halo and other light nuclei[7],[12]. In these calculations selected co-ordinates are treated adiabatically and the eikonal approximation is used to describe the scattering of the frozen object. In his development of a microscopic theory of the nucleon optical potential, Glauber goes further and treats *all* the internal co-ordinates of the target nucleus as frozen during the scattering. The resulting 2-body problem for a set of frozen internal nuclear co-ordinates is then solved using the eikonal approximation. It is one of the purposes of the present paper to advertise the fact that the adiabatic approximation can be useful even when the eikonal approximation breaks down.

2. DEUTERON-NUCLEUS COLLISIONS

This section discusses a 3-body model of the system $n + p + A$, where A is a heavy nucleus in its ground state. The theory we describe can, and has been, applied to other systems and we shall mention several in passing. Some cases, for which the adiabatic approximation is potentially useful, have special problems, e.g., Coulomb and antisymmetrisation effects. These are best discussed separately and will be ignored here.

In a 3-body model of deuteron-nucleus collisions, for example, channels corresponding to elastic deuteron scattering and elastic deuteron break-up in which the target is left in its ground state are all included in a unified way. Excited states of the target A do not appear explicitly. The relation between this model and the underlying many-body problem will be discussed below.

In a widely used notation we use $\vec{r}$ for the position of the neutron relative to the proton and $\vec{R}$ for the position of the centre-of-mass of n and p relative to A. The Hamiltonian of the model is

$$H = T_R + H_{np} + V(\vec{R},\vec{r}), \; V(\vec{R},\vec{r}) \equiv V_{nA}(\vec{R}+\vec{r}/2) + V_{pA}(\vec{R}-\vec{r}/2) \tag{1}$$

where $H_{np} = T_r + V_{np}$ is the Hamitonian for relative motion of the $n-p$ system. The T's are kinetic energy operators. For the purpose of this talk we assume that all Coulomb interactions are screened at large distances. We use $\phi_0(\vec{r})$ for the ground state of the deuteron with energy $-\varepsilon_0 < 0$, and $\phi_{\vec{k}}^{(+)}(\vec{r})$ for the continuum of $n-p$ scattering states which are eigenstates of H_{np} with energy $\varepsilon_k > 0$ and satisfy outgoing wave boundary conditions. It is the purpose of the adiabatic approximation to treat the coupling between the deuteron and the scattering state continuum in as accurate and transparent a way as possible. In the 3-body model this coupling comes from the tidal forces generated by the fact that, over the volume of the deuteron, $V_{nA}(\vec{R}+\vec{r}/2)$ and $V_{pA}(\vec{R}-\vec{r}/2)$ generate forces on the nucleons which differ in magnitude and direction.

2.1. Time dependent picture

Under the transformation

$$\Psi^{trans} = \exp(-\frac{iH_{np}t}{\hbar})\Psi. \tag{2}$$

the time dependent Schrodinger equation for $\Psi(\vec{R},\vec{r},t)$ becomes

$$(T_R + V(\vec{R},\vec{r}(t)))\Psi^{trans}(\vec{R},\vec{r},t) = i\hbar\frac{\partial\Psi^{trans}}{\partial t}. \tag{3}$$

In eq.(3) the $n-p$ relative co-ordinate $\vec{r}(t)$ has acquired a time dependence through the relation

$$\vec{r}(t) = \exp(\frac{iH_{np}t}{\hbar})\vec{r}\exp(-\frac{iH_{np}t}{\hbar}). \tag{4}$$

The adiabatic approximation assumes that the collision time T is so short that we can replace $\vec{r}(t)$ by $\vec{r}(0) = \vec{r}$. A sufficient condition for this step to be valid is that T satisfy

$$\left|\frac{<H_{np}>T}{\hbar}\right| \ll 1. \tag{5}$$

where $<H_{np}>$ is the maximum eigenvalue of H_{np} excited in the collision through the tidal forces. For the strong interaction this maximum is related to the shape of the nuclear surface and is insensitive to the incident deuteron energy. Hence for sufficiently high energy the collision time will always become small enough that the condition (5) is satisfied.

Implementation of the adiabatic approximation for a stationary state requires the solution of the adiabatic equation[13]

$$(T_R + V(\vec{R},\vec{r}) - E_d)\Psi^{ad}(\vec{R},\vec{r}) = 0, \qquad (6)$$

which for fixed $\vec{r}$ is a 2-body problem. Note that even for central V_{nA} and V_{pA} the potential in eq. (6) is not central when considered as a function of $\vec{R}$ for fixed $\vec{r}$ so coupled equations still have to be solved in general.

2.2. Solution of the adiabatic equation

The solution of the adiabatic equation can be reduced to a manageable set of coupled equations by either of two methods which are based on different truncation schemes. Both methods assume the nucleon potentials V_{nA} and V_{pA} are central.

(i) In this method[3],[4],[14][15] the adiabatic wavefunction is expanded in the basis $[Y_l(\hat{r}) \times Y_L(\hat{R})]_{JM}$ and uses the fact that $V_{nA} + V_{pA}$ is diagonal in J, M, although not in l and L. Truncation in these angular momenta is required.

(ii) The method used by Barrett[1] uses the fact that $V_{nA} + V_{pA}$ is diagonal in $\vec{L}.\hat{r}$ and proceeds by making a truncated multipole expansion of the potentials.

Method (ii) is well adapted to the case of scattering by a deformed nucleus when a natural truncation of the multipole expansion occurs. In the present context, convergence in l is found to be very rapid and truncation is linked with the adiabatic assumption of low excitations in the $\vec{r}$ co-ordinate.

The adiabatic wavefunction corresponding to a deuteron incident with momentum $\vec{K}_d$ has the structure

$$\Psi^{ad}_{\vec{K}_d}(\vec{R},\vec{r}) = \phi_0(\vec{r}) \chi^{ad(+)}_{\vec{K}_d}(\vec{R},\vec{r}), \qquad (7)$$

where $\chi^{ad(+)}_{\vec{K}_d}$ satisfies

$$(T_R + V(\vec{R},\vec{r}) - E_d)\chi^{ad(+)}_{\vec{K}_d}(\vec{R},\vec{r}) = 0, \qquad (8)$$

with the boundary condition ($\vec{r}$ fixed)

$$\chi^{ad(+)}_{\vec{K}_d}(\vec{R},\vec{r}) \stackrel{R \to \infty}{\to} \exp(i\vec{K}_d.\vec{R}) + f(\hat{R},\vec{r}) \frac{\exp(iK_d R)}{R}. \qquad (9)$$

The function $\chi^{ad(+)}_{\vec{K}_d}$ and the scattering amplitude $f(\hat{R},\vec{r})$ both depend on $\vec{r}$. The physical meaning of $f(\hat{R},\vec{r})$ is that it describes the elastic scattering in the direction $\hat{R}$ of an $n-p$ pair with fixed separation $\vec{r}$ by the potential $V(\vec{R},\vec{r})$. The factor ϕ_0 in (7) ensures that the coefficient of the plane wave in eq.(7) and the exact 3-body wave function $\Psi^{(+)}_{\vec{K}_d}(\vec{R},\vec{r})$ coincide. The formula (9) can be derived by taking the adiabatic limit of formal expressions for the exact three-body wavefunction[16].

The adiabatic equation must be solved for as many values of $\vec{r}$ as is required for the application. For example, for elastic deuteron scattering we have to evaluate the scattering amplitude $f(\hat{R},\vec{r})$ for as many values of $\vec{r}$ as is required for the accurate evaluation of the integral

$$f_{elastic}(\hat{R}) = \int d\vec{r} \phi_o^*(\vec{r}) f(\hat{R},\vec{r}) \phi_o(\vec{r}), \qquad (10)$$

i.e., for $0 < r < r_d$, where r_d is a measure of the size of the deuteron.

This formalism can be simplified considerably in several interesting cases. When the conditions are such that eq. (6) can be solved in the eikonal approximation an explicit formula for $f(\hat{R},\vec{r})$ in terms of a path integral of $V_{nA} + V_{pA}$ can be obtained. This is Glauber's theory of deuteron-nucleus scattering. The method has been extensively applied to the calculation of elastic scattering and reaction cross-sections of halo nuclei with 2 or more clusters[7],[17]. The integral in (10) is carried out numerically without further approximation of the expression using the best available models for the ground state wave function ϕ_0.

There are also considerable simplifications in the zero range limit of deuteron stripping or when one of the potentials V_{nA} and V_{pA} vanishes. These cases will also be discussed below.

The non-eikonal adiabatic method has been applied to the elastic scattering of ^{6}Li in an $\alpha + d$ model[18], ^{7}Li in an $\alpha + t$ model[19] and to the scattering of ^{11}Li in a $n + n + ^9$Li 3-body model[10].

It is shown in[16] that estimates based on eq.(5) are too conservative. For short range forces the collision time decreases as the impact parameter increases. Hence for a given range of excitation energies the worst violations of the inequality (5) tend to occur at low impact parameters. Under conditions of strong absorption these are just the impact parameters whose contributions to the excitation are suppressed. A revised adiabatic condition which includes this effect is given in ref.[16]. By comparison with CDCC calculations in a special case the revised condition is shown to give an excellent idea of the accuracy of the adiabatic approximation for a model of ^{11}Be scattering.

2.3. An instructive special case

We show the result for deuteron scattering as an example. When one of the interactions V_{nA}, V_{pA} vanishes (or is a constant) the adiabatic equation can be solved exactly in a very simple way. We take $V_{nA} = 0$ for definiteness. This is obviously not a very realistic model of elastic deuteron scattering in general, but it is very relevant to Coulomb break-up of the deuteron[23]. Eq.(6) becomes

$$(T_R + V_{pA}(\vec{R} - \vec{r}/2) - E_d)\Psi^{ad}(\vec{R},\vec{r}) = 0. \tag{11}$$

For a deuteron incident with momentum $\vec{K}_d$ this has the exact solution[20][21][22]

$$\Psi^{ad}_{\vec{K}_d}(\vec{R},\vec{r}) = \phi_0(\vec{r})\exp(i\vec{K}_d.\vec{r}/2)\chi^{(+)}_{\vec{K}_d}(\vec{R} - \vec{r}/2), \tag{12}$$

where $\chi^{(+)}_{\vec{K}_d}$ is the distorted wave for a particle with the mass of the deuteron by the potential V_{pA} and evaluated at the argument $\vec{R} - \vec{r}/2$ i.e., at the $p - A$ relative co-ordinate.

In this limit the elastic deuteron cross-section is simply expressed in terms of the deuteron ground state form factor and the deuteron elastic cross-section generated by V_{pA}. In the generalisation to the case of a projectile with unequal mass clusters the factors of $1/2$ are replaced by ratios involving the masses of the clusters [20]. The generalisation gives a good account of some features of ^{11}Be scattering[20].

The explicit form (12) also makes a deficiency of the adiabatic wavefunction very clear. It predicts that for any $\vec{R}$, $\Psi^{ad}_{\vec{K}_d} \to 0$ exponentially for $r \to \infty$, i.e, in regions of space where we look for outgoing waves in the stripping and break-up channels. We therefore cannot expect the adiabatic wavefunction to be accurate for large r even though it may be perfectly adequate for finite values of r. The reason for this shortcoming can be traced to the treatment of the break-up continuum as degenerate. It is then not possible for the 3-body wavefunction to carry the phase relations between the R and r dependence which are essential to generate the correct asymptotic form in rearrangement and break-up channels. We must use the adiabatic wavefunction in ways which respect this observation. We do this by using it as the basis for an iterative solution to the Schrödinger equation.

2.4. Iteration of the adiabatic solution

We re-write the 3-body Schrödinger equation in the form

$$(E - T_R - T_{np} - V_{pA} - V_{nA})\Psi_{\vec{K}_d} = V_{np}\Psi_{\vec{K}_d}, \tag{13}$$

where we have transferred the V_{np} term to the right-hand-side. This term requires $\Psi_{\vec{K}_d}$ only within the range of V_{np}, and for which we can therefore use the adiabatic wavefunction if the adiabatic conditions are satisfied[13].

The method proceeds by treating the equation

$$(E - T_R - T_{np} - V_{pA} - V_{nA})\Psi_{\vec{K}_d} = V_{np}\Psi^{ad}_{\vec{K}_d}, \tag{14}$$

as an inhomogeneous equation for $\Psi_{\vec{K}_d}$ with the right-hand-side given. In particular, by examining the outgoing Green's function for the operator $E - T_R - T_{np} - V_{pA} - V_{nA}$ it is found that the iterated solution has the correct asymptotic form in the stripping and break-up channels with outgoing waves with momenta correctly given by the conservation of energy, i.e., without the assumption of degenerate break-up channels used in the adiabatic wavefunction. In the exact solution of the inhomogeneous equation the coefficients of the outgoing waves in the stripping and break-up channels (transition amplitudes) are given by

$$T_{d,p} = \langle \chi_p^{(-)} \phi_n | V_{np} | \Psi_{\vec{K}_d}^{ad} \rangle, \qquad (15)$$

$$T_{d,np} = \langle \chi_p^{(-)} \chi_n^{(-)} | V_{np} | \Psi_{\vec{K}_d}^{ad} \rangle, \qquad (16)$$

where the $\chi_p^{(-)}$ and $\chi_n^{(-)}$ are distorted waves generated by V_{pA} and V_{nA}, respectively, and ϕ_n is a neutron bound state, all evaluated at the correct energies predicted by energy conservation.

Strictly speaking, to deduce (15) and (16) from eq.(14) requires the additional assumption that the target has infinite mass, $A \to \infty$. It is only then that the kinetic energy separates into n and p terms and solutions of the homogeneous equation have a product form. Recoil terms of order $1/A$ can mix in terms in the final state in which the neutron is excited out of the state ϕ_n. These corrections (Recoil Excitation and Break-up (REB) effects) can be very significant for light nuclei[24].

All the quantities in (15) and (16) are solutions of 2-body problems and are readily calculated. The evaluation of the amplitudes requires techniques similar to those used in the evaluation of DWBA amplitudes. We emphasise that this iterated theory goes far beyond DWBA. No Born approximation is involved. Couplings between 3-body channels are included to all orders in $\Psi_{\vec{K}_d}^{ad}$. It should be clear that if $\Psi_{\vec{K}_d}^{ad}$ were replaced in equations (15) and (16) by the exact three-body wavefunction $\Psi_{\vec{K}_d}$ then these expressions would give the exact reaction amplitudes. The approximations (15) and (16) assume that all coupling effects in the wavefunction for r less than the range of V_{np} can be adequately accounted for by the adiabatic wavefunction. In principle this whole procedure could be iterated by calculating the complete solution of the inhomgeneous equation (13) (not just the asymptotic form as explained above) and then using the solution to re-calculate an improved inhomogeneous term. As far as I know this has never been done.

In the zero range limit for V_{np} the evaluation of equations (15) and (16) becomes particularly simple because then we can use[13]

$$V_{np} \Psi_{\vec{K}_d}^{ad}(\vec{R},\vec{r}) = V_{np} \phi_0(\vec{r}) \chi_{\vec{K}_d}^{ad(+)}(\vec{R},\vec{r}) = V_{np} \phi_0(\vec{r}) \chi_{\vec{K}_d}^{ad(+)}(\vec{R},0).. \qquad (17)$$

From eq.(8) we see that $\chi_{\vec{K}_d}^{ad(+)}(\vec{R},0)$ satisfies

$$(T_R + V(\vec{R},0) - E_d) \chi_{\vec{K}_d}^{ad(+)}(\vec{R},0) = 0, \qquad (18)$$

and is, therefore, simply a distorted wave generated by the central potential $V_{nA}(R) + V_{pA}(R)$.

We see that in this limit the adiabatic theory of stripping looks even more like a DWBA theory, but this is misleading because the function $\chi_{\vec{K}_d}^{ad(+)}(\vec{R},0)$ includes outgoing waves in break-up channels and the potential $V_{nA}(R)+V_{pA}(R)$ may have very little to do with elastic deuteron scattering.

When s-wave break-up dominates at small r the formalism can be modified to correct for a finite range V_{np}[13]. We expand the three-body wavefunction in terms of the complete set of $n-p$ relative motion states $\{\phi_0, \phi_{\vec{k}}^{(+)}\}$ introduced earlier. Provided the continuum states which contribute do not have very high energy we can safely assume that only s-wave states will overlap V_{np} and we can write

$$\Psi_{\vec{K}_d}(\vec{R},\vec{r}) = \phi_0(r) \chi_o(\vec{R}) + \int d\vec{k} \phi_k^{(+)}(r) \chi_k(\vec{R}), \ r < r_{np}, \qquad (19)$$

where r_{np} is the range of V_{np}. For continuum energies less than roughly 10 MeV the shape of the s-wave state $\phi_k^{(+)}(r)$ does not depend strongly on energy for $r < r_{np}$ and we can write

$$\phi_k(r) \simeq g(k) \phi_0(r), \ r < r_{np}, \qquad (20)$$

where $g(k)$ is independent of r. Inserting this approximation into (19) gives

$$V_{np}\Psi_{\vec{K}_d}(\vec{R},\vec{r}) \simeq V_{np}\phi_0(r)\bar{\chi}_{\vec{K}_d}(\vec{R}), \qquad (21)$$

where

$$\bar{\chi}_{\vec{K}_d}(\vec{R}) = \chi_o(\vec{R}) + \int d\vec{k} g(k)\chi_k(\vec{R}). \qquad (22)$$

We emphasise that the form implied by (21) for the 3-body wavefunction is generally only valid inside the range of V_{np}. The r and R dependence of the adiabatic wavefunction does not factorise in this way in general (see, e.g., eq.(12)).

In the next subsection we will show how these qualitative ideas can be systematically exploited to give an equation for $\bar{\chi}_{\vec{K}_d}(\vec{R})$.

2.5. The method of Johnson and Tandy

A more general approach to obtaining $V_{np}\Psi$, the projection of the 3-body wavefunction which is most relevant to the transition amplitude for stripping, and break-up according to (15) and (16), is to expand in terms of a set of functions which are complete within the range of V_{np}.

A convenient set for this purpose is the set of Weinberg states, or Sturmians, $\bar{\phi}_i(r)$, $i = 0\ldots\infty$, used by Johnson and Tandy[25]. They satisfy

$$(T_r + \alpha_i V_{np})\bar{\phi}_i = -\varepsilon_0 \bar{\phi}_i \qquad (23)$$

where the α_i's are Sturmian eigenvalues. For $i = 0$, $\alpha_0 = 1$ and $\bar{\phi}_0$ is proportional to the deuteron ground state ϕ_0. These states all look like the deuteron asymptotically, but as i and α_i increase they oscillate more and more rapidly at short distances.

An expansion in terms of this set converges rapidly if the dependence on r of the three-body wave function inside the range of V_{np} is similar to ϕ_0. Coupled equations for the coefficients are readily derived using the orthogonality property

$$\langle \bar{\phi}_i \mid V_{np} \mid \bar{\phi}_j \rangle = -\delta_{i,j}. \qquad (24)$$

The first term in the expansion has the form $\phi_0 \bar{\chi}$ with $\bar{\chi}$ defined by

$$(E_d - T_R - \bar{V}(R))\bar{\chi}_{\vec{K}_d}(\vec{R}) = 0, \qquad (25)$$

where the potential $\bar{V}$ is given by

$$\bar{V}(R) = \frac{\langle \phi_0 \mid V_{np}(V_{nA} + V_{pA}) \mid \phi_0 \rangle}{\langle \phi_0 \mid V_{np} \mid \phi_0 \rangle}. \qquad (26)$$

The bra and ket in this equation imply an integration over $\vec{r}$ with fixed $\vec{R}$. $\bar{V}$ reduces to the zero-range result $V(R,0)$ of eq.(18) if the variation of the nucleon optical potentials over a distance of the order of r_{np} can be neglected. For nucleon potentials with a Wood-Saxon shape the effect of the finite range correction in eq.(26) is to increase their diffuseness slightly. A simple way of incorporating these modifications, which can be important for light nuclei, can be found in refs.[26],[27]. Results that take into account terms in the expansion in $\bar{\phi}_i$'s beyond $i = 0$ are given in refs.[25],[28] in specific cases. They show that this is a promising approach to the calculation of break-up effects in stripping which go beyond the adiabatic approximation.

An interesting feature of this derivation is that it makes no reference to the incident energy but only the assumption that the break-up states excited have low enough excitation that the 3-body wave function is well approximated by the form $\phi_0(r)\bar{\chi}(\vec{R})$ inside the range of V_{np}. This suggests that a stripping theory which takes into account break-up effects can be based on the use of $\bar{\chi}$ as a distorted wave even at low energies where the adiabatic condition is not well satisfied.

The situation for elastic deuteron scattering is quite different because there the adiabatic wave function is needed out to distances of the order of the size of the deuteron where the form $\phi_0(r)\bar{\chi}(\vec{R})$ has no justification. The use of Sturmians is not then appropriate.

There have been many comparisons between theory based on eqs. (15) with $V_{np}\Psi^{ad}$ given by (21), (25), (26) and stripping experiments. We call this the Adiabatic Distorted Wave Approximation (ADWA). Over a wide range of energies the ADWA has gives angular distributions for differential crosssections and polarization observables which agree with stripping and pick-up experiments more convincingly and consistently than the DWBA and without the extra ambiguities associated with the use of a deuteron optical potential in the DWBA. Everything in an ADWA calculation is determined by *nucleon* optical potentials for the appropriate energy and target. Some early examples are given in ref.[2], pages 732-734 but there have been many others since, *e.g.*[45]. The method has also been successfully used for (p,d*) [29] and (d,^{2}He)[30] charge exchange reactions.

The study of Cadmus and Haeberli[31] is particularly noteworthy. These authors measured a large number of deuteron and proton elastic scattering observables to pin down optical model parameters so that the DWBA could be applied unambiguously. It was found to fail badly. They were able to use their measurements of deuteron and proton polarization parameters to identify the source of the failures and how these were remedied by the ADWA.

A more recent example of how the ADWA can be used to give an improved account of the systematics of a particular (d,p) transition as function of energy is ref.[32].

Although the ADWA goes well beyond the DWBA and includes effects due to coupling between the elastic deuteron channel and other 3-body channels to all orders it is nevertheless an approximate theory which is expected to need correction at some level. An important example of a clear indication from experiment of the need to go beyond the ADWA theory and the nature of those corrections is the work of the Indiana-Surrey collaboration[33].

One way of going beyond the ADWA for stripping and pick-up is to use the Sturmian expansion method of refs.[25],[28]. An alternative is to use the CDCC wavefunction in (15). For deuteron stripping this is done in refs.[44],[46]. Ref.[47] reports a surprisingly large discrepancy between measured proton polarisation and CDCC predictions for ^{208}Pb(d,p)^{209}Pb at 20 MeV incident energy. Other observables are well reproduced.

2.6. Elastic Coulomb break-up

An interesting application of the expression (16) is the case of Coulomb break-up of a 2-body projectile where one body is uncharged and we can neglect its interaction with the target. In the deuteron case, for example, we can then use eq.(12) and the matrix element factorises[23] as

$$T_{d,np}^{ADWA} = \left(\int d\vec{r}\exp(i(\vec{k}_n - \frac{1}{2}\vec{K}_d).\vec{r})V_{np}\phi_0(\vec{r})\right)\int d\vec{r}_p \chi_{\vec{k}_p}^{(-)*}(\vec{r}_p)\exp(-i\vec{k}_n.\vec{r}_p)\chi_{\vec{K}_d}^{(+)}(\vec{r}_p). \quad (27)$$

where $\chi_{\vec{K}_d}^{(+)}$ and $\chi_{\vec{k}_p}^{(-)}$ are distorted waves describing the scattering of a point deuteron and a proton by the Coulomb field of the target. For a very large screening radius the second factor has the form of an unobservable phase factor which goes to infinity with the screening radius, multiplied by an integral which is similar to that which occurs in the theory of Bremsstrahlung and can be evaluated analytically for a point target. The first factor is easily evaluated for any V_{np}. The restriction to $A \to \infty$ is easily lifted[23].

This theory has been applied successfully to Coulomb break-up of the deuteron[23], ^{11}Be[34], ^{6}He [35] and ^{19}C[34],[36]

We emphasise that the theory which leads to (27) is not perturbation theory. Terms of all orders in V_{pA} and V_{np} are included. The effects of coupling between Coulomb break-up channels are included in all orders with the 2 assumptions that the adiabatic approximation is valid and that the nuclear interaction between the neutron and the target can be neglected.

A DWBA theory which is often used for Coulomb break-up starts from the expression

$$T_{d,np}^{DWBA} = \langle \chi_p^{(-)\vec{k}_n} | V_{np} | \Psi_{\vec{K}_d}^{elast}\rangle, \quad (28)$$

where the elastic deuteron wavefunction $\Psi_{\vec{K}_d}^{elast}$ has the form

$$\Psi_{\vec{K}_d}^{elast}(\vec{R},\vec{r}) = \phi_o(\vec{r})\chi_{\vec{K}_d}^{(+)}(\vec{R}). \quad (29)$$

For the case of Coulomb break-up $\chi_{\vec{K}_d}^{(+)}(\vec{R})$ is a Coulomb wavefunction describing the scattering of a point deuteron in the Coulomb field of the target.

The input data required for the 2 expressions (27) and (28) are identical, i.e., V_{np}, and the Coulomb potential of the target, although they are based on very different physical assumptions. The DWBA expression assumes that the coupling between deuteron elastic and break-up channels is small and can be treated in first order. The adiabatic expression makes no such approximation but instead makes the assumption[1] that any break-up channels that are relevant for the 3-body scattering wavefunction inside the range of V_{np} have low energy compared with the incident deuteron energy.

The DWBA amplitude for Coulomb break-up involves a 6-dimensional integration. Various approximations, including the zero-range approximation for V_{np}, have invariably been made to simplify its evaluation. Recently Zadro[36],[37] has published a momentum-space technique for the exact evaluation of the DWBA amplitude with a finite range V_{np}. This has enabled a meaningful comparison to be made with the ADWA theory. He studied ^{11}Be→^{10}Be+n and ^{19}C→^{18}C+n elastic break-up on a ^{208}Pb target at energies near 70 MeV/nucleon. Both theories give very similar projectile-fragment relative energy distributions in quite good agreement with experiment[38] but the predicted DWBA crosssection magnitudes for a 2-body projectile model are up to 50% bigger. Crosssections for break-up into states of more than a few MeV are very small. The adiabatic approximation therefore ought to be excellent at 70MeV/nucleon[16]. This suggests that spectroscopic factors obtained by comparison of predicted DWBA crosssections with break-up data may be significantly underestimated.

3. THE UNDERLYING MANY-BODY THEORY

Our presentation so far is based on the 3-body Hamiltonian (1) in which V_{nA} and V_{pA} are optical potentials. These are usually taken at $\frac{1}{2}$ of the incident deuteron kinetic energy E_d. This is reasonable if any break-up components in the 3-body wave function have a small fraction of E_d and is certainly consistent with the adiabatic assumption. More generally the $\frac{1}{2}E_d$ prescription can be justified[39] by detailed calculation if the energy dependence arises purely from non-locality and break-up effects are negligible. A deeper question is why the effective interaction in the 3-body model should have anything to do with the nucleon optical potential.

To study this further we recall that the 3-body wavefunction in eq.(3) is the projection of the full many-body $A+2$ wavefunction onto the target ground state. In a standard fashion[40] we can show that the effective Hamiltonian which governs this component when the target is in its ground state in the incident channel is

$$H_{eff} = T_R + H_{np} + \langle \phi_A \mid U \mid \phi_A \rangle, \tag{30}$$

where the bra-ket notation implies integration over the target nucleus co-ordinates to leave an operator in n and p co-ordinates only. The complicated many-body operator U satisfies

$$U = (v_{nA} + v_{pA}) + (v_{nA} + v_{pA})\frac{Q_A}{e}U, \quad v_{NA} = \sum_{i=1}^{A} v(N,i). \tag{31}$$

The v_{NA}'s, $N = p, n$, are the sums of the 2-body interactions between the incident p and n and the target nucleons $1...A$. The operator Q_A projects on to excited states of the target. U sums up all processes via excited states which begin and end on the target ground state.

U can be separated into its p and n contributions by using manipulations from multiple scattering theory. We obtain

$$U = (U_{nA} + U_{pA}) + U_{nA}\frac{Q_A}{e}U_{pA} + U_{pA}\frac{Q_A}{e}U_{nA} + ..., \tag{32}$$

where

$$U_{nA} = v_{nA} + v_{nA}\frac{Q_A}{e}U_{nA}, \quad U_{pA} = v_{pA} + v_{pA}\frac{Q_A}{e}U_{pA}, \tag{33}$$

and the dots in (32) are terms of 3rd or higher order in U_{nA} and/or U_{pA}, always with an excited target (though not necessarily excited deuteron) as intermediate state.

---

[1] Note that because we use the solution (12) in this case we have no need for the extra assumptions about the break-up spectrum which lead to (21).

The above expressions for U_{nA} and U_{pA} are strongly reminiscent of Feshbach's[40] expressions for the operator which gives the nucleon optical potential when sandwiched between the target ground state. Note however that the energy denominator e which appears everywhere is not the denominator one expects to see in the nucleon operator. It is given (for an infinitely massive target) by $e = E + i0 - T_n - T_p - V_{np} - H_A$ where H_A is the target Hamiltonian. It is plausible that if low energy weakly correlated break-up states dominate, then in U_{nA}, for example, we can neglect V_{np} and replace $E - T_p$ by $\frac{1}{2}E_d$ on the average. $\langle \phi_A | U_{nA} | \phi_A \rangle$ then reduces to a formal expression for the neutron optical potential at energy $\frac{1}{2}E_d$.

The higher order terms in (32) still have to be dealt with, however. The second order terms describe a process in which the neutron excites the nucleus and the proton subsequently de-excites it, and vice versa. The magnitude of such effects will be small for weakly correlated $n - p$ configurations such as in the deuteron or low energy break-up configurations. Their neglect is consistent with approximations already made.

We learn from this analysis that the validity of the 3-body model as usually assumed is intimately bound up with the assumption of dominance of low energy break-up configurations. However, all the arguments given above are very qualitative. Very little work has been done to give substance to them and estimate any corrections to the usual model. It would seem to be hardly worth while to go much beyond the adiabatic or CDCC treatments of three-body effects, both of which assume that break-up excitations can be truncated, without investigating many-body corrections to the 3-body Hamiltonian, eq.(1), more thoroughly.

Finally in this section, note that even in the lowest order version of the effective interaction one expects to see corrections arising from the identity of n and p and the target nucleons. There are essentially two distinct approaches to these antisymmetrization effects. The RGM methods starts from a many-body Hamiltonian with an assumed N-N interaction and puts in antisymmetrization right from the start, treating the nucleons in the deuteron and in the target on the same footing. On the other hand it is difficult in practice to treat all possible open channels and absorption has to be inserted by hand. RGM calculations have been published[41] which include deuteron break-up effects using discretisation methods similar to the CDCC (next section). Antisymmetrisation effects are very important in this approach[41].

The alternative approach of refs.[13],[42],[43] is based on the idea that because of the loosely bound extended spatial nature of the deuteron, the nucleons in the deuteron see the target nucleus much as if they were completely independent and so much of the effects of antisymmetry and coupling to excited target states are contained in the complex optical potentials of the 3-body model.In this way one automatically generates a total deuteron reaction cross section which is close to that observed even when deuteron elastic break-up is neglected. New effects arise for deuteron collisions only because the nucleons in the deuteron may scatter off each other into occupied target states (Pauli blocking). These effects are included through a generalisation of the Bethe-Goldstone equation. The role of break-up channels is to re-adjust the flow of flux into inelastic channels involving excited target states as well as transferring flux into break-up channels. For some impact parameters the effect of the break-up channel may even be to *decrease* the partial reaction crosssection because in the break-up configuration the nucleons may overlap spatially less well with the imaginary parts of the nucleon optical potentials. Tostevin, et al[43] found that absorption tends to suppress Pauli blocking and no new major qualitative effects were found. Aoki[44], using a formula for these effects due to Pong and Austern[42], reported that Pauli blocking effects gave a 10% repulsive correction to the deuteron optical potential and improved CDCC fits to elastic deuteron scattering on ^{208}Pb at 20 MeV. The effect on (d,p) crosssections was negligible.

3.1. Many-body theory of stripping

One advantage of a theory of stripping and elastic break-up based on the matrix elements (15) and (16), which we have here derived within a 3-body model, is that they generalise easily to include important many body effects. One starts from formulae analogous to these but with target nucleus co-ordinates still explicit. The wavefunction $\Psi^{ad}_{\vec{K}_d}$ is replaced by the full many-body scattering state corresponding to a deuteron incident on the target ground state. The final state wavefunction describes scattering of a proton incident on the residual nucleus and scattered by the *target* nucleons. There is no V_{np} involved in this state (see [49], pages 838-839, and [24]). The effect of the identity of the neutron in the deuteron and the N target neutrons is included exactly by multiplying the T-matrix by the factor $\sqrt{(N+1)}$ and using properly antisymmetrised wavefunctions for the target and residual nucleus (see [49], pages 836-838).

To obtain the connection with the 3-body theory we

(i) Ignore explicit contributions from channels in which the target is excited by the incident $n-p$ pair. [2]

(ii) Ignore Recoil Excitation and Break-up of the final nucleus by elastic scattering of the proton by the target nucleons in the final state [3].

Because the operator V_{np} in the matrix element is independent of the target internal co-ordinates these two steps automatically give a matrix element which involves the projection of the final nucleus state onto an un-excited target state.

(iii) Ignore the identity of the protons in the target and incident deuteron.

Very little is known about the validity of step (iii). The usual qualitative argument is that proton exchange terms in a (d,p) reaction involve the overlap of bound and continuum proton states instead of the continuum-continuum overlap which contributes to the direct term. The exchange term is therefore expected to be small.

The result of these steps is simply that in the expressions (15) ϕ_n is replaced by the *overlap function* ([2], page 710)

$$\phi_n^{BA}(\vec{r}_n) = \sqrt{N+1} \int d\xi_A \phi_B^*(\xi_A,\vec{r}_n)\phi_A(\xi_A), \tag{34}$$

where ϕ_A, ϕ_B are wavefunctions of the initial and final states in the stripping reaction and $\vec{r}_n$ is the coordinate of the neutron relative to the centre-of-mass of the target. By definition the spectroscopic factor S_{AB} is the norm of the overlap function.

If approximation (ii) is relaxed the (d,p) transition matrix will be a linear combination of terms involving the overlap function for different states B, A, but always with the target ground state. If approximation (i) is relaxed overlaps and spectroscopic factors for many different states A and B will enter, reflecting the many different paths between the initial and final states which then become possible. The corresponding generalisation of the DWBA is referred to as Coupled Channels Born Approximation (CCBA) but the theory described here is not Born approximation because of the way couplings in the n-p space are treated.

4. LINK WITH THE CDCC METHOD

In the CDCC method the 3-body wave function for deuteron-nucleus scattering, $\Psi(\vec{R},\vec{r})$, is expanded in a set of orthonormal functions $\phi_s(\vec{r})$, $s = 0,1,2......$, which diagonalise H_{np} with eigenvalues ε_s and discretise the $n-p$ continuum. The set are usually defined so that $\phi_{s=0}$ is the deuteron ground state. Coupled equations are then derived as a technique for solving the 3-body Schrödinger equation. See the talk by I J Thompson for further details of the CDCC method.

We can expand the adiabatic wavefunction $\Psi^{ad}(\vec{R},\vec{r})$ in a volume $\mathscr{V}$ in $\vec{r}$ space in terms of an orthonormal set $\psi_s(\vec{r})$, $s = 0,1,2......$ which is complete in $\mathscr{V}$:

$$\Psi^{ad}(\vec{R},\vec{r}) = \sum_{s=0}^{\infty} \psi_s(\vec{r})\chi_s(\vec{R}), \tag{35}$$

and derive coupled equations for the χ_s's of the form

$$(E_d - T_R)\chi_s(\vec{R}) = \sum_{s'} \langle \psi_s | V | \psi_{s'} \rangle \chi_{s'}(\vec{R}), \tag{36}$$

where V is defined in eq.(1) and the coupling matrix elements involve an integration over $\mathscr{V}$.

If we identify the ψ_i's and ϕ_i's (to obtain the CDCC equations the ϕ_i's must diagonalise H_{np}) these equations are similar to the CDCC equations with all the channel energies set equal to $-\varepsilon_0$. We can put this another way. If the set $\psi_s(\vec{r})$, $s = 0,1,2......$, is complete in $\mathscr{V}$, and the functions $\chi_s(\vec{R})$ satisfy the coupled equations (36) then these equations show that, for $\vec{r}$ in $\mathscr{V}$, $\sum_s \psi_s(\vec{r})\chi_s(\vec{R})$ satisfies

$$\begin{aligned}(E_d - T_R)\sum_s \psi_s(\vec{r})\chi_s(\vec{R}) &= \sum_s \psi_s(\vec{r}) \int d\vec{r}' \psi_s^*(\vec{r}') \sum_{s'} V(\vec{R},\vec{r}')\psi_{s'}(\vec{r}')\chi_{s'}(\vec{R}) \\ &= V(\vec{R},\vec{r})\sum_{s'} \psi_{s'}(\vec{r})\chi_{s'}(\vec{R}),\end{aligned} \tag{37}$$

---

[2] The implicit effects of target excitation are, of course included in the nucleon optical potentials.
[3] See [24] for an estimate of these effects for some light targets.

where the completeness of the ψ_s's has been used. Eq.(37) is just the adiabatic equation. Hence $\sum_s \psi_s(\vec{r})\chi_s(\vec{R})$ is the adiabatic solution $\Psi^{ad}(\vec{R},\vec{r})$.

We see that the Adiabatic approach can be regarded as an approximation to the CDCC method. Thompson[8] describes how this result can be used as a check of CDCC calculations by talking the limit when all the channel energies (the ε_s's) are set equal. In making these comparisons note that the adiabatic method does not take into account any restrictions imposed by the Pauli Principle on the states which should be included in the set ϕ_s. For example, in the case of ^{11}Be scattering the adiabatic calculations include transitions into a state in which the neutron is in a nodeless s-state with respect to the ^{10}Be core. Such contributions are easily excluded in the CDCC calculation or in the Johnson-Tandy approach[25], but it is not obvious how to do this in an adiabatic calculation without introducing non-local projection operators with the consequential loss of some of the characteristic simplicity of the adiabatic equation.

At first sight it is puzzling that the adiabatic calculation can take into account effects due to excited deuteron states when only the deuteron ground state wave function appears explicitly. In CDCC calculations the wave functions of all excited states deemed to be important must be inserted explicitly into the calculation of the coupling matrix elements. Our derivation above shows how this puzzle can be resolved but it is also helpful to note that the ground state wavefunction ϕ_0 determines the Hamiltonian through the identity (see the Appendix to [16])

$$H_{np} = -\varepsilon_0 - \frac{\hbar^2}{2\mu_{np}} \phi_0^{-1} \nabla_r \phi_0^2 . \nabla_r \phi_0^{-1}, \qquad (38)$$

where the ∇_r operators act on everything to the right of them.

We note that the CDCC method uses a basis which is complete in large volumes of $\vec{r}$ space. As we have seen the stripping and break-up matrix elements (15) and (16) explore a very restricted part of this space, i.e., within the range of V_{np}. For this purpose the complete set used by Johnson and Tandy[25] and its generalisations may be more efficient.

In their exploration of the adiabatic approximation Johnson and Soper[13] proposed an approximation to the CDCC method which replaced the deuteron continuum by a single pseudo state. In their method the component $V_{np}\Psi$ of the 3-body wave function is still governed by equations (21), (25) and (26), but the deuteron elastic scattering wavefunction and the pseudo break-up state satisfy a pair of coupled equations. This method was critically examined in great detail by Rawitscher[3],[4] within the CDCC framework. A more sophisticated version of the single pseudostate method was developed by Amakawa, Austern and Vincent[48] and is known as the quasi-adiabatic method.

5. CONCLUDING REMARKS

Some of the clearest evidence for the importance of deuteron break-up effects and the failure of the DWBA for (d,p) and (p,d) reactions has been obtained by using the adiabatic approximation as implemented in the ADWA. However, we have seen that the adiabatic approximation can be regarded as an approximation to the CDCC, so it might be argued that the adiabatic approximation no longer has a role. It is only recently, however, that the CDCC method has become available for projectiles with more than 2 clusters and, when coupled with the eikonal approximation where applicable, the adiabatic approximation is a powerful tool for the analysis of reactions with composite projectiles.

An attractive feature of the adiabatic approach which it shares with CDCC is that it provides a framework for inserting the systematics of the interaction of the constituents of the projectile with the target into reaction analyses. This means that the need for optical potentials for unstable projectiles can often be avoided, but it still requires reliable information about the constituents' optical potentials and hence good elastic scattering data for appropriate energies and targets.

An advantage of the adiabatic method over CDCC is that its implementation does not need detailed wavefunctions of strongly coupled excited bound and continuum states of the projectile. The construction of these states may introduce considerable uncertainties into a CDCC calculation. It is important therefore to understand the limitations of the adiabatic approximation.

Perhaps the most important feature of the adiabatic approximation is its ability to provide insights into the mechanism of complex reactions. It can be used to provide checks of CDCC and other theories as well as being a relatively easy and transparent way to take into account complicated effects of channel coupling in some important special cases.

REFERENCES

1. R. C. Barrett, Nucl. Phys.**51**(1964)27

2. G. R. Satchler, *Direct Nuclear Reactions*, Oxford University Press, Oxford, 1983.
3. G. H. Rawitscher, Phys.Rev.**C9**(1974)2210.
4. G. H. Rawitscher, Nucl. Phys. **A241**(1975)365.
5. M. Kamimura, M. Yahiro, Y. Iseri, H. Kameyama, Y. Sakuragi and M. Kawai, Prog.Theor.Phys.Suppl.**89**(1986)1.
6. N. Austern, Y. Iseri, M. Kamimura, M. Kawai, G. Rawitscher and M. Yahiro, Phys.Rep.**154**(1987)125.
7. J. S. Al-Khalili and J. A. Tostevin, *Few-body models of nuclear reactions*, Chapter 3.1.3 in "Scattering" eds. R. Pike and P. Sabatier, Academic Press, London and San Diego, 2002, pages 1373-1392.
8. I. J. Thompson, *this conference*
9. T. Matsumoto, E. Hiyama, K. Ogata, Y. Iseri, M. Kamimura, S. Chiba and M. Yahiro, Phys.Rev. **C70**(2004)061601(R).
10. J.A. Christley, J.S. Al-Khalili, J.A. Tostevin and R.C. Johnson, Nucl. Phys. **A624**(1997) 275 .
11. R. J. Glauber, in Lectures in Theoretical Physics, edited by W E Brittin (Interscience, New York, 1959), Vol. 1, pp 315-414.
12. J. Al-Khalili and F. Nunes, J.Phys.G: Nucl. Part. Phys.**29**(2003)R89.
13. R. C. Johnson and P. J. R Soper Phys. Rev.**C1** (1970)976.
14. H. Amakawa, S. Yamaji, A. Mori and K. Yazaki, Phys Lett, **B82**(1979) 13.
15. I.J. Thompson, Computer Programme ADIA, Daresbury Laboratory Report, 1984, unpublished.
16. N. C. Summers, J. S. Al-Khalili and R. C. Johnson , Phys. Rev. **C66**(2002) 014614 .
17. K. Varga, S. C. Pieper, Y. Suzuki, and R. B. Wiringa Phys. Rev. **C66**(2002) 034611.
18. I. J. Thompson and M. A. Nagarajan, Phys. Letts., **106B** (1981)163.
19. M. A. Nagarajan, I. J. Thompson and R. C Johnson, Nucl. Phys. **A385**(1982)525.
20. R. C. Johnson , J. S. Al-Khalili, J. A. Tostevin, Phys. Rev. Letts. 79 (1997) 2771.
21. R. C. Johnson, J. Phys G: Nucl. Part. Phys. 24 (1998) 1583.
22. R. C. Johnson , *Elastic scattering and elastic break-up of halo nuclei in a special model* in Proc. Eur. Conf. On Advances in Nuclear Physics and Related Areas, Thessoloniki, Greece, July 8-12, 1997, (eds. DM Brink, ME Grypeos and SE Massen, Giahoudi-Giapouli, Thessaloniki, 1999) 156.
23. J. A. Tostevin, S. Rugmai and R. C. Johnson, Phys. Rev. **C57** (1998) 3225.
24. N. K. Timofeyuk and R. C.Johnson, Phys. Rev.**C59** (1999) 1545.
25. R. C. Johnson, P. C. Tandy, Nucl. Phys.**A235** (1974) 56.
26. J. D. Harvey and R. C. Johnson, Phys.Rev.**C3**(1971)636-645.
27. G. L. Wales and R. C. Johnson, Nucl. Phys. **A274** (1976) 168.
28. A. Laid, J. A. Tostevin and R. C. Johnson, Phys. Rev. **C48** (1993) 1307.
29. B. Gönül and J. A. Tostevin, Phys. Rev. **C53** (1996) 2949.
30. S. Rugmai J. S.Al-Khalili, R. C. Johnson, and J. A. Tostevin, Phys. Rev. **C60** (1999) 027002.
31. R. R. Cadmus and W. Haeberli, Nucl.Phys.A327(1979)419, ibid., **A349**(1980)103.
32. X. D. Liu, M. A. Famiano, W. G. Lynch, M. B.Tsang, and J. A. Tostevin, Phys.Rev. **C69**(2004)064313.
33. E. J. Stephenson, *et al.* Phys.Rev. **C42**(1990)2562.
34. P. Banerjee, I. J. Thompson and J. A. Tostevin, Phys.Rev.**C58**(1998)1042.
35. P. Banerjee, I. J. Thompson and J. A. Tostevin, Phys.Rev.**C58**(1998)1337.
36. M. Zadro, Phys.Rev. **C70**(2004)044605.
37. M. Zadro, Phys.Rev. **C66**(2002)034603.
38. T. Nakamura *et al.* Nucl. Phys. **734**(2004)319.
39. R. C. Johnson, P. J. R Soper, Nucl. Phys. **A182** (1972) 619.
40. H. Feshbach, Ann.Phys.(N.Y.)**5** (1958) 357.
41. T. Kaneko and Y. C. Tang, Nucl. Phys. **A612**(1997)204
42. W. S. Pong and N. Austern, Ann.of Phys.(N.Y.) **93**(1975)369
43. J. A. Tostevin, M-H Lopes and R. C. Johnson, , Nucl. Phys. **A465** (1987)83.
44. M.Masaki, Y. Aoki, K. Katoh, S. Nakagawa, N. Nakamoto and Y.Tagashi, Nucl.Phys. **A573**(1994)1
45. Y. Aoki, H. IIda, K. Nagano, Y. Toba and K. Yagi, Nucl.Phys. **A393** (1983)52.
46. K. Hirota, Y. Aoki, N. Okamura and Y. Tagishi, Nucl.Phys. **A628**(1998)547.
47. M.Yamaguchi, Y. Tagishi, Y. Aoki, N. Kawachi, N. Okamura and N. Yoshimaru, Nucl.Phys. **A747**(2005)3.
48. H. Amakawa, N. Austern, and C. M. Vincent, Phys. Rev. **C 29** (1984)699
49. M. L. Goldberger and K. M. Watson, *Collision Theory*, Wiley, New York, 1964.

Near-Barrier Elastic Scattering of Weakly Bound Nuclei and the Threshold Anomaly

M. S. Hussein*, L. C. Chamon* and P. R. S. Gomes[†]

*Instituto de Física, Universidade de São Paulo, Caixa Postal 66318, 05315-970 São Paulo, SP, Brazil
[†]Instituto de Física, Universidade Federal Fluminense, Av. Litoranea s/n, Gragoatá, Niterói, RJ, 24210-340, Brazil

Abstract. It is pointed out that the usual threshold anomaly, found operative in the energy behavior of the imaginary and real parts of the optical potential representing the elastic scattering of tightly bound nuclei at near- and below-barrier energies, suffers a drastic qualitative change in the case of the elastic scattering of weakly bound projectile nuclei. Owing to the strong coupling to the breakup channel even at sub-barrier energies, the imaginary potential strength seems to increase as the energy is lowered down to below the natural, barrier, threshold, accompanied by a decrease in the real potential strength. This feature is consistent with the dispersion relation. It also clearly indicates the effective increase of the barrier height. The systems ^{9}Be,^{16}O + ^{64}Zn and ^{6,7}Li + ^{138}Ba are analyzed to illustrate this new phenomenon.

Keywords: nuclear reaction, optical potential, dispersion relations
PACS: 25.70.Bc,25.70.Mn

The by now well known Threshold Anomaly (TA), seen in the behavior of the real and imaginary parts of the optical potential as a function of decreasing energy in the elastic scattering of tightly bound nuclei at near-barrier energies, has been reviewed and discussed by several authors [1, 2]. The phenomenon is a direct consequence of the dispersion relation which quantifies the concept of causality in scattering: no scattered wave emerges before the incident wave reaches the target. Recently, the TA has been looked for in the case of the elastic scattering of weakly bound stable and radioactive nuclei [3–7]. Careful analyses of the data show that what happens in these systems is a new manifestation of the dispersion relation unique for the breakup coupling dynamical polarization potential. Because the coupling to the breakup in these systems continues to be important even at energies below the barrier, the "threshold" ceases to be the barrier itself. Thus, the imaginary part of the potential could increase at lower energies and, as the dispersion relation dictates, the real part of the dynamic potential would show a decrease, implying an overall decrease in the real part of the optical potential that fits the elastic scattering. This is indeed what is found [3–7]. The purpose of this paper is to give a detailed account of this phenomenon, which we call the Breakup Threshold Anomaly (BTA).

We first present the analyses for the "normal" system ^{16}O + ^{64}Zn, where elastic scattering angular distributions at several energies have been measured [8]. We have re-analyzed the data using the São Paulo (SP) optical potential. The SP interaction [9–12] is based on the double folding potential which accounts for exchange through an effective energy dependence:

$$V_{SP}(R,E) = (1 + i\,0.78)F(R,E), \qquad (1)$$

where $F(R,E)$ is the double folding potential whose energy dependence results from the local equivalence of the otherwise non-local interaction [13, 14]. This energy dependence is not dispersive. In the present analysis, we have assumed for the optical potential a normalized version of the SP potential:

$$V_{SP}(R,E) = [N_R(E) + iN_I(E)]F(R,E). \qquad (2)$$

The coefficients $N_R(E)$ and $N_I(E)$ are energy dependent normalization factors that take into account the effects of the dynamic polarization potentials (DPP) arising from direct channel couplings. It is worth mentioning here that all DPPs are dispersive with their real and imaginary parts being connected through a dispersion relation. One important exception to this is the elastic transfer DPP [15].

From the properties of the Green function, that enters in the definition of the DPP, one can immediately derive the dispersion relation between $N_R(E)$ and $N_I(E)$, represented by:

$$N_R(E) = N_{R0} + \Delta N_R(E), \qquad (3)$$

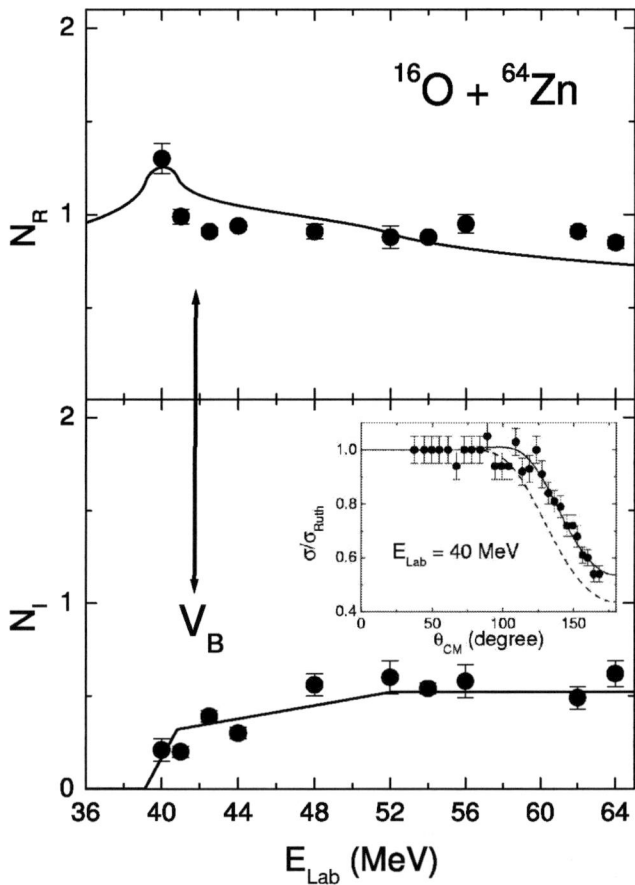

FIGURE 1. Energy dependence of the normalization factors N_R and N_I of the SP potential for the $^{16}O + ^{64}Zn$ system. The lines represent possible behaviors of N_R and N_I that are compatible with the dispersion relation, where N_R was obtained through Eqs. 3-10 with N_{R0}=1.2. The approximate position of the barrier height (V_B) is indicated in the figure. The elastic scattering angular distribution for the lowest energy is presented in the inset. The dashed line represents predictions with the standard values N_R=1 and N_I=0.78, while the solid line corresponds to the results obtained with the best N_R and N_I values for this energy.

$$\Delta N_R(E) = \frac{P}{\pi} \int \frac{N_I(E')}{E' - E} dE', \qquad (4)$$

and its subtracted form

$$\Delta N_R(E) = \Delta N_R(E_s) + (E - E_s)\frac{P}{\pi} \int \frac{N_I(E')}{(E' - E_s)(E' - E)} dE', \qquad (5)$$

where E_s is some sufficiently high energy at which information about both N_I and ΔN_R are known [2]. These equations are the analog of the Kramer-Kronig dispersion relation in optics, and arise from general principle of causality: no scattered wave emerges before the incident wave reaches the scattering center.

In figure 1, we show the behavior of $N_R(E)$ and $N_I(E)$ obtained from elastic scattering data fits for the normal system mentioned above. Because the threshold for non-elastic channel coupling effects is the Coulomb barrier, one observes the TA as expected: a decrease in $N_I(E)$ as the energy is lowered below the barrier, accompanied by an increase in $N_R(E)$, implying an increase in the attraction and accordingly a reduction of the barrier height. This favors an enhancement in the complete fusion cross section as has been demonstrated for a wide range of systems involving

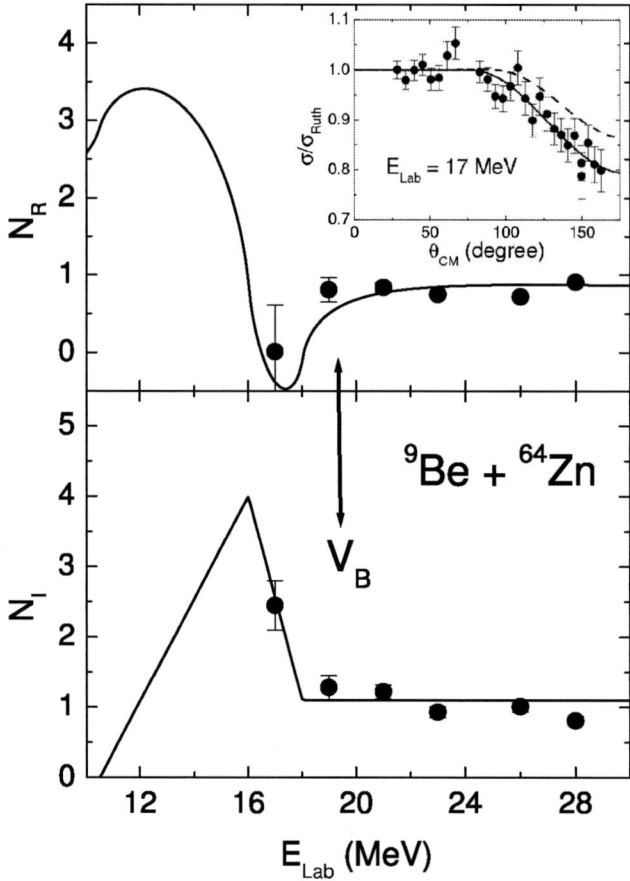

FIGURE 2. The same as Figure 1, for the ^{9}Be + ^{64}Zn system ($N_{R0}=0.9$).

normal, tightly bound nuclear projectiles [16]. Just to illustrate the dispersion relation, we have assumed a schematic description for N_I

$$N_I = 0 \text{ for } E \leq E_1, \tag{6}$$

$$N_I = a(E - E_1) \text{ for } E_1 \leq E \leq E_2, \tag{7}$$

$$N_I = a(E_2 - E_1) + b(E - E_2) \text{ for } E_2 \leq E \leq E_3, \tag{8}$$

$$N_I = a(E_2 - E_1) + b(E_3 - E_2) = N_{I\infty} \text{ for } E \geq E_3. \tag{9}$$

Within this assumption, and owing to the constancy of N_I at $E > E_3$, the subtracted dispersion relation, Eq. (5), gives the same result as the non-subtracted one, Eq. (4). One obtains an analytical expression for ΔN_R [1, 2]

$$\begin{aligned}\Delta N_R(E) &= a(E_2 - E_1)\left[\varepsilon_1 ln|\varepsilon_1| - \varepsilon_2 ln|\varepsilon_2|\right] \\ &+ \left[b(E_3 - E_2) - a(E_2 - E_1)\right] \\ &\times \left[\varepsilon_2' ln|\varepsilon_2'| - \varepsilon_3' ln|\varepsilon_3'|\right],\end{aligned} \tag{10}$$

with $\varepsilon_i = (E - E_i)/(E_2 - E_1)$ and $\varepsilon_i' = (E - E_i)/(E_3 - E_2)$. The solid lines in figure 1 represent a behavior of $N_R(E)$ and $N_I(E)$ within this schematic picture. The elastic scattering angular distribution for the lowest energy of the ^{16}O + ^{64}Zn system is presented in the inset of Fig. (1). The dashed line in this inset represents predictions obtained with the standard values $N_R=1$ and $N_I=0.78$, while the solid line corresponds to the results obtained with the best N_R and N_I values for this energy. A comparison between these dashed and solid lines shows that the data fit is quite sensitive to the N_R and N_I values even at this low energy region.

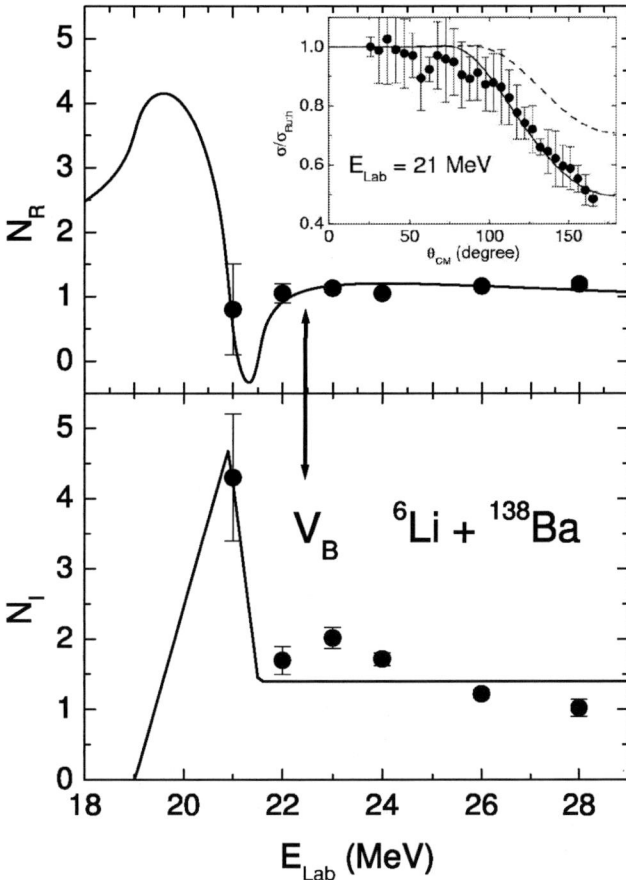

FIGURE 3. The same as Figure 1, for the ^{6}Li + ^{138}Ba system (N_{R0}=1.15).

We turn now to the weakly bound nucleus ^{9}Be and study its elastic scattering on the same target ^{64}Zn. We have re-analyzed elastic scattering angular distributions at several bombarding energies [17]. The corresponding $N_R(E)$ and $N_I(E)$ are shown in figure 2. A striking difference in the energy dependence of these normalization coefficients from the one shown in the previous figure (figure 1) is clearly seen. As the energy is lowered below the barrier $N_R(E)$ decreases, while $N_I(E)$ increases. This implies an effective reduction of the nuclear attraction leading to an increase in the barrier height. Accordingly, the complete fusion cross section may be hindered in comparison with the opposite effect expected for tightly bound stable nuclei, giving more credence to the findings of several authors in theory [18–20] and experiment [21–24]. The solid lines in figure 2 represent behaviors of N_R and N_I that are compatible with the dispersion relation. We have also performed similar re-analyses for the ^{6,7}Li + ^{138}Ba systems (elastic scattering data from Ref. [25]). The corresponding results (see Figs. 3 and 4) are very similar to those for ^{9}Be + ^{64}Zn. It seems therefore that the threshold for non elastic processes of these systems is several MeV below the barrier, as the solid lines in figures (2)-(4) indicate, in agreement with the schematic calculations of Ref. [19].

A further evidence of the consistency of our analyses are the values obtained for N_{R0} and $N_{I\infty}$. As the SP potential, equation (1), has been successful in describing the elastic scattering for a large number of different systems in energies above the barrier [12], one should expect that both N_{R0} and $N_{I\infty}$ to be close to unity. Of course, the values for N_{R0} and $N_{I\infty}$ found in the present work are not identical to the standard (from high energies) N_R=1 and N_I=0.78. However, these standard values in fact represent mean values obtained from data analyses of several systems [12] and one can expect variations around the average values for particular systems. Indeed, structure effects on the nuclear densities involved in the folding calculations may affect N_R while different degrees of absorption from particular reaction channels could affect N_I. An inspection of Figs. 1-4 shows that $0.9 \leq N_{R0} \leq 1.15$ and $0.5 < N_{I\infty} < 1.5$ have been found in the present work.

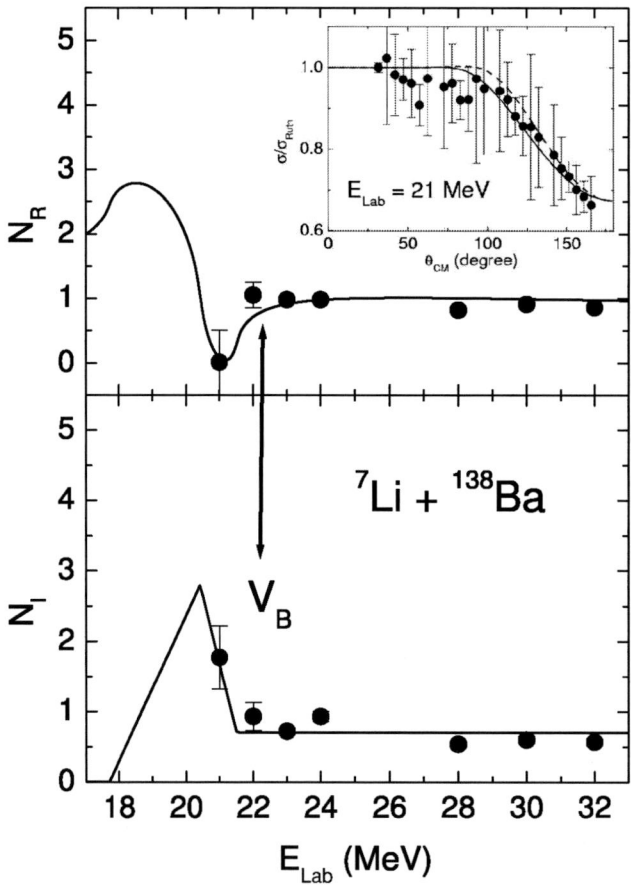

FIGURE 4. The same as Figure 1, for the ^{7}Li + ^{138}Ba system (N_{R0}=1.05).

As already commented, the solid lines in Figs. 1-4 represent behaviors of N_R and N_I compatibles with the dispersion relation. In fact, different behaviors, that also could follow the "data", can be found even within the schematic picture of Eqs. 6-10 just by slightly modifying the values of the parameters involved in these equations. Therefore, clearly our findings are mainly based on the N_R and N_I "data" themselves. The detected difference between the normal and weakly-bound systems is mostly based on the results from the corresponding lowest energies (see Figs. 1-4). Even so, we consider that significant evidence for the proposed BTA has been obtained, because the data fits for these low energies, except in the ^{7}Li + ^{138}Ba case, are still quite sensitive to N_R and N_I, as illustrated by the dashed and solid lines in the insets of Figs. 1-4. We mention that quite similar behavior was already observed earlier for other systems [3–7].

Traditionally, the threshold anomaly is formally displayed in terms of complex renormalization factors of the double-folding potential, as we have done in the present work. This assumption, which corresponds to assuming that the radial dependence of the optical potential is the same as the bare one, is well established for normal tightly bound stable nucleus systems. The situation is more complicated in the case of weakly bound nuclei, where the effect of the couplings give rise to a polarization potential with different radial shape. In particular, an important role is played by the tail of the polarization potential which is much longer than that of the folding potential [26]. Nevertheless, our simple approach of considering renormalization factors has provided good fits, not only for the data shown in the insets of figures 1-4, but for all angular distributions analyzed in the present work

In conclusion, we have found some evidence of differences in the energy-dependence of the optical potential for tightly and weakly bound nucleus systems. The Break-up Threshold Anomaly implies an increase of the imaginary part and a decrease of the real part of the optical potential that fits the elastic scattering at low energies. This reduction in the attraction results in an increase of the barrier height. Consequently, a possible hindrance in the complete fusion may erase any enhancement due to the larger size of the weakly bound nucleus. In fact this behavior has recently been clearly seen in the reaction induced by the radioactive, very weakly bound, halo nucleus ^{6}He on a ^{238}U target [27]. Our findings have mostly been based on the results obtained from elastic scattering data analyses for the lowest energies of four systems, and also on earlier data analyses for other similar systems [3–7]. Clearly more work is required to further our understanding of the BTA and its connection with the hindrance of complete fusion of both weakly bound, stable, nuclei and more so of radioactive nuclei.

ACKNOWLEDGMENTS

This work was partially supported by Financiadora de Estudos e Projetos (FINEP), Fundação de Amparo à Pesquisa do Estado de São Paulo (FAPESP), Fundação de Amparo à Pesquisa do Estado do Rio de Janeiro (FAPERJ), Instituto de Milênio de Informação Quântica (MCT), and Conselho Nacional de Desenvolvimento Científico e Tecnológico (CNPq).

REFERENCES

1. M. E. Brandan and G. R. Satchler, Phys. Rep. **285**, 143 (1997).
2. G. R. Satchler, Phys. Rep. **199**, 147 (1991).
3. C. Signorini, Europ. Phys. Journal **A13**, 129 (2002).
4. C. Signorini et al., Phys. Rev. **C61** 0601603 (2000).
5. N. Keeley, S. J. Bennett, N. M. Clarke, B. R. Fulton, G. Tungate, P. V. Drumm, M. A. Nagarajan, J. S. Lilley, Nucl. Phys. **A571**, 326 (1994).
6. A. Pakou et al., Phys. Rev. **C69**, 054602 (2004).
7. I. Martel, J. Gomez-Camacho, K. Rusek, G. Tungate, Nucl. Phys. **A605**, 417 (1996).
8. C. Tenreiro et al., Phys. Rev. **C53**, 2870 (1996).
9. L. C. Chamon, D. Pereira, M. S. Hussein, M. A. Candido Ribeiro and D. Galetti, Phys. Rev. Lett. **79**, 5218 (1997).
10. L. C. Chamon, D. Pereira and M. S. Hussein, Phys. Rev. **C58**, 576 (1998).
11. L. C. Chamon et al., Phys. Rev. **C66**, 014610 (2002).
12. M. A. G. Alvarez et al., Nucl. Phys. **A723**, 93 (2003).
13. M. A. Candido Ribeiro, L. C. Chamon, D. Pereira, M. S. Hussein and D. Galetti, Phys. Rev. Lett. **78**, 3270 (1997).
14. D. T. Khoa, G. R. Satchler and W. von Oertzen, Phys. Rev. **C56**, 954 (1997).
15. A. Lepine-Szily, M. S. Hussein, R. Lichtenthaler, J. Cseh and G. Levai, Phys. Rev. Lett. **82**, 3972 (1999).
16. A. B. Balantekin and N. Takigawa, Rev. Mod. Phys. **70**, 77 (1998).
17. S. B. Moraes et al., Phys. Rev. **C61**, 64608 (2000).
18. M. S. Hussein, M. P. Pato, L. F. Canto and R. Donangelo, Phys. Rev. **C46**, 377 (1992).
19. M. S. Hussein, M. P. Pato, L. F. Canto and R. Donangelo, Phys. Rev. **C47**, 2398 (1993).
20. N. Takigawa, M. Kuratani and H. Sagawa, Phys. Rev. **C47**, R2470 (1993).
21. M. Dasgupta et al., Phys. Rev. Lett. **82**, 1395 (1999).
22. M. Dasgupta et al., Phys. Rev. **C66**, 041602(R) (2002).
23. M. Dasgupta et al., Phys. Rev. **C70**, 024606 (2004).
24. V. Tripathi et al., Phys. Rev. Lett. **88**, 172701 (2002).
25. A. M. M. Maciel et al., Phys. Rev. **C59**, 2103 (1999).
26. R. S. Mackintosh and N. Keeley, Phys. Rev. **C70**, 024604 (2004).
27. R. Raabe et al., Nature **431**, 823 (2004).

Is the optical model valid for the scattering of exotic nuclei?

J. Gómez-Camacho*, M. Alvarez*, A. Moro*, I. Martel[†], A. Sánchez-Benítez[†], D. Escrig** and M. J. G. Borge**

*Departamento de FAMN, Universidad de Sevilla, Aptdo 1065, 41080 Sevilla, Spain
[†]Departamento de FA, Universidad de Huelva, E-21819 Huelva, Spain
**Instituto de Estructura de la Materia, CSIC, Serrano 113bis 28006 Madrid, Spain

Abstract. The optical model, and its applications to elastic, inelastic and transfer reactions is reviewed. The applicability of the optical model to the collisions of exotic nuclei is discussed. The sensitivity of recent experimental data of elastic scattering of ^{6}He on ^{208}Pb to the characteristics of the optical potential is investigated.

Keywords: Optical model, Exotic nuclei, Coupled Channels, Continuum discretization, Elastic scattering, Break-up
PACS: 24.10.Ht, 24.10.Eq, 25.60.Bx, 25.70.Bc

INTRODUCTION

In the study of nuclei, the structure properties are often inferred from reaction measurements. This can only be done if the dynamics of the collision of the projectile with the target is properly described. For that, one would need to solve the many body problem, including all degrees of freedom. In practise, this is not possible, and one has to select certain states which are relevant, and define the proper effective interactions. From formal point of view, the optical model defines the expression of these effective interactions. However, from the practical point of view, the optical model introduces a phenomenological object, the optical potential, which is fitted to some experimental observables (usually the elastic scattering cross sections), and then it is used to describe other scattering observables (inelastic, transfer, etc). In this way, important structure parameters from the colliding nuclei, such as deformation lengths and spectroscopic factors, can be obtained. This procedure has been broadly successful for the description of the structure of stable nuclei. However, as we shall dicuss throughout this paper, it is not trivial that it should also be valid for the scattering of weakly bound exotic nuclei.

Let us recall the main features of the optical model. When two nuclei collide, the scattering wavefunction will have, in general, components of all possible reaction channels. The complete solution of this scattering wavefunction would require to solve a many body Schrödinger equation, containing the interaction $\mathcal{V}$, that will depend in general on the relative as well as on the internal coordinates of the colliding nuclei. This is not possible in general. However, if we restrict our description of the collision to some specific channels, characterized by the projector P, and $Q = 1 - P$ as the complementary projection operator, then the "model" wavefunction $P\Psi$ is the solution of a Schrödinger equation which contains an interaction given by

$$U(E) = P\mathcal{V}P + P\mathcal{V}Q\frac{1}{E^+ - QHQ}Q\mathcal{V}P. \qquad (1)$$

The formal derivation of this optical potential was due to Feshbach [1]. In this work, we will make use of Feshbach formalism, as presented by Satchler [2].

The optical model in elastic scattering

Let us first consider elastic scattering. Then, P projects on the ground state of projectile and target. The optical potential U is a function of the variables associated to the relative motion ($\vec{r}$ and $\vec{p}$, in general). The expression (1) of the optical potential is, in principle, of little practical use, because it would require to solve the scattering problem with all the relevant degrees of freedom. The success of the optical model relies on the fact that one is able to approximate the complicated non-local operator U by a local, L-independent complex function which can be written as the sum of a real and an imaginary potential

$$U(E) \simeq V(r,E) + iW(r,E). \tag{2}$$

It is by no means trivial that this can be achieved. However, the experience of several decades of studies of nuclear collisions indicates that, for stable nuclei, local optical potentials, obtained taking some usual analytic forms (usually Woods-Saxon), and fitting the parameters to reproduce elastic scattering data, give a sensible approach to the complicated object $U(E)$ of eq. (1).

The real part of the optical potential $V(r,E)$ is obtained in some cases by double folding of an effective nucleon-nucleon interaction over the densities of projectile and target. This procedure is justified when the "direct" term PVP dominates in eq. (1). The imaginary potential $W(r,E)$ comes from the "dynamic polarization" term $P\mathcal{V}Q\frac{1}{E^+ - QHQ}Q\mathcal{V}P$ and cannot, in general, be obtained by folding. The imaginary optical potential can be parameterized in terms of some analytic form (Woods-Saxon). The study of the fenomenology of nuclear collisions indicates that the imaginary potential vanishes at distances in which there is not a significant overlap of the nuclear densities. However, it has been found that there is an enhanced surface absorptions in the case of weakly bound projectiles [3].

One important aspect about the description of elastic scattering is the presence of strong absorption [2]. In the nuclear collisions of heavy ions, it is found that, when the angular momentum of the relative motion is below a certain critical value, which is called grazing angular momentum, there is a strong damping of outgoing waves. That means that the S matrix, for angular momenta well below the grazing, has very small values. This makes that the elastic scattering is not really sensitive to the values of the optical potential at distances well below the strong absorption radius, which is the distance of closest approach corresponding to the classical trajectory associated to the grazing angular momentum.

In strong absorption situations, the elastic scattering is mostly sensitive to the value of the optical potential at the strong absorption radius. For smaller distances, the optical potential may be very strong, but the S matrix tends to vanish. For larger distances the optical potential is very small (apart from the monopole Coulomb potential), and both direct and dynamic polarization potentials tend to vanish.

The real and imaginary parts of the optical potential (1) are, in general, energy dependent. Then, it can be shown that the energy dependence of the real and imaginary parts are linked by the expression

$$\mathfrak{R}(U(E) - U(E_0)) = \frac{\mathcal{P}}{\pi} \int dE' \frac{\mathfrak{I}(U(E') - U(E_0))(E_0 - E)}{(E - E')(E_0 - E')}. \tag{3}$$

Assuming that the complicated operator $U(E)$ can be substituted by the local, L-independent potentials given by eq. (2), the dispersion relation become

$$V(E,r) - V(E_0,r) = \frac{\mathcal{P}}{\pi} \int dE' \frac{(W(E',r) - W(E_0,r))(E_0 - E)}{(E - E')(E_0 - E')} \tag{4}$$

The validity of this relation has been investigated in detail, within the context of the study of the "threshold anomaly". This refers to a sharp maximum in the real potential, at energies around the Coulomb barrier, associated to a sharp increase imaginary potential. This effect has been observed in the scattering of ^{16}O+^{208}Pb [4], and in many other systems. In some systems, however, there is not a threshold anomaly, in the sense that there is not any sharp energy dependence of the optical potential. In any case, the dispersion relations (4) are fulfilled.

The optical model in inelastic scattering

The optical model can be applied to the description of inelastic scattering. In this case, the projector operator P includes the relevant states of the projectile and target. The only change is that both real and imaginary potentials in eq. (2) will become matrices $\mathbf{V}(\vec{r},E) + i\mathbf{W}(\vec{r},E)$, and they will depend explicitly on the direction of $\vec{r}$. A multipole expansion gives rise to the usual formfactors:

$$U(E) \simeq \mathbf{V}(\vec{r},E) + i\mathbf{W}(\vec{r},E) = \sum_{\lambda\mu}(\mathbf{V}_{\lambda\mu}(r,E) + i\mathbf{W}_{\lambda\mu}(r,E))Y^*_{\lambda\mu}(\hat{r}) \tag{5}$$

The determination of the real and imaginary formfactors $\mathbf{V}_{\lambda\mu}(r,E) + i\mathbf{W}_{\lambda\mu}(r,E)$ is not trivial. In the collective model, when the excited states included in the calculation can be considered in terms of surface oscillations, the formfactors can be obtained deforming the phenomenologic optical potentials. This procedure, plus some additional assumptions, leads to formfactors given by the derivatives of the optical potential multiplied by the deformation length operator $\delta_{\lambda\mu}$.

$$\mathbf{V}_{\lambda\mu}(r,E) + i\mathbf{W}_{\lambda\mu}(r,E) \simeq \delta_{\lambda\mu}\left(\frac{dV(r,E)}{dr} + i\frac{dW(r,E)}{dr}\right) \quad (6)$$

This expression may be justified for the real formfactors, assuming that the direct term $P\mathcal{V}P$ is dominant, but it is not at all justified for the imaginary formfactors. However, in general, one assumes that the deformation lengths of the real and imaginary potentials are the same, and in this way one obtains in general a reasonable agreement with inelastic scattering data. This procedure allows to extract the values of the matrix elements of the deformation length operator from the inelastic cross section data.

Alternatively, the formfactors can be obtained by folding nucleon-nucleon interactions with transition densities. Again, this procedure is reasonable for the real formfactors, assuming that the direct term is dominant, but it does not apply to the imaginary formfactor. Usually, what it is done is to multiply the real formfactors obtained from the folding procedure by a complex factor $N_r(E) + iN_i(E)$, and in this way all the formfactors in the coupled channels systems are expressed in terms of two parameters, which are a function of the energy, and should be linked by dispersion relations. This procedure is purely empirical, but it has been used successfully by several authors [5, 6].

The optical model in transfer reactions

The measurement of transfer cross sections plays a crucial role in the description of the single particle structure of nuclei. It allows, in principle, to measure the spectroscopic factors of the colliding nuclei. However, this requires a precise understanding of the reaction mechanism. The description of a transfer process implies two logical steps. First, one approximates the complicated many body problem of all the participant nucleons in terms of a three-body problem,

$$(A+C) + B \rightarrow A + (C+B) \quad (7)$$

where one has two "cores", A and B and a transferred particle C. Indeed, the fact that one reduced the many body problem to a three body problem implies that the interactions become complex, and will be given in general by an operator U given by eq. (1), where P projects in the internal states considered of the clusters A, B and C. This operator would be, in general, a three body potential, which is a function of the relative coordinates and momenta of the clusters. Note that there is a priori no reason for which this effective interaction should be written as the sum of two-body potentials which depend only on the relative coordinates or momenta of each pair of clusters $A+C$, $B+C$ and $A+B$. This will only be the case if the dynamic polarization terms in (1) is neglected. Even for the direct term $P\mathcal{V}P$, writing it as a sum of $A+C$, $B+C$ and $A+B$ potentials implies that one neglects the effects of the Pauli principle, because the presence of, say, particle A affects the states to which $B+C$ can scatter. Despite of these problems, what it is usually done is to take for the interaction the sum of the two-body optical potentials of $A+C$, $B+C$ and $A+B$, obtained (when possible) from the fit of elastic scattering data at some average energies E_{AC}, E_{BC}, E_{AB}.

$$U(E) \simeq U_{AC}(\vec{r}_{AC},E_{AC}) + U_{BC}(\vec{r}_{BC},E_{BC}) + U_{AB}(\vec{r}_{AB},E_{AB}) \quad (8)$$

Note that this approach has some consistency problems. In the incident channel, the system $A+C$ is bound. So, the energy E_{AC} is negative, and the interaction U_{AC} should be purely real, to be able to maintain an stationary bound state. In the incident channel, it would be reasonable to take the same energy per nucleon for the fragments as the one for the projectile, so that E_{BC} and E_{BA} are positive. Hence, the interactions U_{BC} and U_{BA} will be complex. However, in the final channel, $B+C$ is bound, E_{BC} is negative, and the interaction U_{BC} should be purely real, while the others are complex.

The second step is to solve the three body problem. This could be done, in principle, solving the Faddeev equations. However in practise this is more commonly done using a DWBA approach, the validity of which depends on how accurately is the exact three-body wave function described in terms of the distorted waves in the incident or outgoing channels. Thus, the DWBA approach, and also its extensions as the CCBA or the CRC approach, depend on whether the break-up components of the exact three-body wavefunctions affect significantly the transfer process.

Even assuming that the break-up components are not important for the transfer process, it is clear that the integrals that appear in the DWBA expressions depend on the values of the wavefunctions in the incident and outgoing channels at all distances, and not only the asymptotic part. The use of optical potentials to generate distorted waves ensures, at best, that the asymptotic behavior of the wavefunctions is correct, but it does not say much about the wavefunctions in the interior. Nevertheless, in situations of strong absorption, it could be argued that the only configurations that affect transfer are those in which the colliding nuclei just touch each other, and these should be well described if the asympotics is correct.

The success of the optical model

Despite the limitations discussed above, the optical model, with optical potentials obtained from elastic cross sections, has been very successful in allowing us to understand the mechanisms of nuclear collisions, as well as extracting structure information from the scattering data of stable nuclei. Some of the highlights of the optical model are the following:

- Elastic cross sections are reproduced with optical potentials which depend smoothly on the energy and the target mass.
- Deformation parameters are obtained from the fits to inelastic scattering.
- Spectroscopic factors are obtained from the fits to transfer reactions.
- Sub-barrier fusion enhancement is explained through coupled channels effects in the entrance channels.
- The sharp energy dependence in the elastic optical potential, where present (threshold anomaly), is explained as the effect of coupling to collective states or transfer channels.

Why does the optical model work so well? There is no clear answer to this question. However, the success could be related to the fact that the relative short range of the nuclear interaction, plus the presence of strong absorption makes that the relevant scattering observables are sensitive to a peripheral region of the configuration space. Hence, the optical model wavefunctions describing the reaction channels may be sufficiently accurate in the peripheral region.

THE OPTICAL MODEL IN THE COLLISIONS OF EXOTIC NUCLEI

The optical model, which is very successful to describe the collisions of stable nuclei, as previously discussed, may not be adequate to describe exotic nuclei. Exotic nuclei are, in general, less bound than stable nuclei. Hence, the dynamics of the collision may be much more affected by the excitation of the internal degrees of freedom.

In general, the optical potential defined by eq. (1) is applicable to all nuclei, stable or exotic. The difference relies on the fact that the role of the dynamic polarization component will be more important in the case of exotic nuclei, because the excited states not included in the calculation (described by the projector Q) have lower energies, and are more easily accessible. This means that non-local effects, as well as the energy and L-dependence of the optical potential can be more important in the case of exotic nuclei.

The success of the optical model for the description of elastic scattering relies on the fact that one is able to express the complicated non-local operator U by a local, L-independent complex function which can be written as the sum of a real and an imaginary potential. This is by no means trivial in the case of scattering of exotic nuclei. One can indeed fit a local, L independent potential to reproduce the elastic scattering of exotic nuclei at a given energy. However, these potentials may display unexpected radial and energy dependences.

The use of double folding interactions to obtain the real potential is adequate, assuming that the direct term $P\mathscr{V}P$ dominates the interaction. This assumption will not be adequate for exotic nuclei. Dynamic polarization effects should be taken into account. In the case of dipole coulomb excitation, an analytic expression of the polarization potential can be used to describe this effect [7, 8].

The relevance of strong absorption for exotic nuclei is also, a priori, less important for exotic nuclei than for stable nuclei. Exotic nuclei do not have a compact structure. It should be not expected that they will abandon the elastic channel completely when the distance of closest approach between the centers of the projectile and target gets below a certain value. Exotic nuclei, in their ground state, may be described as a combination of several configurations of their fragments. Some of these configurations will be strongly absorbed for a certain distance of the centers of mass, but not others. Thus, one should expect that absorption does not appear at some fixed distance. Instead, it could be spread over a range of distances, and hence of orbital angular momenta.

Let us consider inelastic phenomena. The use of the collective model, which is based on the existence of a well defined nuclear surface, may not be adequate to describe neutron-rich exotic nuclei, with extended neutron skins or halos. Folding potentials may not be adequate, for the same reasons described for elastic scattering. The main uncertainty, however, is related to the values of imaginary formfactors. There is no reason for which the imaginary deformation parameter of the optical potential should be equal to the real one, and neither of these should be simply related to the structure properties of exotic nuclei.

Microscopic approaches to the scattering of exotic nuclei

Break-up channels play an important role in the scattering of weakly bound nuclei. The explicit description of break-up channels can be done by means of some continuum discretization procedure. Let us consider some exotic nucleus P, which can be described as a weakly bound system $A+C$. It collides with a target B, and as a result many different processes may occur. The exotic nucleus may survive in the elastic channel, or go to some other bound state of the system $A+C$ (inelastic). The exotic nucleus may be broken, and then the fragments A and C will come out, with some angular and energy distribution (break-up). The fragment C may be transferred to the target, forming a compound system $B+C$.

One can consider explicitly the channels in which the fragments A, C and B remain in their ground state. In this situation, one explicitly considers the elastic and inelastic channels, transfer channels and break-up states. The optical potential will be given by an operator U which is given by eq. (1), where P projects on the ground state of the clusters A, C and B, and Q projects into the states in which these clusters suffer any excitation. It should be noticed that the optical potential is a three-body operator, depending on the three relative co-ordinates, and it may not be written, a priori, as a sum of optical potentials. However, as previously discussed, one might, with many reservations, assume that the interactions are given as a sum of optical potentials between the fragments (8).

When the projectile is weakly bound, it is important to consider explicitly the break-up channels. One possibility is to expand the complete three-body wavefunction in terms of the bound and continuum states of the $A+C$ subsystem. We will refer to this treatment of the continuum, which emphasizes on the $A+C$ continuum states, as "direct break-up". To do this consistently requires to consider that the interaction between the fragments $A+C$ is purely real $U_{AC}=V_{AC}$. The continuum discretization procedure would substitute the infinite range, non-normalizable continuum wavefunction of $A+C$ by normalizable "bins" or "pseudostates", which would in general have a large spatial extension. From these wavefunctions, one obtains the transition densities, which can be bound-bound, bound-continuum or continuum-continuum. Note that the bound-continuum, and specially the continuum-continuum transition densities have a very long range.

The usual approach to obtain the coupling formfactors between the ground state and the continuum states is called the cluster folding. It consists in folding the cluster optical potentials U_{AB} and U_{CB}, obtained from the elastic scattering of the fragments of the projectile with the target, with the very extense transition densities. Note that this procedure can lead to important uncertainties in the transition potentials, associated to the fact that the optical potentials U_{AB} and U_{CB} are very uncertain at small distances r_{AB}, r_{CB}. These short range contributions affect the coupling formfactors at larger projectile-target separations, due to the long range of the transition densities.

The direct break-up plus cluster folding approach has been used for deuteron scattering [9], and also for ^{6,7}Li scattering [5]. This approach may be adequate when the scattering energy is relatively high, and the relative momenta of the break-up fragments is small, compared with the momentum of the projectile. This approach is in principle inadequate to describe transfer, because the complex potential U_{CB} cannot hold bound states [9].

An alternative treatment of the continuum is to expand the complete three-body wavefunction in terms of the bound and continuum states of the $B+C$ subsystem. We will refer to this treatment of the continuum, which emphasizes the $B+C$ continuum states, as "transfer to the continuum". This treatment requires that the interaction between the fragments $B+C$ is purely real $U_{BC}=V_{BC}$. The bound and continuum states would be obtained solving the Schrödinger equation for the $B+C$ system. Then, the continuum states can be discretized. The interactions U_{AB} and U_{AC} can be taken as the corresponding optical potentials, and then transfer and break-up can be consistently calculated [10].

It is an open question whether "transfer to the continuum" is more adequate than "direct break-up", for the purpose of evaluating break-up cross sections. One would expect that when the fragment C is left with a small relative energy with respect to the target B, then transfer to the continuum would be a more adequate approach than direct break-up.

OPTICAL POTENTIALS OF ^{6}HE ON ^{208}PB

In this section we will present the preliminary analysis of experimental data taken at the CRC at Louvain la Neuve. We performed measurements of the elastic scattering of ^{6}He + ^{208}Pb at energies ranging from 14 to 22 MeV. We also measured the alpha particles coming from break-up. We have also considered the elastic scattering of this same system at 27 MeV, which was recently published [11].

We used a detector array consisting of CD Silicon telescopes, covering forward as well as backward scattering angles. Details of the experiment as well as the measured cross sections will be published separately.

The most remarkable qualitative feature of the elastic scattering data, at all the energies measures, was the absence of a rainbow, this is, of a maximum in the ratio of the elastic differential cross section to the Rutherford value. This feature, which was also seen at 27 MeV [11], is a clear indication of the presence of long range absorption. Note that the rainbow occurs because the attractive nuclear force starts to overcome the deflection produced by the repulsive Coulomb force. If there is long range absorption, the trajectories for which the rainbow occurs get partly absorbed, and the rainbow dissapears.

Part of the long range absorption is due to dipole Coulomb excitation. This effect can be explicitly taken into account including the dipole polarization potential [7, 8], which is completely determined from the $B(E1)$ distribution of the projectile. This polarization potential accounts for part of the long range absorption required to fit the data. However, as it is shown in Figure 1, an additional long range absorption is needed to fit the data. To interpret this figure, it should be considered that the strong absorption radius is around 11.0 fm. So, one can see that the imaginary potentials that fit the data extend well beyond this distance.

We have performed several fits to the data, using different families of optical potentials. The real potentials used in the calculations had a radius of 7.86 fm, and a diffuseness of 0.811 fm. The depth depends on the energy, and it is shown in the figures. Our calculations, which will be reported in detail elsewhere, indicate clearly that the real potentials that fit the data are well determined in the vicinity of the strong absorption radius. However, as we see in the figure, the imaginary potentials are well determined around 13.0 to 13.5 fm, at which the real potential is negligible. This fact should be taken into account when investigating the energy dependence of the optical potentials. The question about the presence of a threshold anomaly in the optical potential cannot be investigated, as for stable nuclei, looking at the energy dependence of the real and imaginary potentials at the strong absorption radius.

The nature of this long range absorption has to be investigated. The imaginary potential which arises of (1) is a non-local operator. It is possible that the long range found in the phenomenologic imaginary potentials shown in figure 1 is an indication of this non-locality. It seems reasonable to consider that the long range absorption is associated to the effect of break-up channels. A proper description of the break-up of ^{6}He requires to consider four-body continuum states, which is very difficult, although there is rapid progress in this field. We have performed preliminary calculations making use of a di-neutron model for ^{6}He, which is considered as an α-particle plus a di-neutron cluster. The calculations indicate that the coupling to break-up channels indeed generates the long range absorption shown in the elastic data.

We have also investigated the distribution of α-particles coming from break-up. Data seem to indicate that the neutrons are left with a very small energy with respect to the target. This indicates that the transfer to the continuum approach may be more adequate to describe some break-up observables.

SUMMARY AND CONCLUSIONS

We have reviewed the applicability of the optical model for the scattering of exotic nuclei. Following Feshbach formalism, we expect that an optical potential (1) should describe consistently all the relevant channels (elastic, inelastic, break-up, transfer) in the collision of exotic nuclei, as it was the case for the collisions of stable nuclei. However, we expect that the difference will be that, in the scattering of exotic nuclei, dynamic polarization effects will play a much more important role than for stable nuclei.

The crucial aspect of the practical use of the optical model in the collisions of stable nuclei is the use of local imaginary potentials, which are obtained from the fit to elastic scattering data, and are then used in reaction calculations. It is not clear that the same procedure will be valid for exotic nuclei. Non-locality effects, break up effects, as well as long range absorption produced by the loosely bound structure of exotic nuclei, can modify the relation between elastic and reaction observables which is implicitly assumed when local imaginary potentials are used in reaction observables.

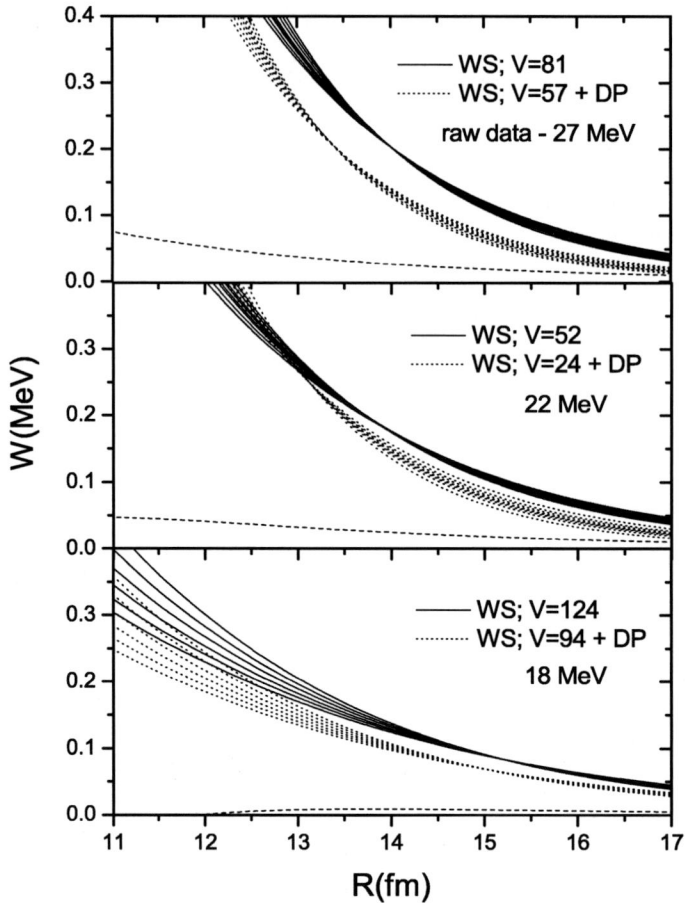

FIGURE 1. Imaginary potentials obtained from several fits to the data. The dashed lines correspond to the imaginary dipole coulomb polarization potential. Dotted lines correspond to the nuclear potentials needed to fit the data, once that the dipole polarization potentials are included. Full lines are the potentials needed to fit the data, without the explicit inclusion of dipole polarization effects.

The strong absorption argument, by which scattering observables are sensitive only to the optical potential around the strong absorption radius, is not applicable in the case of loosely bound exotic nuclei.

A preliminary analysis of experimental data of ^{6}He on ^{208}Pb at energies around the Coulomb barrier indicates that the optical potentials that fit the elastic scattering data show a long range absorption. This absorption is partly due to the Coulomb dipole polarizability, but there are other mechanisms generating absorption at relatively large separations of ^{6}He and ^{208}Pb. Microscopic calculations including break-up states indicate that the absorption mechanisms are associated to processes in which ^{6}He looses its weakly bound neutrons.

We consider that it is necessary to improve our understanding of the reaction mechanisms which are relevant in the collisions of weakly bound nuclei. For that purpose, one needs to have detailed and precise experimental data of elastic scattering and reaction cross sections of weakly bound systems, such as ^{6}He, for which the structure is reasonably well understood. The comparison of experimental data with the scattering calculation will indicate if the present approximation of the optical model, which is based on the use of local optical potentials between the fragments of the projectile and the target, is adequate for exotic nuclei. If this is not the case, then new concepts should be applied to correlate the different scattering observables, which do not rely on the use of local optical potentials. One such approach is the Uncorrelated Scattering Approximation [12, 13], in which elastic and break-up S matrices are obtained from the S-matrices of the fragments with the targets, and no optical potentials have to be used.

A better understanding of the relation between structure properties and reaction observables for exotic nuclei, for which the structure is reasonably well known, will facilitate to obtain reliable structure information for more exotic nuclei, from the measurement of reaction cross sections.

ACKNOWLEDGMENTS

This work has been supported by the spanish CICyT projects FPA2002-04181-C04-02, FPA2002-04181-C04-04, and FPA2003-05958.

REFERENCES

1. H. Feshbach, *Ann. of Phys.* **19**, *287* (1962).
2. G. R. Satchler, *Direct Nuclear Reactions*, Oxford University Press, New York, 1983.
3. G. Baur, R. Shyam, F. Rosel, and D. Trautman, *Phys. Rev.* **C21**, *2668* (1980).
4. J. S. Lilley, M. A. Nagarajan, D. W. Banes, B. R. Fulton, and I. J. Thompson, *Nucl. Phys.*, **A 4632**, 170 (1987).
5. Y. Sakuragi, M. Yahiro, and M. Kamimura, *Prog. Theor. Phys. Suppl.* **89**, *136* (1986).
6. J. Gómez-Camacho, M. Lozano, and M. A. Nagarajan, *Nucl. Phys.* **A440**, *543-556* (1985).
7. M. V. Andrés, J. Gómez-Camacho, and M. A. Nagarajan, *Nucl. Phys.* **A579**, *273* (1994).
8. M. V. Andrés, J. Gómez-Camacho, and M. A. Nagarajan, *Nucl. Phys.* **A583**, *817* (1995).
9. N. Austern, Y. Iseri, M. Kamimura, M. Kawai, G. Rawitscher, and M. Yahiro, *Phys. Rep.*, **154**, 125 (1987).
10. A. M. Moro, R. Crespo, H. García-Martínez, E. F. Aguilera, E. Martínez-Quiroz, J. Gómez-Camacho, and F. M. Nunes, *Phys. Rev.*, **C 68**, 034614 (2003).
11. O. R. Kakuee, J. Rahighi, A. M. Sánchez-Benítez, M. Andrés, S. Cherubini, T. Davinson, W. Galster, J. Gómez-Camacho, A. M. Laird, M. Lamehi-Rachti, I. Martel, A. Shotter, W. Smith, and J. Vervier, *Nucl. Phys.* **A728**, *339* (2003).
12. A. M. Moro, J. A. Caballero, and J. Gómez-Camacho, *Nucl. Phys.* **A695**, *143* (2001).
13. A. Moro, J. A. Caballero, and J. Gómez-Camacho, *Nucl. Phys.*, **A689**, 547c–550c (2001).

Coulomb Breakup for Spectroscopy

I.J. Thompson

Department of Physics, University of Surrey, Guildford GU2 7XH, U.K.

Abstract. In order to use Coulomb breakup as a spectroscopic tool, the accuracy of breakup model calculations needs to be examined. The coupled discretised continuum channels (CDCC) method is widely used, and I report on its convergence and accuracy. One confirmation of its convergence is that it reproduces analytic solutions where these are known, such as the elastic scattering from the Johnson special three-body model. The CDCC results are also compared with those from Bremsstrahlung, time-dependent and DWBA models, and comments are made about the correct description of one-step DWBA for Coulomb breakup. The convergence of CDCC at low beam energies is shown with respect to increasing model space, as well as evidence for the accuracy of the converged solution.

Keywords: Coulomb breakup, Bremsstrahlung, Adiabatic, Three-body, CDCC
PACS: 25.60.Gc, 25.70.De

USING COULOMB BREAKUP

For extracting spectroscopic information from exotic or halo nuclei, breakup is one of the important experimental techniques. Nuclear breakup, whether diffraction dissociation or stripping (elastic or inelastic breakup) can produce large excitation energies ε. Coulomb breakup, by contrast, is a more gentle process that is suitable for exciting structures at low lying energies with little momentum transfer. It is more selective in its angular momentum coupling, and hardly couples to the intrinsic spin degrees of freedom, but the precision with which we know the Coulomb potential allows absolute calibration of spectroscopic information obtained in this way. This is subject, of course, to having an accurate and reliable model of the mechanism of Coulomb breakup, along with good estimates of higher order components in the model that come from nuclear processes and/or from multistep effects.

In nuclear physics, however, there are a variety of models and techniques. There is no "universal" model for calculation of reaction rates, so only particular approaches depending on the reaction (beam) energy E, on the target (whether Coulomb or nuclear is dominant), and on the structures involved (whether the important degrees of freedom are deformation or breakup, etc). Calculation of reactions always requires knowledge of structures involved, as well as a model for the reaction. A simple mechanism may allow us to deduce structure (eg at high energy), but in general *both* structure and reaction models must be nontrivially combined to predict a reaction rate. This means that only if we have an *accurate* reaction model can we extract spectroscopic information, and decide whether a given structure model fails or prevails.

At high beam energy E compared with the excitation energies ε, some specific simplifications become possible. If the reaction is 'fast' in the sense that $E \gg \varepsilon$, then *adiabatic* (or *sudden*) approximations become more accurate, and if the adiabatic approach is combined with an *eikonal* approximation for the spatial scattering wave functions, then the Glauber theory may be derived. When the energy is not so high, coupled-channels methods may be needed. When the breakup channels are coupled together, we have what are called 'coupled discretised continuum channels' (CDCC) methods. These are computationally longer and more complex than adiabatic methods, so their accuracy needs to be checked and confirmed wherever possible. One useful check of CDCC methods is that they reproduce the results of simpler models that follow from specific physical approximations, if those same approximations are made equivalently in the CDCC analysis. This contribution examines some of these checks and confirmations, and also the convergence of the CDCC method itself. The CDCC convergence for nuclear mechanisms has already been examined [1], so here we consider especially the convergence for Coulomb mechanisms.

SOLUBLE THREE-BODY MODEL

Johnson [2] describes some of the adiabatic scattering models and applications that exist. Of particular interest now are those that use his Soluble Three-body Model presented in [3]. This soluble model for a three-body problem (of *core*, *valence particle* and *target* with masses m_c, m_v and m_t respectively) uses the adiabatic approximation in conjunction with neglecting the valence-target interaction V_{nt}. Setting $V_{nt} = 0$ is physically plausible for Coulomb breakup of neutron haloes, and for nuclear breakup if $V_{ct} \gg V_{nt}$, but in any case the important result is that the three-body problem now has an exact solution [3].

If $\alpha = m_v/(m_v + m_c)$ and $\chi_K(\mathbf{R})$ is the scattering of a point projectile from the core-target potential V_{ct}, then the exact solution of the three-body model under the adiabatic approximation is

$$\Psi_{\mathbf{K}}^{\mathrm{Ad}}(\mathbf{R},\mathbf{r}) = \phi_0(\mathbf{r})e^{i\alpha \mathbf{K}\cdot\mathbf{r}}\chi_{\mathbf{K}}(\mathbf{R} - \alpha\mathbf{r}) \tag{1}$$

for initial projectile bound state $\phi_0(\mathbf{r})$. (The derivation of this result [3] is valid for finite range forces, or with screened Coulomb forces. If, however, the screening radius is allowed to become very large, then it appears only as an overall phase factor in any matrix element and therefore does not affect any observable except at extremely forward angles). This solution may be used directly to predict elastic scattering, or may be used in a T-matrix integral to predict breakup cross sections.

For elastic scattering, the predicted angular distribution is the scattering cross section from the $\chi_K(\mathbf{R})$ solution multiplied by a form factor $F(Q)$ which is a Fourier transform of the projectile two-body density,

$$F(Q) = \int d\mathbf{r}|\phi_0(\mathbf{r})|^2 exp(i\mathbf{Q}\cdot\mathbf{r}) , \tag{2}$$

where $\mathbf{Q} = \alpha(\mathbf{K} - \mathbf{K}')$ is proportional to the elastic momentum transfer. Predictions of this model are shown in [3], where they are also compared with Glauber and more complete adiabatic solutions, as well as, for the case of ^{11}Be + ^{12}C at 49.3A MeV, with experiment.

For breakup, the $\Psi_{\mathbf{K}}^{\mathrm{Ad}}(R,r)$ cannot be used directly as it goes to zero for $\mathbf{r} \to \infty$ because of the $\phi_0(\mathbf{r})$ factor. However, it can be used iteratively within T-matrix integrals, such as the post-form integral for elastic breakup

$$T = \langle \chi^{(-)}(\mathbf{k}_c,\mathbf{R}_c)e^{i\mathbf{k}_v\cdot\mathbf{R}_v}|V_{cv}|\Psi_{\mathbf{k}_a}^{(+)}(\mathbf{r},\mathbf{R})\rangle , \tag{3}$$

giving

$$T^{\mathrm{AD}} = \langle \chi^{(-)}(\mathbf{k}_c,\mathbf{R}_c)e^{i\mathbf{k}_v\cdot\mathbf{R}_v}|V_{cv}|\phi_0(\mathbf{r})e^{i\alpha\mathbf{k}_a\cdot\mathbf{r}}\chi^{(+)}(\mathbf{k}_a,\mathbf{R}_c)\rangle . \tag{4}$$

which produces the factorised form

$$\begin{aligned} T^{\mathrm{AD}} &= \langle e^{i\mathbf{q}\cdot\mathbf{r}}|V_{cv}|\phi_0(\mathbf{r})\rangle \; \langle \chi^{(-)}(\mathbf{k}_c,\mathbf{R}_c)e^{i\alpha\mathbf{k}_v\cdot\mathbf{R}_c}|\chi^{(+)}(\mathbf{k}_a,\mathbf{R}_c)\rangle \\ &= \langle \mathbf{q}|V_{cv}|\phi_0\rangle \; \langle \chi^{(-)}(\mathbf{k}_c);\alpha\mathbf{k}_v|\chi^{(+)}(\mathbf{k}_a)\rangle . \end{aligned} \tag{5}$$

The momentum $\mathbf{q}$ appearing in the first term is $\mathbf{q} = \mathbf{k}_v - \alpha\mathbf{k}_a$, giving

$$\langle \mathbf{q}|V_{cv}|\phi_0^{l\mu}\rangle = D_l(q)Y_{l\mu}(\hat{\mathbf{q}}) = D(\mathbf{q}) , \tag{6}$$

the Fourier transform of the projectile vertex function $V_{cv}\phi_0$. If $V_{ct}(r_c)$ is a point Coulomb potential, then the second integral has an analytical form in terms of a Bremsstrahlung integral [4]. Note that in the expression (5) for T^{AD}, both $V_{ct}(r_c)$ and $V_{cv}(r)$ are included to *all* orders, so that this *cannot* be called a first order distorted wave approximation, but is of higher order.

The cross sections predicted by the Bremsstrahlung integral (5) of the adiabatic wave function may be directly compared with experiments. Such comparisons have been presented by Tostevin et al [5, 6] for forward-angle deuteron breakup on targets ranging from ^{12}C to ^{208}Pb. Experimental absolute magnitudes are well reproduced for heavier targets but, as expected, this Coulomb-only breakup model deteriorates with light targets when nuclear effects should become relatively more important even at forward angles. The Bremsstrahlung method has also been used for Coulomb breakup of ^{11}Be [7] and ^{6}He [8].

The Bremsstrahlung method also provides a reference case for other methods, including non-adiabatic methods such as CDCC or time-dependent methods. Since the excitation energies produced in Coulomb breakup are small compared with beam energies, we might expect the adiabatic approximation in the Bremsstrahlung method to be good. On the other hand, the long time scale for Coulomb dipole interactions may imply that approximations relying on even small energy differences may yet produce significant phase errors over the long duration of the reaction. Comparisons with experiments and other theories, therefore, are also significant for testing the adiabatic approximation itself.

CONSTRUCTING THE CDCC BASIS

The CDCC method consists of expanding the three-body continuum wave function by means of 'bins' constructed by averaging the continuum states of the cv system over small energy ranges. These bin wave functions are defined as

$$u_{\ell sj,[k_1,k_2]}(r) = \sqrt{\frac{2}{\pi N}} \int_{k_1}^{k_2} w(k) e^{-i\delta_k} u_{\ell sj,k}(r) dk, \qquad (7)$$

with δ_k the scattering phase shift for $u_{\ell sj,k}(r)$, the single-energy scattering wave function in the chosen potential $V_{cv}^\ell(r)$ which may be ℓ-dependent. The normalisation constant is $N = \int_{k_1}^{k_2} |w(k)|^2 dk$ for some weight function $w(k)$. These bin states are normalised $\langle u|u\rangle = 1$ once a sufficiently large maximum radius R_{bin} for r is taken. They are orthogonal to any bound states, and are orthogonal to other bin states if their energy ranges do not overlap. The phase factor $e^{-i\delta_k}$ ensures that they are all real valued for real potentials $V_{cv}^\ell(r)$. The weight function $w(k)$ may be taken as unity in the simplest case, but for resonances the choice of $w(k) = \sin\delta_k$ gives a more succinct summary of the resonance strength within one bin [9], and is much preferable to dividing a resonance into a dozen or more narrow bins [10].

There are two alternative methods of constructing CDCC basis states. One is to use the 'mid-point' method: to take not an average of the cv scattering wave function $u_{\ell sj,k}(r)$ over the bin as in eq. (7), but simply the scattering wave function $u_{\ell sj,k_m}(r)$ at the mid-point $k_m = (k_1 + k_2)/2$. This gives similar solutions for nuclear breakup [1], but we would not expect continuum-continuum couplings to converge with these non-square-integrable basis functions. A second method is to use a basis set the positive energy *pseudo-states* obtained by diagonalisation using extended Gaussian expansion methods. This method naturally puts resonant strength into one (or a few) pseudostates. It should converge to the same result as the standard CDCC method described above, but gives less flexibility to focus the accuracy of the basis on particular regions of the partial-wave continuum. When calculating fragment breakup cross sections, more work is needed to overlap the pseudo-states with the true breakup asymptotic forms $u_{\ell sj,k}(r)$, which overlap is given in the method of eq. (7) by simply the coefficients of $u_{\ell sj,k}(r)$ in that integral.

TESTING CDCC ACCURACY

The rms radius of a bin wave function increases as the bin width $k_2 - k_1$ decreases, approximately as $1/(k_2 - k_1)$, so large radial ranges are needed to include narrow bin states. If the maximum radius R_{bin} is not sufficiently large, then the bin wave functions $u_{[k_1,k_2]}$ will not accurately be normalised to unity by the factors given in equation (7). Since the missing normalisation comes at large distances, it is important to check the convergence of the CDCC basis for the maximum radii used in practice. The first check is that the correct Coulomb $B(E\lambda)$ distributions are generated, and then that the correct pure Coulomb excitation cross sections are obtained. At high energies, with a sufficient number of scattering partial waves, the first order results should agree with the Alder-Winther semiclassical theory [11], and higher-order results should agree with the Bremsstrahlung results of the previous section.

Our first check of the CDCC method consists of reproducing as accurately as possible the Johnson analytic solution (eq. 1) of the three-body problem. That is, in a calculation for example of $d + {}^{208}$Pb elastic scattering, the neutron-target potential is set to zero, and the coupled channels equations have asymptotic kinetic energies set the same as in the elastic channel. This Adiabatic-CDCC calculation *should* reproduce exactly the partial-wave S-matrix elements derived from the analytic solution. The comparisons reported in [12] show that convergence is obtained both for nuclear and Coulomb breakup mechanisms. The left part of Fig. 1 shows the real and imaginary analytic S_L partial wave S matrix elements from nuclear breakup as the solid curves, and then progressively more converged CDCC calculations. The right part shows the absolute errors in nuclear+Coulomb breakup CDCC models. We find in fact slightly more ready convergence for Coulomb breakup, as it is the partial waves $L \sim 15$ just inside grazing that converge the most slowly. Note that convergence in the adiabatic limit is slower than in the realistic case, because in the non-adiabatic realistic case the bins at higher energy have reduced kinetic energies, and therefore become cut off as their kinetic energies tend to zero.

COMPARISON OF BREAKUP FROM CDCC AND BREMSSTRAHLUNG METHODS

Fig. 2 (left) shows a comparison of semiclassical and CDCC Coulomb *breakup* for ^{19}C incident on ^{208}Pb at 1273 MeV lab energy, with a $2s$ ground state at with 0.530 MeV binding energy. Here, the assumptions about the reaction

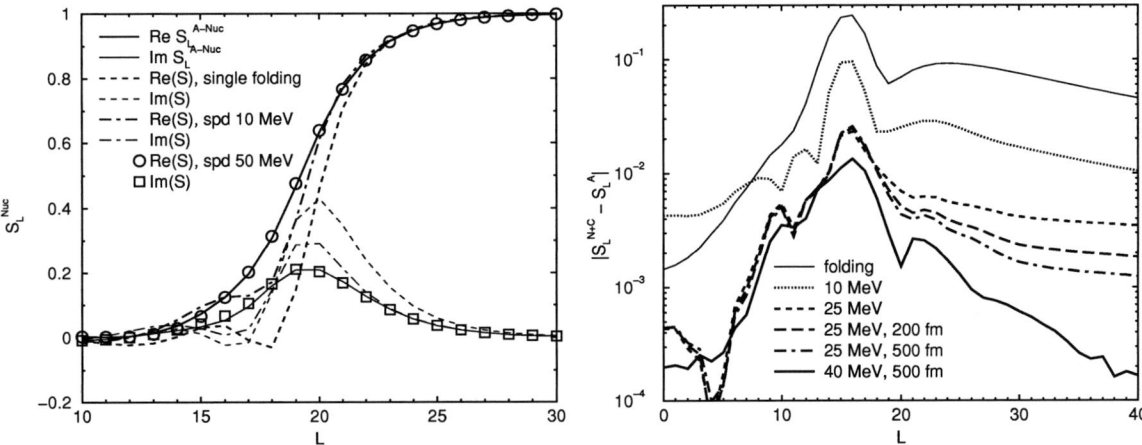

FIGURE 1. Deuteron scattering on ^{208}Pb at 50 MeV. Comparison of (left) the exact nuclear-only adiabatic S-matrix elements S_L^{A-nuc} with those of CDCC calculations including $\ell=0,1,2$ in bins up to 10 and 50 MeV, and also those from a single-folding calculation. TRight: Absolute errors of Coulomb+nuclear CDCC S-matrix elements in comparison with the adiabatic limit. All calculations, except the folding curve, use, unless otherwise indicated, $\ell \leq 2$ and R_m=100 fm.

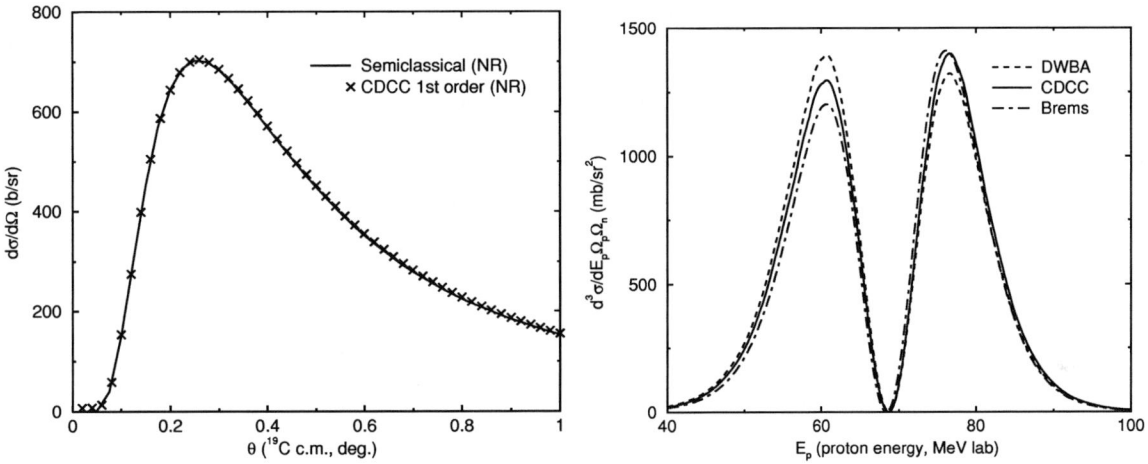

FIGURE 2. (left) Pure Coulomb semiclassical and CDCC breakup for ^{19}C + ^{208}Pb at 1273 MeV to 5 breakup bins over the continuum energy range 0 – 0.5 MeV. (right) Comparison of CDCC, Bremsstrahlung and DWBA calculations for purely Coulomb deuterons on ^{12}C at 140 MeV, and $\theta_p = \theta_n = 0°$.

mechanism have been kept in close agreement as possible in the two models. The CDCC calculation was reduced to a first-order plane-wave Born approximation for the transition, and both models use non-relativistic kinematics: excellent agreement is then achievable.

A second breakup example in Fig. 2 (right) now compares CDCC, Bremsstrahlung and DWBA calculations in the case of pure Coulomb breakup of deuterons on ^{12}C at 140 MeV to forward angle $\theta_p = \theta_n = 0°$ fragment detectors. Here the breakup couplings are expected to be weak, but we still see small differences between the models: the Bremsstrahlung gives a larger higher-order correction to the lower-energy peak than does the CDCC in comparison with the DWBA result. A comparison of ^{11}Be + ^{208}Pb breakup calculations (not shown here, but discussed also in the next section) shows that the Bremsstrahlung calculations systematically predict about 15%-25% less breakup than the CDCC method for this projectile.

A third breakup example is of an even large discrepancy between Bremsstrahlung and other predictions. We consider the breakup of ^{8}Li into a neutron and ^{7}Li, with a separation energy of 2.03 MeV. Fig. 3 shows the parallel momentum distributions from four different theoretical models. The adiabatic Bremsstrahlung prediction is significantly less than

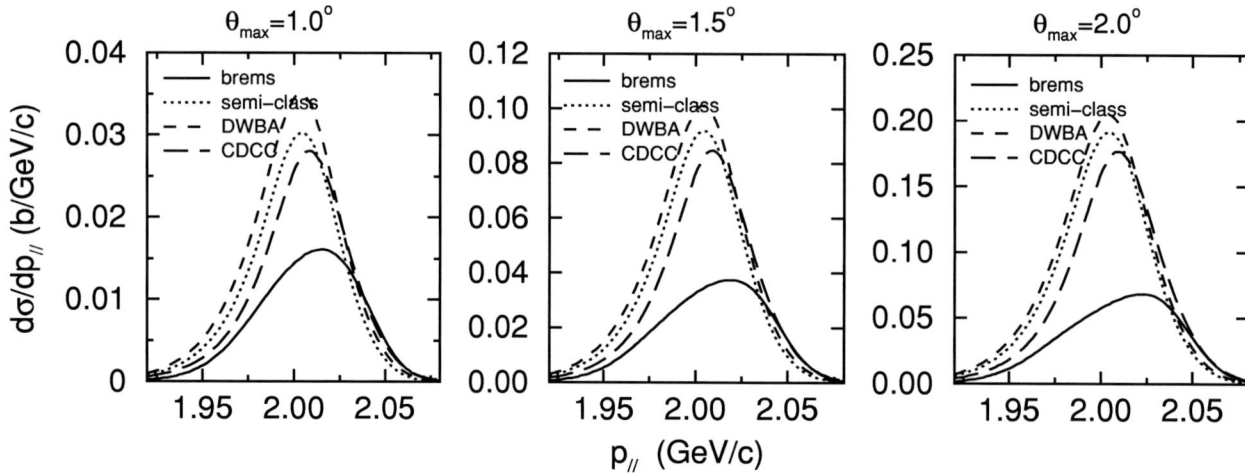

FIGURE 3. Calculated parallel momentum distributions of ^{7}Li from the breakup of ^{8}Li on ^{208}Pb at 44A MeV for different transverse angular apertures.

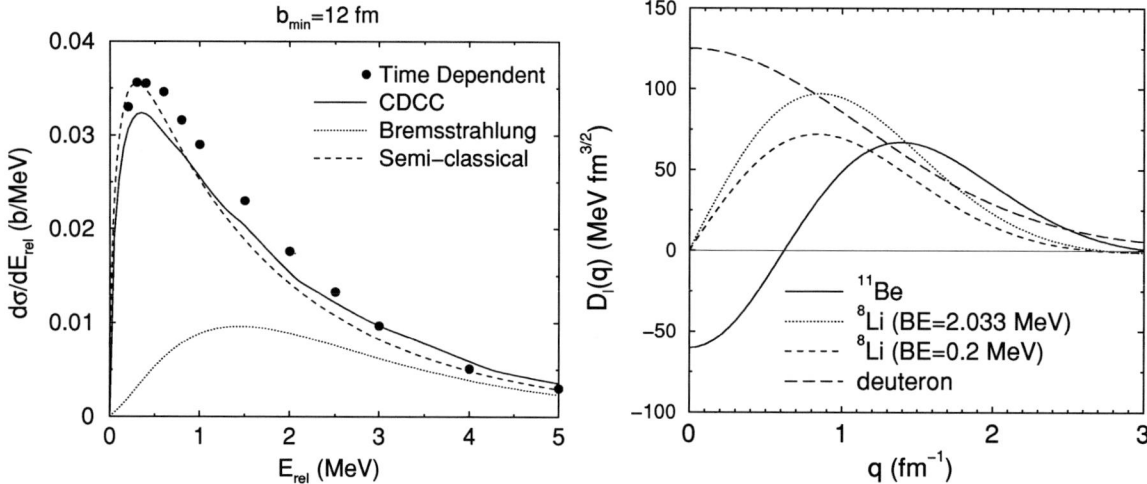

FIGURE 4. Left: Excitation energy spectrum for the breakup of ^{8}Li on ^{208}Pb at 44A MeV with minimum impact parameter $b_{min} = 12$ fm. Right: Vertex functions $D_l(q)$ for ^{11}Be, ^{8}Li and deuterons, and also for a hypothetical ^{8}Li with weaker binding.

the other three models, and the same pattern is seen in the excitation energy spectrum of the left part of Fig. 4. This figure also shows with filled circles the result of a preliminary time-dependent calculation by the Brussels group [13], and this is closer to the semiclassical and CDCC curves than to the Bremsstrahlung result. We do not show angular distributions since these are not yet produced by the time-dependent method. Ideally, some experimental results would enable us to determine which of these models was the more accurate.

DWBA

If we reflect on what is different for the ^{11}Be and ^{8}Li breakup, compared with deuteron breakup where the models all seem to agree reasonably, we focus on the vertex function $D_l(q)$. Fig. 4 (right) shows how the vertex function is different for the three nuclei. The zero-range approximation [14] is equivalent to treating this function as a constant, and this may be a tolerable approximation for deuterons. For wave functions with a node as in ^{11}Be, or wave functions with $l > 0$ as in ^{8}Li, it is clear that $D_l(q)$ varies significantly with q, and zero-range models cannot be accurately used. It appears that more rapid variation of $D_l(q)$ with q is perhaps associated with the discrepancy between the Bremsstrahlung predictions and those of the other methods.

We remark in passing that various approximate treatments have been proposed to allow for the variation of $D_l(q)$. One prescription is in eq. (5) to replace $\mathbf{q} = \mathbf{k}_v - \alpha \mathbf{k}_a$ by $\mathbf{k}_v - \alpha \mathbf{K}_a$ for some fixed momentum $\mathbf{K}_a$ to be determined. Sometimes such prescriptions are called 'local momentum approximations' (LMA) [15, 16, 17], but it is clear that $\mathbf{K}_a$ is a *global* effective momentum and not a *local* momentum at all. Sometimes they are also seen as good approximations to a finite-range calculation of the post form DWBA for breakup [17]. However, we see that setting $\mathbf{K}_a = \mathbf{k}_a$, the incident momentum, the theory becomes identical to a T-matrix integral using the adiabatic analytic three-body wave function which includes potentials to all orders. This implies that the so-called LMA-DWBA models are not first order distorted wave Born approximations either.

For true finite-range post-form DWBA calculations of Coulomb breakup, the only known method is that of Zadro [18, 19]. In [19] he shows that the true Coulomb DWBA is systematically distinct from both the Bremsstrahlung and the so-called LMA-DWBA predictions, which are very close to each other. This reinforces our conclusion of the previous paragraph. When DWBA and Bremsstrahlung results are compared for ^{11}Be breakup, the DWBA is found to be approximately 25% larger than the Bremsstrahlung curve. When CDCC and time-dependent predictions are again compared with these, the CDCC curve is near to the (true) DWBA, and the early time-dependent result of [20] is nearer to the Bremsstrahlung prediction. We cannot by this comparison conclude which is correct, only that both the Bremsstrahlung and time-dependent theories require larger spectroscopic factors to fit experimental breakup data: nearer unity rather than a factor in the range 70%–80% which is expected from most structure models. A replication of the result of [20] with the more recent accurate time-dependent methods would be useful contribution to this debate.

Finally, we remark on the role of three-body asymptotic Coulomb wave functions, for which some of the leading terms in an asymptotic expansion have been found recently [21]. The time-dependent and CDCC methods both take a more systematic approach, in which the dissociation amplitude is calculated taking into account the coupling of the different channels. The results of [21] are useful for an estimation of the possible effects of three-body final-state Coulomb interactions on the breakup cross section at small scattering angles and relatively high energies, by resorting to some approximations that are admissible under such conditions. For more quantitative conclusions, both the coupling of the different channels and an exact, and not only asymptotically valid, final-state three-body wavefunction needs be considered. For reactions at low energies, a more detailed numerical solution is unavoidable.

LOW-ENERGY CDCC CONVERGENCE

We now examine the convergence of the CDCC method for low-energy reactions, at and below the Coulomb barrier. We examine the predictions for the Notre Dame experiment [22, 23, 24], where ^{8}B was incident on ^{58}Ni at 25.8 MeV beam energy. Predictions have been made in DWBA [25, 26] and CDCC methods, the later for two-body [27] and then later three-body [28] kinematics.

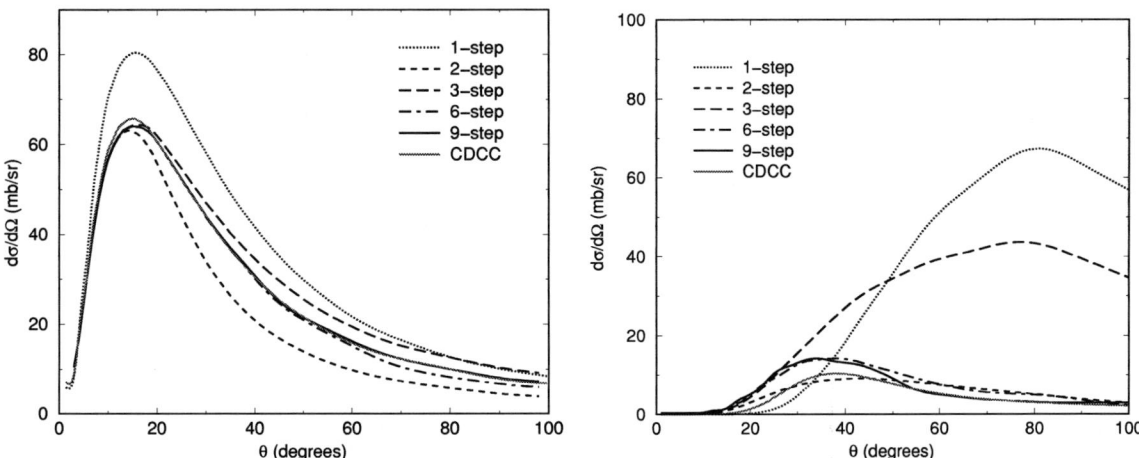

FIGURE 5. The differential cross section obtained for the multi-step breakup of ^{8}B into s and p-wave bins using Padé acceleration. On the left for Coulomb and on the right for nuclear mechanisms.

Fig. 5 shows the convergence of the Coulomb and nuclear mechanisms by themselves, where the graphs show the total breakup cross sections as a function of the centre of mass angle of the fragments ^{8}B*. The left hand Coulomb plot shows a pronounced destructive interference at second order for all angles. This can be traced to the interference between second order E1 and first order E2 transitions, as both lead from a p-wave ground state to the p-wave continuum. This is a manifestation of the Barkas effect [29] known in the stopping of charged particles in matter, and also shown by Esbensen to occur not only there [30] but also [31] in nuclear reactions over a range of breakup energies. The right-hand nuclear plot shows that first-order prior DWBA calculations of nuclear breakup grossly overestimate the breakup at these low energies. This suggests that there is considerable nuclear distortion of the ^{8}B projectile that appears as breakup in first order, but is in fact virtual and reversible so that the all-order CDCC real breakup is much smaller. It would be interesting to compare these results with those from a finite-range post-form nuclear breakup model, if we knew in practice how to calculate the required integrals.

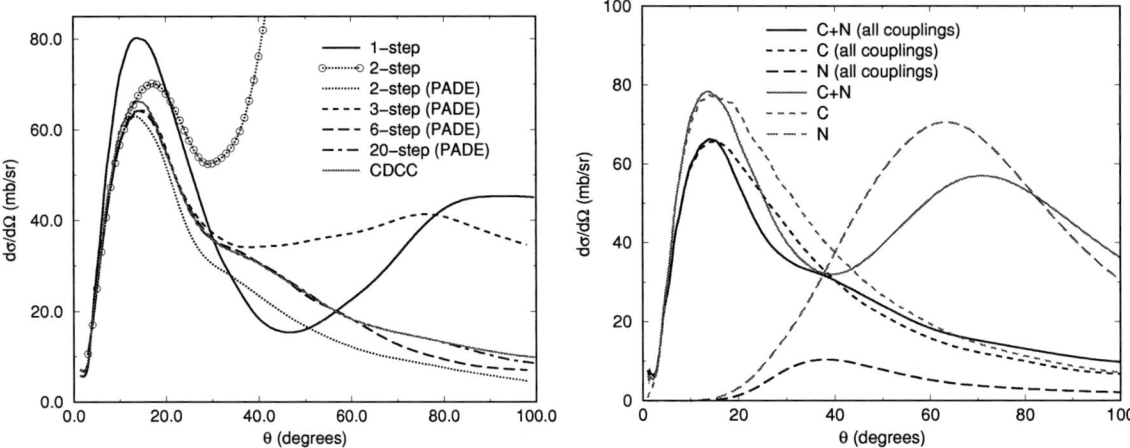

FIGURE 6. Left: The differential cross section obtained for multi-step Coulomb and nuclear breakup of ^{8}B into s and p-wave bins: comparing the full CDCC calculation with Born approximations. Right: Comparing the CDCC results with and without continuum-continuum couplings

When both nuclear and Coulomb mechanisms are included, Fig. 6 left, iterative solutions of the coupled equations become even less likely to converge, but the non-perturbative CDCC solutions are stable and apparently realistic. The right hand plot of Fig. 6 shows the effect of omitting continuum-continuum couplings between bins. The large nuclear peak reappears, and the Barkas interference effect disappears, indicating again that first-order calculations missing important physical processes.

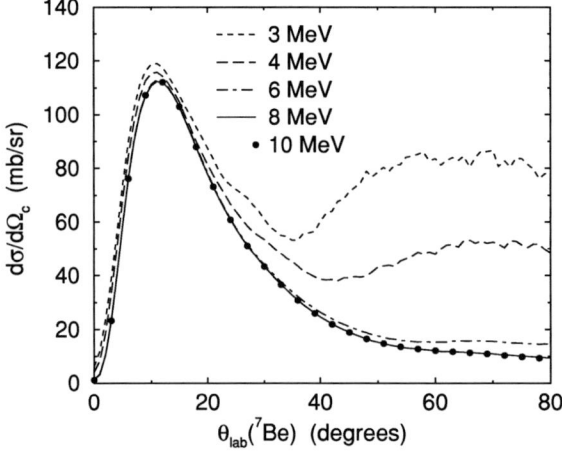

FIGURE 7. Convergence of the calculated laboratory frame ^{7}Be cross section angular distribution following the breakup of ^{8}B on ^{58}Ni at 25.8 MeV as a function of the maximum proton-^{7}Be relative energy included in the calculation.

The CDCC calculations must always be checked for convergence with respect to increasing the maximum energy of the continuum bins. Fig. 7 considers 5 separate calculations with maximum bin energies of 3 to 10 MeV. The total breakup cross section as a function of the centre of mass angle of the fragments ^{8}B* (not shown) are almost identical, but Fig. 7 shows that the *fragment* angular distributions are not at all converged unless a full range of energy bins is included in the calculation. The cause of this is that when there are insufficient bin energies, the total nuclear breakup is approximately constant, but it 'piles up' in the bin of highest energy. This pileup gives a spuriously energetic fragment distribution, one that disappears if more bins are included in the calculation, and the flux allowed to distribute naturally across all bins.

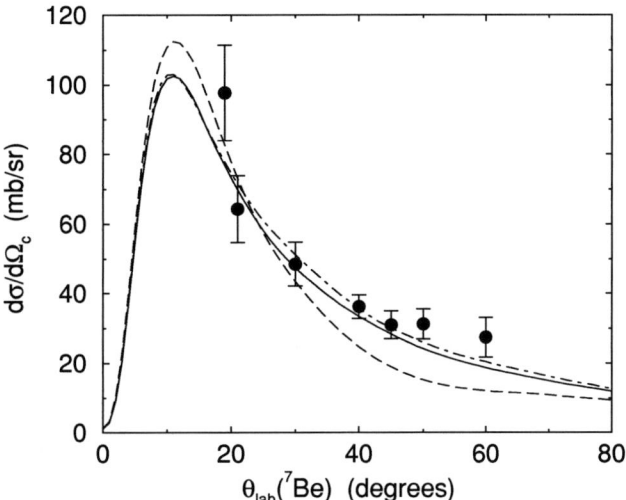

FIGURE 8. The calculated laboratory frame ^{7}Be cross section angular distribution following the breakup of ^{8}B on ^{58}Ni at 25.8 MeV. The long-dashed curve is the $E_{max} = 10$ MeV, $\ell \leq 3$, $q \leq 2$, calculation from Fig. 7. The solid curve includes $q = 3$ multipole terms while the dot-dashed curve includes both $q = 4$ and $\ell = 4$ effects.

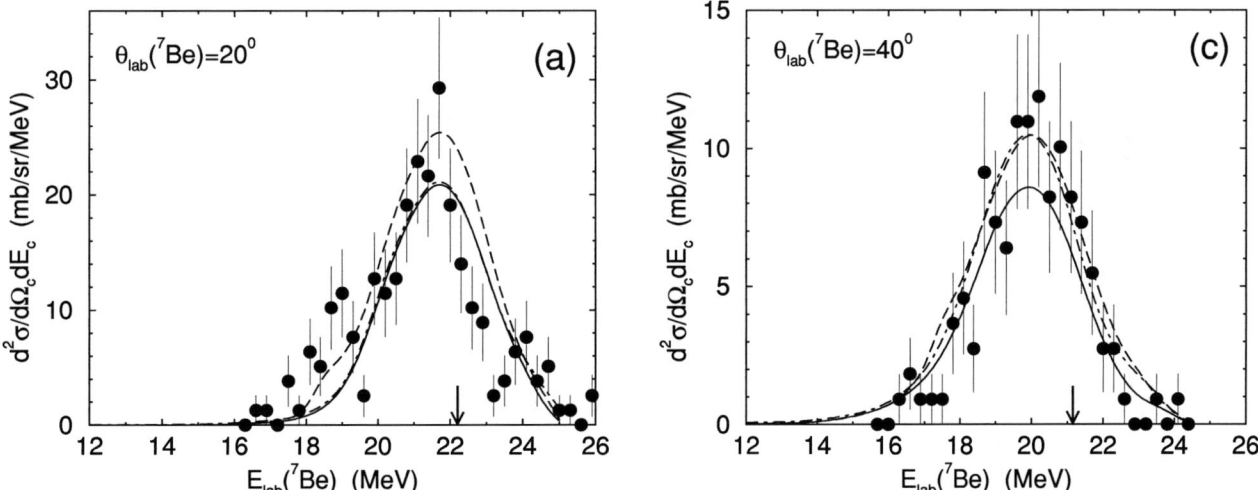

FIGURE 9. Calculated laboratory frame ^{7}Be cross section energy distributions following the breakup of ^{8}B on ^{58}Ni at 25.8 MeV for the laboratory angles indicated. The calculations use the EB (solid) and Kim (dashed) models for the proton-^{7}Be interaction and the BG proton-target interaction. The dot-dashed curves use the EB proton-^{7}Be interaction and the VG proton-target interaction. The arrows on the energy axis indicate 7/8 of the ^{8}B energy for elastic scattering at each laboratory angle. The experimental data are from Ref. [24].

The convergence of the calculations with multipole order, and also with the included continuum partial waves, is shown in Fig. 8. Here the long dashed curve is the result shown in Fig. 7, converged with respect to excitation energy, with $q \leq 2$ and $\ell \leq 3$. The solid curve includes also the effects of the $q = 3$ multipole couplings for $\ell \leq 3$. The dash-dot curve is a calculation where $q = 4$ multipole couplings and the $\ell = 4$ breakup partial wave are included, and we see that the additional effects are small.

Fig. 7 shows experimental breakup data from [24], and more detailed ^{7}Be energy distributions are compared with experimental data in Fig. 9. We conclude that at these low energies, CDCC calculations with all orders of couplings between bound and continuum states can be made to converge, and converge towards predictions in agreement with experiment.

CONCLUSIONS AND FUTURE

The results shown here suggest that the CDCC method is good for both the nuclear and Coulomb breakup of two cluster nuclei, of which halo nuclei are an important subset. The CDCC gives a non-adiabatic treatment of Coulomb breakup, and includes multistep effects caused by all final-state interactions with a certain but large reaction volume. Finite-range and recoil effects are naturally included, and Coulomb and nuclear breakup amplitudes both approach convergence as long as we pay attention to the large radii and partial-wave limits needed. With new computing resources, good calculations become now easily feasible.

Development is currently underway for extensions of the CDCC method. Rotational excitations of one of the projectile fragments are recently being calculated by the MSU theory group, and will be compared with existing Glauber calculations [32] for the same three-body problem.

The Kyushu group [33, 10] has developed CDCC for three-body projectiles, to solve the four-body breakup problem using the basis of positive energy pseudo-states. Another group at Sevilla is developing a similar method using transformed harmonic oscillators to generate the set of pseudostates. These can again be compared with existing four-body Glauber calculations for elastic scattering [34, 35], but will be applicable over a much wider range of beam energies, including energies around the Coulomb barrier.

ACKNOWLEDGMENTS

Support from EPSRC grant GR/T28577 is acknowledged.

REFERENCES

1. R.A.D. Piyadasa, M. Kawai, M. Kamimura and M. Yahiro, Phys. Rev. C **60** (1999) 044611
2. R.C. Johnson, *this conference*.
3. R.C. Johnson et al, PRL **79** (1997) 2771
4. A. Nordsieck, Phys. Rev. **93** (1954) 785.
5. J.A. Tostevin et al., Phys. Lett. **B424** (1998) 219.
6. J.A. Tostevin, S. Rugmai and R.C. Johnson, Phys. Rev. C **57** (1998) 3225.
7. P. Banerjee, I.J. Thompson and J.A. Tostevin, Phys. Rev. C **58** (1998) 1042.
8. P. Banerjee, J.A. Tostevin and I.J. Thompson, Phys. Rev. C **58** (1998) 1337.
9. I.J. Thompson, Comp. Phys. Rep., **7** (1988) 167
10. M. Kamimura, *this conference*.
11. A. Winther and K. Alder, Nucl. Phys. **A319** (1979) 518
12. B. Cross, R.C. Johnson, I.J. Thompson and J.A. Tostevin, to be submitted.
13. D. Baye, private communication.
14. G. Baur, *this conference*.
15. R. Shyam and M. A. Nagarajan, Ann. Phys. **163** (1985) 265.
16. P. Braun-Munzinger, H. L. Harney, and S. Wenneis, Nucl. Phys. **A235** (1974) 190.
17. R. Chatterjee, P. Banerjee and R. Shyam, Nucl. Phys. **A675** (2000) 477
18. M. Zadro, Phys. Rev. C **66** (2002) 034603
19. M. Zadro, Phys. Rev. C **70** (2004) 044605
20. V.S. Melezhik and D. Baye, Phys. Rev. C **59** (1999) 3232
21. E.O. Alt, B.F. Irgaziev, A.M. Mukhamedzhanov, Phys. Rev. C **71** (2005) 024605
22. J. von Schwarzenberg *et al.*, Phys. Rev. **C53** (1996) R2598
23. V. Guimarães *et al.*, Phys. Rev. Lett. **84** (2000) 1862
24. J.J. Kolata *et al.*, Phys. Rev. C **63** (2001) 024616
25. F.M. Nunes and I.J. Thompson, Phys. Rev. C **57** (1998) R2818

26. R. Shyam and I.J. Thompson, Phys. Rev. C **59** (1999) 2645
27. F.M. Nunes and I.J. Thompson, Phys. Rev. C **59** (1999) 2652
28. J.A. Tostevin, F.M. Nunes and I.J. Thompson, Phys. Rev. C **63** (2001) 024617
29. W.H. Barkas, W. Birnbaum, and F.M. Smith, Phys. Rev. **101** (1956) 778
30. H. Esbensen *et al.*, Phys. Rev. B **18** (1978) 1039
31. A. Volya and H. Esbensen, Phys. Rev. C **66** (2002) 044604
32. P. Batham, I.J. Thompson and J.A. Tostevin, submitted to Phys. Rev. C.
33. T. Matsumoto, E. Hiyama, K. Ogata, Y. Iseri, M. Kamimura, S. Chiba, and M. Yahiro, Phys. Rev. C **70** (2004) 061601
34. I.J. Thompson, J.S. Al-Khalili, J.A. Tostevin and J.M. Bang, Phys. Rev. **C47** (1993) R1364
35. J.S. Al-Khalili, J.A. Tostevin, and I.J. Thompson, Phys. Rev. **C54** (1996) 1843

Three-body decay of many-body resonances

A.S. Jensen*, D.V. Fedorov*, H.O.U. Fynbo* and E. Garrido[†]

*Department of Physics and Astronomy, University of Aarhus, DK-8000 Aarhus C, Denmark
[†]Instituto de Estructura de la Materia, CSIC, Serrano 123, E-28006 Madrid, Spain

Abstract. We use the hyperspherical coordinates to describe decay of many-body resonances. Direct and sequential decay are described by different paths in the distances between the particles. We generalize the WKB expression for the α-decay width to decay of three charged particles. Decay mechanisms and resonance structures are computed in coordinate space. The energy distributions of the particles after decay are discussed. Moderate s-wave scattering lengths prefer decay via corresponding virtual state possibly leaving unique fingerprints of this reminiscence of the Efimov effect in the decay of excited states. Numerical illustrations are resonances in ^{6}He, ^{12}C, ^{17}Ne.

Keywords: resonance structure, decay mechanisms, final state energies, Efimov effect
PACS: 21.45.+v, 31.15.Ja, 25.70.Ef

MOTIVATION

Resonances can be populated directly in reactions or for example by beta-decay from an initially populated state in a neighboring nucleus. The subsequent decay of the resonance carries information about the resonance structure and the decay mechanism. This could be interpreted as a reaction mechanism and then fall within the scope of this workshop. In any case a number of reactions do proceed via population of a resonance, and series of measurements are available and many more are on the way [1, 2]

The recent and the planned experiments, where only a few clusters (two, three or four) are present in the final state, are both accurate and kinematically complete. It is a challenge to understand such reactions proceeding via resonance decay. The interpretation must relate the initial structure, the process and the final state properties. We would like to understand the process in details, i.e. in medical terms explain the anatomy of the decay. The dissecting instruments are the tools from theoretical few-body techniques.

We restrict ourselves to three-body decay of a many-body (more or less stable) resonance. The preceding formation process is assumed not to influence the decay and we are left with precisely three particles in the final state. Then at large distances we have inevitably a three-body problem. At smaller distances when the particles are close together forming the resonance structure we have a problem involving more particles except in the special cases of three-body resonances. The attempt in this contribution is to assume that the three-body structure is decisive for the process, the partial three-body width of the resonance, and the relative energy distribution of the particles in the final state.

This is analogous to α-decay where a many-body state emits an α-particle [3]. Preformation before entering the confining barrier is decided at the corresponding small distances but everything else is determined by the "large-distance" properties of the two-body problem of the α-particle and the daughter nucleus. In the second section we briefly describe the method employed, the concept introduced, and which approximations we use. In the third section we generalize the WKB expression for the α-decay width to three-body decay. We also discuss the competition between direct and sequential decay via intermediate resonance structures. In the fourth section we discuss widths and resonance structure obtained from only the lowest adiabatic potential for a number of nuclear states. In section five we give preliminary results for the energy distribution obtained with several of the lowest adiabatic potentials. Finally in section six we study the effect of a low-lying virtual state in one of the two-body subsystems.

METHOD AND CONCEPT

We use the adiabatic hyperspherical expansion method to compute the necessary three-body quantities [4]. The masses, charges, coordinates and momenta are denoted m_i, eZ_i, $\mathbf{r}_i$, and $\mathbf{p}_i$, $i=1,2,3$, where e is the (positive) unit charge. The Jacobi coordinates are $\mathbf{r}_{ij} = \mathbf{r}_i - \mathbf{r}_j$ and $\mathbf{r}_{k,ij} = \mathbf{r}_k - \mathbf{R}_{ij}$, where $\mathbf{R}_{ij} = (m_i \mathbf{r}_i + m_j \mathbf{r}_j)/(m_i + m_j)$. The

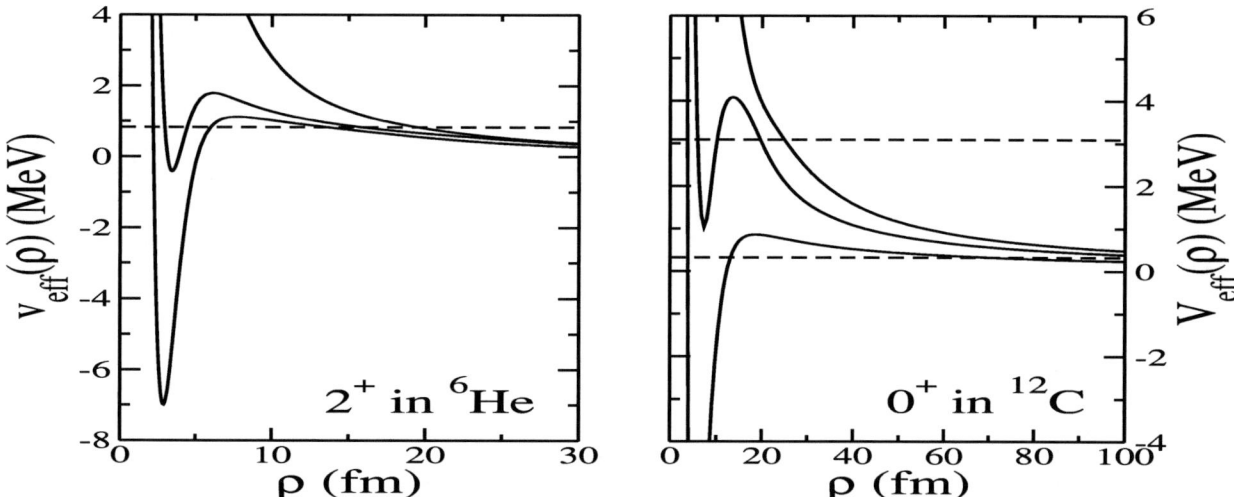

FIGURE 1. The lowest adiabatic potentials as functions of ρ. Left side shows the 2^+ states in ^{6}He (^{4}He + n + n). The resonance energy (horizontal line) and the width are 0.824(25) MeV and 0.113(20) keV. Right shows the 0^+ states in ^{12}C ($\alpha + \alpha + \alpha$). For the lowest two unbound states the energies (horizontal lines) and widths are, 0.369(10) MeV and 8.5(1.0) eV, 3.0(3) MeV and 3.0(7) MeV.

hyperspherical coordinates consist of the hyperradius ρ defined by $\rho^2 = \sum_{i<j} m_i m_j \mathbf{r}_{ij}^2/(mM)$ ($M = m_i + m_j + m_k$, m is an arbitrary mass), and five dimensionless angles collectively denoted Ω. The particles interact pairwise through two-body potentials $V_{ij}(r_{ij})$. If necessary a short-range three-body potential depending on ρ is added.

Energies and wave functions of the resonances are computed by use of the complex scaling method where $\mathbf{r}_j \to \mathbf{r}_j \rho \exp(i\theta)$, i.e. the coordinates are rotated in the complex plane by an angle θ [5]. This is particularly simple for the hyperspherical coordinates, because angles then remain unchanged and only the hyperradius changes, i.s. $\rho \to \exp(i\theta)$.

The procedure to obtain both bound states and resonances is then to fix ρ and solve the Faddeev equations in angular space. The solutions correspond to a discrete set of eigenvalues due to the definition of the angles within finite intervals. The set og angular eigenfunctions is complete and used as a basis for expansion of the total wavefunction. The expansion cofficients, the radial wavefunctions, are functions of ρ obeying a coupled set of radial equations. The radial solutions then provide wavefunctions and complex eigenvalues. Bound states energies are real, and resonance energies have real and imaginary parts where the width is minus half of the imaginary value. All these quantities are independent of the rotation angle θ.

To be specific we show examples in fig.1 for the lowest adiabatic potentials as function of hyperradius. The total wavefunction is then a linear combination of terms corresponding to each of these potentials. All these coefficients depend on ρ. Different decay paths lead from small to large distances. They are quantum mechanically connected by the radial wavefunction. Classical paths can be defined in the underlying multidimensional space. Some lead directly to the three-body continuum where all particles are far apart from each other. Some lead via two-body resonances or perhaps virtual states, where two particles are close and the third is far away, to the final state of three particles all far from each other. The radial wavefunctions contain the information of the weights on the angular wavefunctions where each represent a specific asymptotic behavior. Combining these classical paths with the proper ρ-dependent phases is equivalent to the full quantum mechanical solution as for Feynmann paths.

To focus on the concept we first imagine that only the lowest potential contributes. For a many-body resonance the short-distance behavior is by definition an effective potential. This part is initially obtained within the three-body model and with the related two-body interactions. However, it may be totally wrong especially clearly seen when the resonance differs very much from three-body structure, i.e. has N-body characteristics. This possibility is analogous to the description of α-decay in terms of an α-nucleus potential which also is unreliable at short distances. The α-decay width is nevertheless computed by knocking rate times tunneling probability modified by a preformation factor which describes the probability of having an α-particle at the edge of the barrier [3].

We generalize this concept to three-body decay where the distance between α-particle and daughter nucleus is replaced by the hyperradius [9]. The underlying structure of the three-body system could be well described by the angular wavefunction for the lowest adiababtic potential but many potentials could as well be necessary. Although one radial dimension and one potential look very similar to two-body decay we should appreciate that the concept is very different for three particles due to the underlying angular structure.

Three quantities characterize the resonance. The position, the real part of the eigenvalue, is mostly sensitive to the depth of the minimum times the square of the radius. For a many-body resonance there may not even be a minimum in the three-body hyperradial potential. Then the three-body potential is added. In the extreme we can use an attractive square well potential creating a minimum within a reasonable hyperradius and producing a three-body resonance with the desired real part of the energy. From then on everything is technically three-body computations.

The second charcateristic quantity is the width of the resonance which is most sensitive to the barrier shape (height and width). The width is approximately given by the tunneling probability as for α-decay. Adjusting the real part of the energy by use of a three-body potential could also modify the barrier for example by changing the range of this interaction. In principle both real and imaginary part of the eigenvalue could in this way be adjusted to desired values. This would only leave the third characteristic quantity, i.e. the energy distribution of the particles after the decay. The large distance properties of the potential and the corresponding resonance wavefunction are then crucial.

We shall discuss results from three different approximations. First we generalize the α-particle model and give analytical estimates for the partial decay width of three charged particles. Second we give results from the use of only the lowest adiababtic potential, and finally we use several adiabatic potentials.

SCHEMATIC MODEL

The path from small to large ρ-values can be defined by specifying how the distance between pairs of particles, r_{ij}, increase as function of ρ. Let us here assume scaling, s_{ij}, independent of ρ corresponding to a proportional increase of an initial triangle with the three particles at the corners, i.e. $r_{ij} = \rho s_{ij}$. The definition of ρ then implies the constraint

$$mM \equiv \sum_{i<k} m_i m_k s_{ik}^2 \,. \tag{1}$$

The total Coulomb potential arising from the interaction of the three pairs is then

$$V(\rho) = \sum_{i<k} \frac{Z_i Z_k e^2}{r_{ik}} = \frac{1}{\rho} \sum_{i<k} \frac{Z_i Z_k e^2}{s_{ik}} \tag{2}$$

for constant scaling. For sequential decay, where particle 3 is emitted and particles 1 and 2 are left in an intermediate configuration of energy E_{12}, the Coulomb potential is given by

$$V(r_{13}) = E_{12} + \frac{Z_1 Z_3 e^2}{r_{13}} + \frac{Z_2 Z_3 e^2}{r_{23}} = E_{12} + \frac{(Z_1 + Z_2) Z_3 e^2}{r_{13}}, \tag{3}$$

where the distances are related by $r_{13} \approx r_{12}$. The intermediate two-body structure could correspond to a resonance or maybe results from an s-wave attraction keeping the particles together.

As soon as one effective potential is responsible for the decay we can derive simple estimates by use of the WKB tunneling transmission T, i.e.

$$T = \exp(-2S) \,, \quad S = \frac{1}{\hbar} \int_{\rho_0}^{\rho_t} d\rho \, \sqrt{2m(V(\rho) - E)}, \tag{4}$$

where E is both the total energy and the kinetic energy of the particles after separation, ρ_0 and ρ_t are the classical turning points where $V(\rho_0) = V(\rho_t) = E$.

The results for the above Coulomb potentials are [9, 7]

$$S = \frac{\pi}{2} \sum_{i<k} \frac{Z_i Z_k e^2}{s_{ik}} \sqrt{\frac{2m}{\hbar^2 E}} \geq \frac{\pi e^2}{2} \sqrt{\frac{2}{\hbar^2 EM}} \left(\sum_{i<k} (Z_i Z_k)^{2/3} (m_i m_k)^{1/3} \right)^{3/2} \equiv S_{min}, \tag{5}$$

$$S_{12,3} = \frac{\pi}{2} (Z_1 + Z_2) Z_3 e^2 \sqrt{\frac{2\mu_{12,3}}{\hbar^2 (E - E_{12})}}, \tag{6}$$

TABLE 1. Energies and widths for the two lowest 0^+ resonances in ^{12}C, and the $3/2^-$ and $5/2^-$ resonances in ^{17}Ne obtained from the schematic model. Experimental results are compared with direct, sequential and WKB (for one potential) estimates. In the third row, the number in brackets corresponds to sequential decay through the 2^+ state in ^{8}Be. All the energies and widths are given in MeV.

	J^π	Γ_{WKB}	$\Gamma_{sch.}^{(Dir.)}$	$\Gamma_{sch.}^{(Seq.)}$	$E_{exp.}$	$\Gamma_{exp.}$
^{12}C	0_1^+	$6 \cdot 10^{-5}$	$\sim 10^{-6}$	$\sim 10^{-4}$	0.37	$8 \cdot 10^{-5}$
	0_2^+	0.4	0.2	0.7 (0.3)	3.0 ± 0.3	3 ± 0.7
^{17}Ne	$\frac{3}{2}^-$	$3.6 \cdot 10^{-12}$	$\sim 10^{-9}$	—	0.34	$< 2.5 \cdot 10^{-11}$
	$\frac{5}{2}^-$	$1.3 \cdot 10^{-10}$	$\sim 10^{-5}$	$\sim 10^{-5}$	0.82	$> 3 \cdot 10^{-10}$

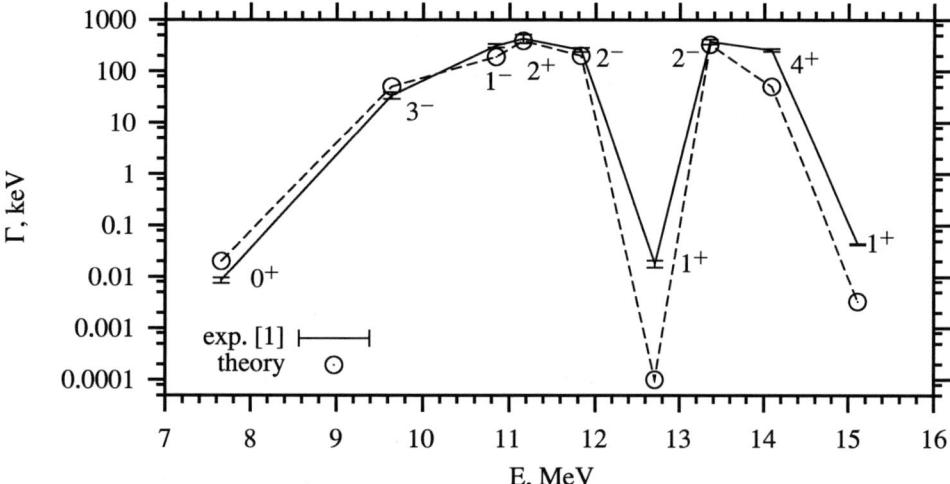

FIGURE 2. The widths of the excited states in ^{12}C. The angular momentum and parity are given at the points. The lines are only to connect experimental and computed points, respectively. The theoretical results are for one adiabatic potential.

where the relevant reduced mass for sequential decay is denoted $\mu_{12,3}$. Each of the three different particles could be emitted first, and possibly even via different two-body stepping stones. The lower limit for S is obtained by minimizing with respect to s_{ik}. The corresponding optimum path is defined for the minimizing scaling parameters $s_{ik}^3 m_i Z_j = s_{jk}^3 m_j Z_i$. The energies of the particles are determined for this optimum path to be

$$E_k = E_{total} \left(1 + \left(\frac{m_k Z_k^2}{m_i Z_i^2}\right)^{1/3} + \left(\frac{m_k Z_k^2}{m_j Z_j^2}\right)^{1/3} \right)^{-1}, \qquad (7)$$

where E_{total} is the total energy distributed among the three particles.

Comparison of S_{min} and $S_{12,3}$ indicates whether direct or sequential decay is preferred, i.e. if $S_{12,3}$ is smaller or larger than S_{min} then sequential or direct decay is preferred, respectively. If $\frac{E - E_{12}}{E}$ is sufficiently small then direct decay wins while sequential decay is favored by large $E_3 \equiv E - E_{12}$, small m_3 and small Z_3.

To test the estimates we show in table 1 the results for a few measured resonances. The calculated results are within the experimental limits for ^{17}Ne. The sequential estimate for the lowest ^{12}C state is close to the measured value while the second state is found to be too small. This may be related to the apparent inconsistency in the measured position and width [9].

A more systematic test case is the set of measured excited states in ^{12}C. They are shown in fig. 2. The non-monotonous behavior can never be reproduced by the schematic model, which only depends on energy. The neglect of the short-range interactions and possibly also angular momentum effects are crucial in these cases.

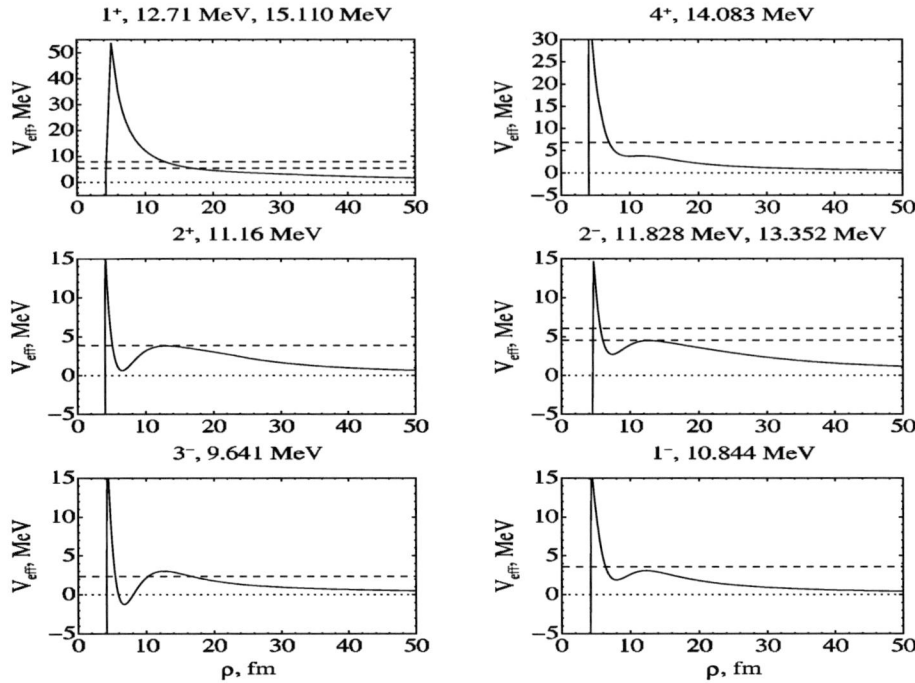

FIGURE 3. The lowest adiabatic potentials for a number of excited states in ^{12}C. Angular momentum, parity and energy are given above the figures. The horizontal lines are (apart from zero) the energies of the resonances. The vertical line is at the inner turning point.

RESONANCE STRUCTURE

The schematic model can clearly not reproduce the series of ^{12}C data. The better approximation is to use the lowest adiabatic potential corresponding to the given angular momentum and parity as the examples shown in fig. 3. The hyperradial dependence varies a lot with angular momentum and parity. Some barriers even have double hump structure due to the long-range effect of the short-range interaction. This comes about because two particles can remain close together for a while even with increasing ρ. This effect would be enhanced by two-body resonances. The width is estimated by the WKB tunneling probability through the barriers. With the same inner turning point for all states we compare in fig. 2 the results with the measured values. The agreement is rather striking although closer inspection reveals substantial deviations.

The structure of the dominating components (lowest angular eigenfunction) of two 0^+ resonances is seen in fig. 4. At small distances the lowest resonance is essentially only s-waves whereas the second resonance is an equal mixture of s and d-waves. At larger distances, below and outside the barrier, the second resonance becomes more and more dominated by s-waves while the first resonance mixes in more and more of the higher angular momenta. This "dynamic" behavior could reflect different decay mechanisms.

The structure of these resonances are shown in fig. 5. Already at moderate ρ the lowest wavefunction has a strong peak at very small distance between two of the α-particles. Another peak at larger distances is also seen but it is entirely due to the symmetry of the total wavefunction. This peak is for two other close-lying α-particles. The decay seems to pass a configuration similar to ^{8}Be in the 0^+ ground state, i.e. sequential decay. The other 0^+ resonance has only one broad peak for intermediate distances between the α-particles. This could be interpreted as direct decay to the continuum without passing an intermediate ^{8}Be state.

The 2^+ resonance in ^{6}He can be approximated as a three-body problem with only short-range interactions. The partial wave decomposition of the lowest angular eigenfunction is shown in fig. 6 in the two different Jacobi coordinate systems. At small distance within the attractive pocket of the lowest adiabatic potential the dominating component is two $p_{3/2}$ neutron-α states coupled to 2^+. As ρ increases this structure quickly changes in this lowest angular eigenfunction reaching an equal distribution of the three different possible p-wave couplings. This behavior is mirrored in the other Jacobi coordinates. At small distance the two neutrons are distributed in relatively small amounts of s,

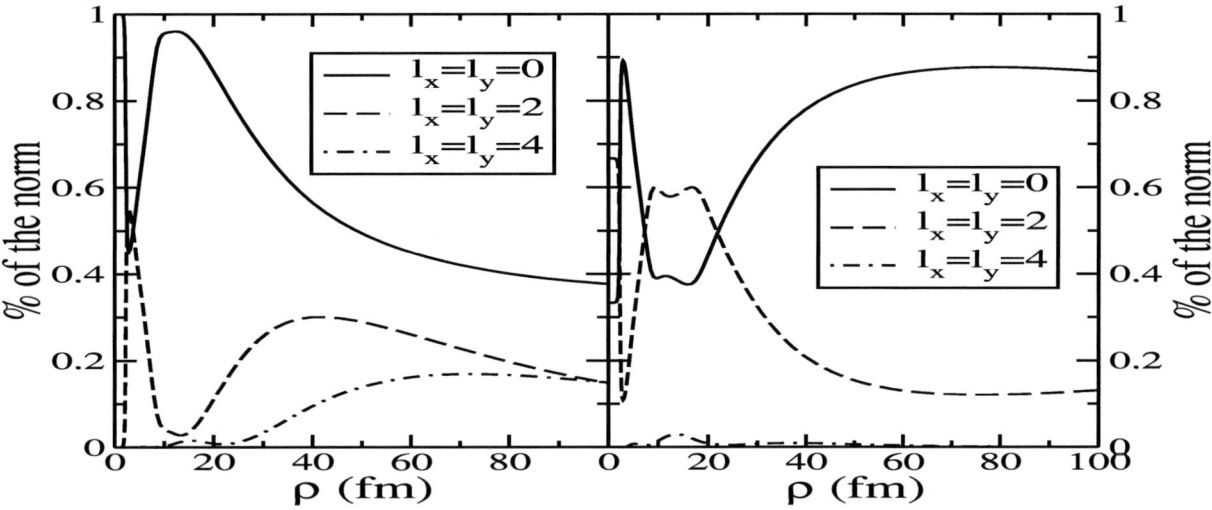

FIGURE 4. The percentage of the norm for the two lowest 0^+ resonances for different components in the first (left) and second (right) adiabatic potential as function of ρ for ^{12}C(0^+).

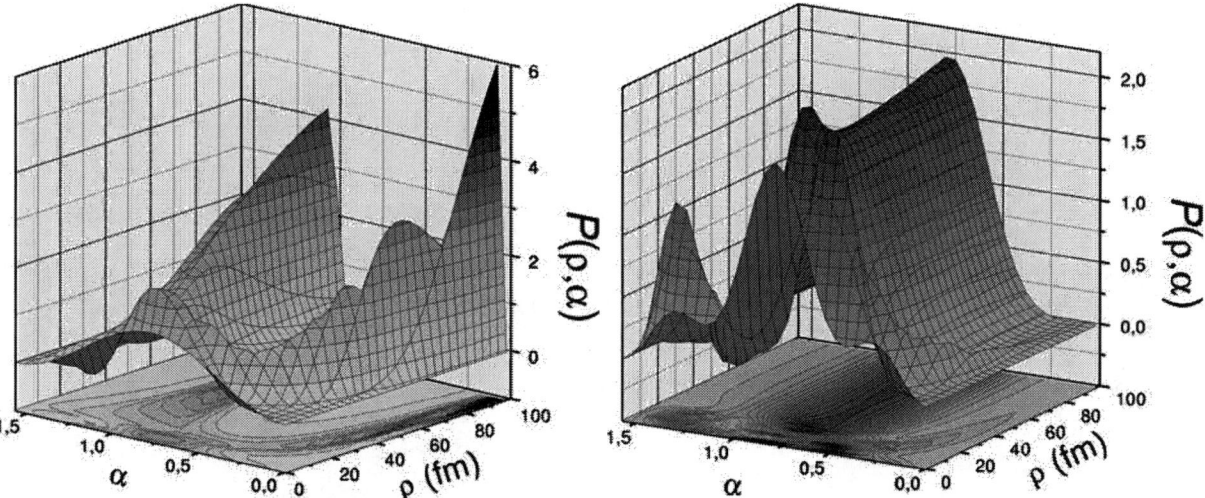

FIGURE 5. The probability distribution for the cases in fig. 4 as function of hyperradius ρ and α related to the distance between two α-particles by $r_{\alpha\alpha} \propto \rho \sin\alpha$.

p and d-waves. The largest components are $s_{1/2}$ and $p_{3/2}$-states. As ρ increases the s-wave takes over completely corresponding to the qual distribution of all p-waves in the other coordinate system.

The probability distribution for ^{6}He(2^+) is shown in fig. 7 in both Jacobi coordinates. Only a broad peak at intermediate distances between neutron and α-particle is seen. This structure is reflected in a relatively broad peak at intermediate distances beteen the two neutrons. If anything this indicates a direct decay mechanism into the continuum.

ENERGY DISTRIBUTION

The resonance structures must eventually produce an energy distribution of the particles after the decay. This observable has to be established as a large distance property of the resonance wavefunction. Several asymptotic structures are possible simultaneously, each with an amplitude corresponding to the relative probability. In computations the resonance wavefunction is obtained as a linear combination of adiabatic angular wavefunctions with weights depending on ρ.

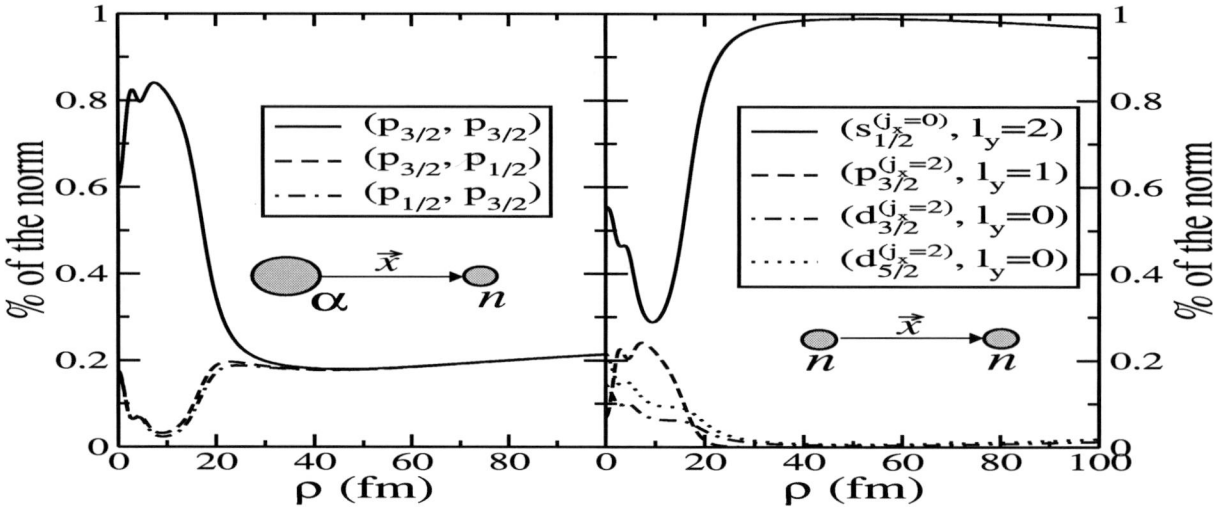

FIGURE 6. The same as fig. 4 for ^{6}He(2^+) in two different Jacobi coordinate systems. Left and right x refer to the neutron-α and the two-neutron systems and y to the corresponding center of masses relative to the third particle. We give the angular momenta as $\ell_j^{(j_x)}$, where j_x is the angular momentum quantum number obtained by coupling of ℓ_x and the neutron spin of $1/2$.

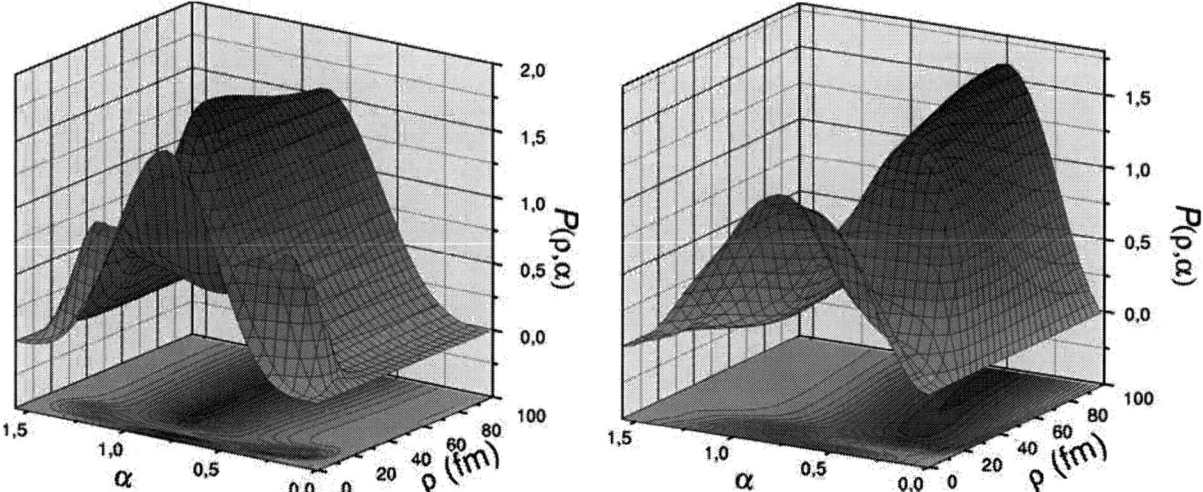

FIGURE 7. The same as fig. 5 for the cases in fig. 6. The neutron-α (r_{nc}, left) or and the neutron-neutron (r_{nn}, right) distances are described by α, i.e. $r_{ik} \propto \rho \sin \alpha$.

The energy of the outgoing particle is proportional to $\cos^2 \alpha$ where $\rho \sin \alpha / \sqrt{2}$ is the distance between the two remaining particles. To be more precise these relations should be stated in momentum space corresponding to the Fourier transform of the wavefunction. The probability of finding the system with a given energy (a given $\cos^2 \alpha$) is then the probability divided by the derivative of $\cos^2 \alpha$, i.e. $\sin(2\alpha)$.

For bound states and resonances at small and intermediate distances only the lowest adiabatic potentials contribute. The asymptotic limit of the adiabatic eigenvalues is the hyperharmonic spectrum for short-range interactions. The lowest of these wavefunctions could then play an important role. We show in fig. 8 the energy distribution obtained from each of the three lowest ($K = 2$) rotated wavefunctions corresponding to ^{6}He(2^+). For s or d-waves between the two neutrons the peak is at relatively large or small α-particle energy, respectively. When the angular momentum 2 is evenly distributed the peak is in the middle.

The radial solution for several adiabatic potentials provide the relative weights. In fig. 9 we show the results for a large ρ-value of 100 fm when the three lowest adiabatic potentials are included in a full computation. We recognize the schematic picture from the $K = 2$ contributions in fig. 8. The lowest adiabatic potential with neutron-neutron s-waves

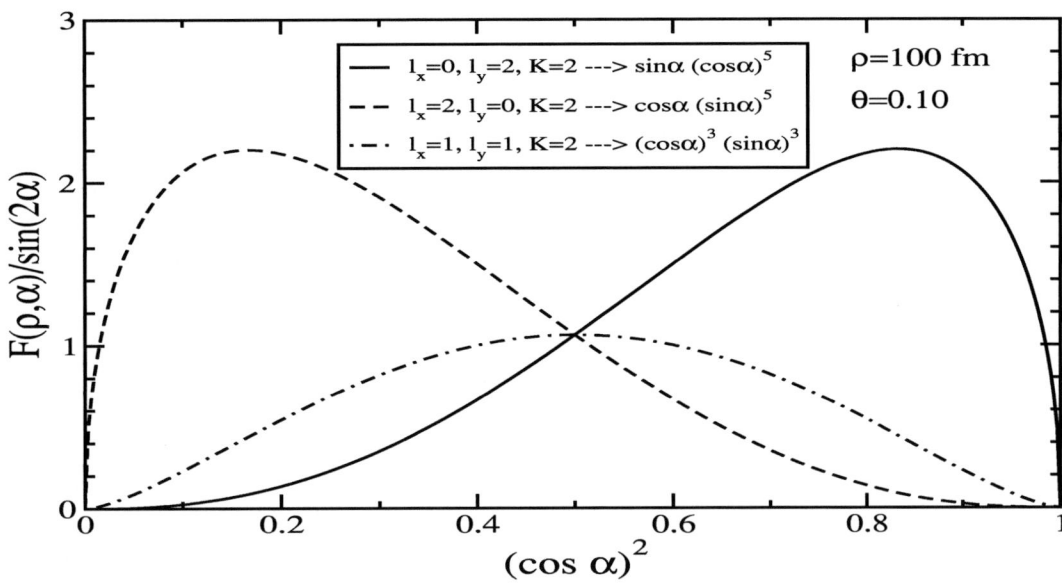

FIGURE 8. The energy distribution for the lowest hyperharmonic wavefunctions related to ^{6}He(2^+), i.e. relative angular momenta between neutrons (l_x) and their center of mass and the α-particle (l_y). The rotation angle is 0.10.

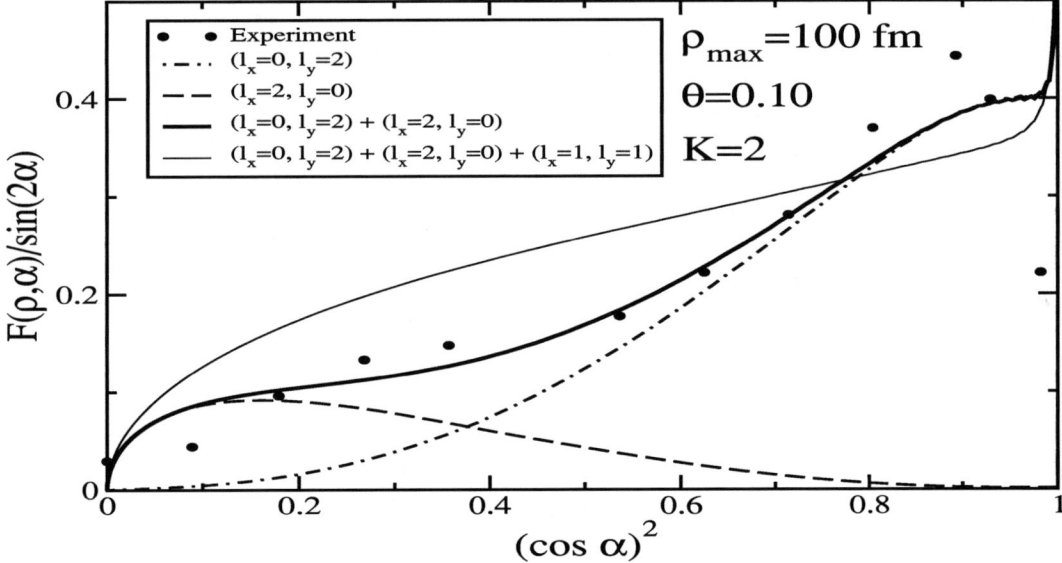

FIGURE 9. The energy distribution for the lowest adiabatic angular wavefunctions. The points are measurements from [8].

gives the peak at large α-energy and with the largest weight. The small weight on the d-wave contribution again gives the peak at small energy. Together they add up to a distribution resembling the measured one. Unfortunately further addition of the contribution from the remaining states approaching $K = 2$ for large ρ produces a distribution far from the measurement. This indicates that a coupling is included wrongly or a peculiar cancellation from the so far not included higher-lying states.

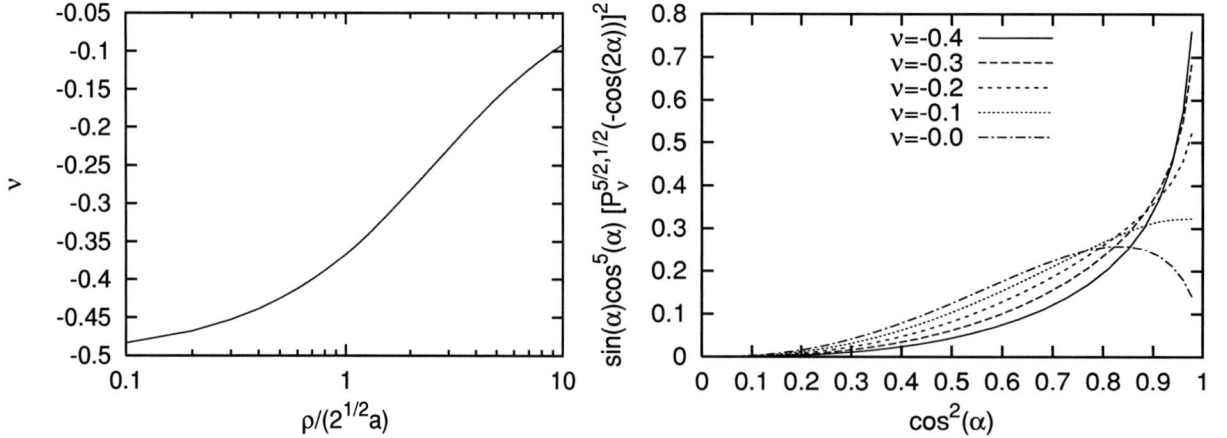

FIGURE 10. The Jacobi index v as function of $\rho/(a\sqrt{2})$ (left) and the energy distribution (right) for the angular wavefunctions corresponding to the different v-values given for each curve.

EFFECT OF VIRTUAL STATE

The peak in the energy distribution of the α-particle at a value close to the maximum possible for the decay of ^{6}He(2^+) corresponds to two neutron leaving together in the same direction. This is certainly not sequential decay through the $p_{3/2}$-resonance of ^{5}He but rather sequential decay through a di-neutron type of structure. Since the di-neutron is unstable the decay configuration must be due to the neutron-neutron attraction expressed as low-lying virtual state. The essence of that configuration can be found by neglecting the neutron-α interaction and look for the large distance angular solution. As ρ increases more and more of the coordinate space is interaction free where the angular solution is obtained solely from the angular kinetic energy operator, i.e. given as a Jacobi function $P_v^{(L+1/2,1/2)}(-\cos(2\alpha))$ where $\rho\sin\alpha/\sqrt{2}$ is the distance between the two neutrons and L is the orbital angular momentum. The index v is determined by matching this wavefunction to the behavior obtained from the smaller α-values where the neutron-neutron interaction still is finite. The zero-range approximation replaces this by a boundary condition, i.e.

$$\frac{1}{\alpha P_v}\frac{\partial}{\partial \alpha}\left(\alpha P_v^{(L+1/2,1/2)}(-\cos(2\alpha))\right)\Big|_{\alpha=0} = \frac{\rho}{a_{nn}\sqrt{2}}, \tag{8}$$

where a_{nn} is the neutron-neutron scattering length. When $\rho \gg a_{nn}$ we find that v is an integer corresponding precisely to the hyperharmonic solutions. When $\rho \ll a_{nn}$ we find that v is half integer which can be interpreted as the solution for the virtual state. The continuous connection between these two extremes can be seen in in fig. 10.

The energy distribution for these wavefunctions is given as function of the α-particle energy by

$$\sin(2\alpha)|P_v^{(l_y+1/2,1/2)}(-\cos(2\alpha))|^2, \tag{9}$$

which is shown in fig. 10 for several v-values. The energy distribution for the virtual state, and in fact for all non-integer v-values, diverge as $1/\alpha$ when $\alpha \to 0$. The distribution resembles qualitatively both the experimental result and the computation with several adiabatic potentials. Thus the ^{6}He(2^+) decay is strongly influenced by the virtual neutron-neutron state and visible in the observable energy distribution.

The neutron-α interaction is essentially restricted to p-waves and the s-wave interaction is repulsive. Imagine another two-neutron halo system with a substantial attraction between neutron and core, i.e. a large scattering length or equivalently a low-lying virtual state. The lowest adiabatic potential at large distance always prefers the s-wave as they have the slowest convergence to the hyperharmonic spectrum [4]. The decay of a resonance in this system would now try to establish a large-distance configuration with a component of this lowest adiabatic eigenvalue. The precise distribution of probabilities on the different adiababtic wavefunctions is determined by the couplings between these potentials. It is a trade off between maintaining the structure or minimizing the energy. In a sense it is a dynamic effect.

With more than one contributing virtual state the lowest adiabatic potential would converge even slower to the hyperharmonic value. Furthermore the structure of the angular wavefunction would be a coherent superposition of these virtual states in the different two-body subsystems. This is precisely as the Efimov effect. The resulting energy distribution, say of the core, would now show a peak close to maximum energy but also another peak at some intermediate energy corresponding to the contribution from the other two (identical) virtual states. This could be a signal of the Efimov effect in decay of excited states of finite angular momentum. The Efimov effect resulting in more than one of these states is not expected in nuclei [6]. However, even for moderate values of the scattering lengths it seems possible to observe a reminiscence of the effect in the energy distribution observed after decay of a nuclear resonance.

SUMMARY AND CONCLUSION

For three particles we use the concept of effective potentials as functions of the average size of the system. This generalized radial coordinate is the hyperradius. The different potentials provide a coupled set of radial equations. For bound states and small distance structure of resonances the lowest of these adiabatic potentials often carries the largest part of the probability. We first approximate this potential by Coulomb and centrifugal barrier contributions for hyperradii corresponding to all three particles just outside the range of their short-range interactions. The WKB tunneling transmission is then a measure of the partial decay width into the three particles. With only Coulomb potential an analytical expression is derived for arbitrary charges and masses. This constitutes a generalization of the WKB decay probability for α-emission.

We then discuss the decay mechanisms and the resonance structures for various nuclear excited states. The decay can be direct (proportional scaling of all distances between the particles), sequential (emission of one particle followed by decay of the remaining two particles), virtual sequential (emission of one particle through an energetically forbidden two-body resonance), sequential through a virtual s-state (like sequential but with only an attractive s-wave to hold the two particles together), Efimov like (through a coherent superposition of more than one virtual s-state. The latter mechanism may bring in traces of the Efimov effect in nuclear decay. All these mechanisms can be described by use of the hyperspherical adiabatic expansion method.

The structure of the resonance wavefunction corresponding to the lowest adiabatic eigenvalue changes sometimes rather dramatically from small to large distances. This change is accompanied by the reverse change of the structure from higher-lying angular eigenvalues. The distribution over the different eigenfunctions may therefore also depend on the hyperradius. The combination of changing structure of the different angular eigenfunctions and varying relative weights in total produce the "dynamic evolution" of the resonance wavefunction. Eventually, at large hyperradii, the particles are free and the decay completed. The energy distribution is then established carrying the signature of the various decay mechanisms.

REFERENCES

1. B. Blank et al., C.R. Physiques **4** (2003)
2. A.A. Korshenennikov, Yad.Fiz. **52**, 1304 (1990), [Sov. J. Nucl. Phys. **52**, 827 (1990)]
3. P.J. Siemens and A.S. Jensen, Elements of nuclei. Many-body physics with the strong interaction, Addison-Wesley, California 1987.
4. E. Nielsen, D. V. Fedorov, A. S. Jensen, and E. Garrido, Phys. Rep. **347**, 373 (2001).
5. D.V. Fedorov, E. Garrido, and A.S. Jensen, Few-body systems, **33**, 153-171 (2003).
6. A.S. Jensen, K. Riisager, D.V. Fedorov and E. Garrido, Rev. Mod. Phys. **76** (2004) 215.
7. O.I. Kartavtsev, Few-Body Systems **34** (2004) 39.
8. B.V. Danilin, M.V. Zhukov, A.A. Korsheninnikov, L.V. Chulkov V.D. Efros, Sov. J. Nucl. Phys. **46** (1987) 225.
9. E. Garrido, D.V. Fedorov, A.S. Jensen and H.O.U. Fynbo, Nucl.Phys. **A 748**, 27-38 (2005), and Nucl.Phys. **A 748**, 39-58 (2005).

Continuum-discretized coupled-channels method for four-body breakup reactions

M. Kamimura*, T. Matsumoto*, E. Hiyama†, K. Ogata*, Y. Iseri** and M. Yahiro*

Department of Physics, Kyushu University, Fukuoka 812-8581, Japan
†*Department of Physics, Nara Women's University, Nara 630-8506, Japan*
**Department of Physics, Chiba-Keizai College, Chiba 263-0021, Japan*

Abstract. Development of the method of CDCC (Continuum-Discretized Coupled-Channels) from the level of three-body CDCC to that of four-body CDCC is reviewed. Introduction of the pseudo-state method based on the Gaussian expansion method for discretizing the continuum states of two-body and three-body projectiles plays an essential role in the development. Furthermore, introduction of the complex-range Gaussian basis functions is important to improve the CDCC for nuclear breakup so as to accomplish that for Coulomb and nuclear breakup. A successful application of the four-body CDCC to ^{6}He+^{12}C scattering at 18 and 229.8 MeV is reported.

Keywords: CDCC, breakup, continuum, unstable nuclei
PACS: 21.45.+v,21.60.Gx,24.10.Eq,25.60.Gc,27.10.+h

INTRODUCTION

In the study of reactions induced by unstable nuclei, analysis of the case where the projectile is considered to be composed of three-clusters such as ^{6}He and ^{11}Li becomes quite important. For this purpose, along the diagram in Fig. 1, we have developed the three-body CDCC (Continuum-Discretized Coupled-Channels) for nuclear breakup of two-body projectiles [1] into the four-body CDCC for Coulomb and nuclear breakup of three-body projectiles.

The momentum-bin method to discretize the continuum states of the two-body projectiles (such as ^{6}Li = $\alpha + d$, ^{8}B + p, etc.) is not practically available to the case of three-body projectiles. On the basis of the Gaussian expansion method (GEM)[2], we proposed, in Ref.[3], the pseudo-state (PS) method to discretize the continuum states and examined it in the case of two-body projectiles (three-body CDCC); this is Step A in Fig.1. In the PS method we diagonalized the two-body Hamiltonian of the internal motion of the projectile using the Gaussian basis functions [2] and obtained dense distribution of the pseudo-states, namely discretized continuum states. An advantage of this method is that it can easily be extended to the case of three-body projectiles by using the GEM. Another advantage of the PS method is that we can derive continuous S-matrix elements as a smooth function of the momentum of the projectile breakup states. We found [3] that the S-matrix elements obtained by the PS method agrees well with the S-matrix elements by the momentum-bin method with very precise bins.

As Step B in Fig.1, we extended the three-body CDCC (for nuclear breakup) to the four-body CDCC (for nuclear breakup) using the three-body Gaussian basis functions of GEM to obtain bound and pseudo-states of the three-body projectiles [4]. The GEM is very suitable for describing bound and pseudo-states of three- and four-body systems; it is extensively reviewed in Ref.[2]. The four-body CDCC was applied to the ^{6}He+^{12}C at 18 and 229.8 MeV. The differential cross sections of the elastic scattering were well reproduced by using the double-folding CC potentials.

In Step C, we improved the three-body CDCC for nuclear breakup to that for Coulomb and nuclear breakup [5] by using the PS method with the complex-range Gaussian basis functions [2] instead of the (real-range) Gaussian basis functions adopted in the previous Steps. Due to the long-ranged Coulomb coupling-potentials, the modelspace required for CDCC is very large. Particularly, one must prepare the internal wave functions of the projectile, both in bound and continuum states, for a wide range of the internal coordinate, say 0–100 fm, which is in general difficult for PS methods. This can easily be achieved by using the complex-range Gaussian basis in the case of two-body projectile.

In order to treat both Coulomb and nuclear breakup processes at *intermediate energies* with high accuracy and computational speed, a new method was proposed in Ref. [6]; namely, a hyprid calculation with the three-body CDCC method and the eikonal-CDCC (E-CDCC) method. E-CDCC describes the center-of-mass motion of the projectile relative to the target by straight-line approximation (or by using Coulomb wave functions instead of plane waves) and treats the excitation of the projectile explicitly by CDCC with the momentum-bin method or the PS method. E-

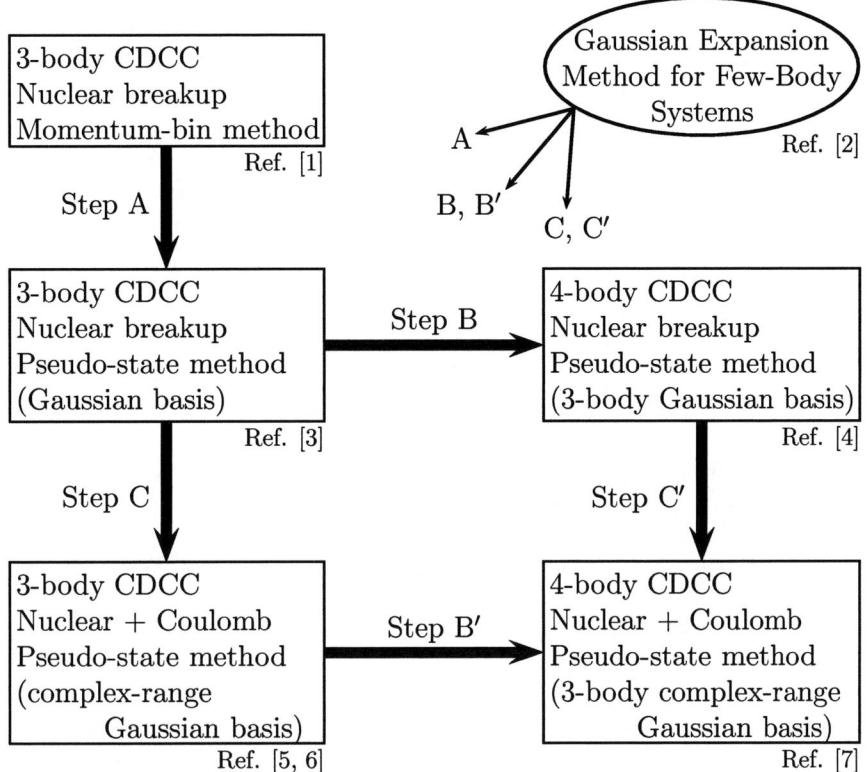

FIGURE 1. A flow of the improvement of the method of CDCC starting from the three-body CDCC for nuclear breakup to the four-body CDCC for Coulomb and nuclear breakup with the aid of the Gaussian expansion method for few-body systems.

CDCC drastically reduces computation time and eliminates many problems concerned with huge angular momentum in solving coupled-channel equations. Thus, the hybrid calculation is expected to be opening the door to the systematic analysis of Coulomb (plus nuclear) dissociation of projectiles in the wide range of beam energies.

Finaly, by Step B' (or by Step C') we can reach the four-body CDCC for Coulomb and nuclear breakup. This step was not reported in the time of the RIA workshop but was recently accomplished and successfully applied to the ^{6}He+^{209}Bi scattering at 19.0 and 22.5 MeV [7].

METHOD OF PSEUDO-STATE CDCC FOR TWO-BODY PROJECTILES

In the method of CDCC, the total wave function of the scattering state Ψ_{JM} is expanded in terms of a finite number of internal wave functions $\Psi_{nIm}(\xi)$ of the projectile:

$$\Psi^{JM}(\xi,\mathbf{R}) = \sum_{nI,L} [\Phi_{nI}(\xi) \otimes \chi^{J}_{nI,L}(\mathbf{R})]_{JM}, \tag{1}$$

where $\mathbf{R}$ is the coordinate of the center-of-mass of the projectile relative to the target, and ξ is the internal coordinates of the projectile. I is the total spin of the projectile and n stands for the nth eigenstate. $\chi^{J}_{nI,L}$ represents the relative motion between the projectile and the target; L is the orbital angular momentum regarding $\mathbf{R}$. The unknown function $\chi^{J}_{nI,L}(\mathbf{R})$ are solved using the usual framework of the coupled-channel method for discrete excited states.

The projectile internal wave functions $\Phi_{nI}(\xi)$ include both bound states and discretized continuum states. To calculate the wave functions of the latter states the momentum-bin method has widely been utilized in the usual three-body CDCC calculations. In the method the exact scattering wave functions are averaged within each narrow intervals of momentum between the two constituents in the projectile. But, this method is not practically suitable for discretizing the breakup states of the three-body projectile.

In the pseudo-state (PS) method [1, 8, 9], on the other hand, wave functions of the discretized breakup states are obtained by diagonalizing the internal Hamiltonian of the projectile, which describes the relative motion of the two constituents, using L^2-type basis functions. Since the wave functions of such pseudo breakup states have wrong asymptotic forms, the PS method was mainly used in the past to describe virtual breakup processes in the intermediate stage of elastic scattering [9] and (d,p) reactions [1].

In the work of Ref.[3], however, we proposed the new method of pseudo-state (PS) discretization for two-body projectiles. It can be used not only for virtual breakup processes in elastic scattering but also for breakup reactions. In order to diagonalize the Hamiltonian of the two-body projectile, we employed two types of basis functions. One is the conventional real-range Gaussian functions

$$\phi_{j\ell}(r) = r^\ell \exp\left[-(r/a_j)^2\right], \qquad (j = 1\text{--}n) \tag{2}$$

where $\{a_j\}$ are assumed to increase in a geometric progression [10, 2]:

$$a_j = a_1 (a_n/a_1)^{(j-1)/(n-1)}. \tag{3}$$

The other is an extension of (2) introduced in Ref. [2], i.e., the following pairs of functions:

$$\begin{aligned}
\phi_{j\ell}^{C}(r) &= r^\ell \exp\left[-(r/a_j)^2\right] \cos\left[b(r/a_j)^2\right], \\
\phi_{j\ell}^{S}(r) &= r^\ell \exp\left[-(r/a_j)^2\right] \sin\left[b(r/a_j)^2\right], \quad (j=1\text{--}n).
\end{aligned} \tag{4}$$

Here, b is a free parameter, in principle, but numerical test showed that $b = \pi/2$ is recommendable. Both $\phi_{j\ell}^C$ and $\phi_{j\ell}^S$ are to be used simultaneously; the total number of basis is thus $2n$. The basis functions (4) can also be expressed as

$$\begin{aligned}
\phi_{j\ell}^{C}(r) &= \{\psi_{j\ell}^*(r) + \psi_{j\ell}(r)\}/2, \\
\phi_{j\ell}^{S}(r) &= \{\psi_{j\ell}^*(r) - \psi_{j\ell}(r)\}/(2i),
\end{aligned} \tag{5}$$

with

$$\psi_{j\ell}(r) = r^\ell \exp[-\eta_j r^2], \quad \eta_j = (1+ib)/a_j^2, \tag{6}$$

i.e., Gaussian functions with a complex-range parameter. We thus refer to the basis $\phi_{j\ell}^C$ and $\phi_{j\ell}^S$ as the complex-range Gaussian basis.

The complex-range Gaussian basis functions are oscillating with r. They are therefore expected to simulate the oscillating pattern of the continuous breakup state wave functions better than the real-range Gaussian basis functions do. Moreover, numerical calculation with the complex-range Gaussians can be done using essentially the same computer programs as for the real-range Gaussians, just replacing real variables for a_j of Eq. (3) by complex ones. Usefulness of the real- and complex-range Gaussian basis functions in few-body calculations are extensively presented in the review work [2].

Here, we explore a typical example in which the complex-range Gaussian basis functions reproduce highly oscillatory functions with high accuracy. A good test is to calculate the wave functions of highly excited states in a harmonic oscillator potential; note that this potential is not specially advantagious for the Gaussian bases. We take the case of a nucleon with angular momentum $l = 0$ in a potential having $\hbar\omega = 15.0$ MeV. Parameters of the complex-range Gaussian basis functions are $\{2n_{\max} = 28, r_1 = 1.4\,\text{fm}, r_{n_{\max}} = 5.8\,\text{fm}, \alpha = \frac{\pi}{2}\frac{1}{1.2^2} = 1.09\}$. For the sake of comparison, we also tested the Gaussian basis functions with the paramters $\{n_{\max} = 28, r_1 = 0.5\,\text{fm}, r_{n_{\max}} = 11.3\,\text{fm}\}$. Optimized r_1 and $r_{n_{\max}}$ are quite different between the two types of bases though the total numbers of basis functions are the same. In Table 1, we compare the calculated energy eigenvalues with the exact ones. It is evident that the complex-range Gaussians can reproduce the energy up to much more highly excited states than the Gausssians do. For the Gaussian basis, even if the number of basis functions is increased, the result is not significantly improved, because the number of oscillation does not increase. On the other hands, for the complex-range Gaussian functions, as the number is increased, the result becomes better so long as the number of oscillation is not too larger than ~ 20. Figure 2 demonstrates good accuracy of the wave function of the 19-th excited state having 38 quanta. Error is within a few %, much smaller than the thickness of the line. The figure suggests that the basis functions is also suitable for describing pseudo-states used for Coulomb breakup reactions.

We here emphasize that even in the case where the projectile is assumed to be three-body system, the Gaussian basis functions with real and complex ranges are easily utilized in the CDCC calculation with the PS method. We discuss this point in the next section.

TABLE 1. Test of the accuracy of real-range and complex-range Gaussian basis functions for highly excited states ($2n+l \leq 46, l = 0$) of a harmonic oscillator potential for a nucleon. The number of basis functions is 28 for both cases. Eigenenergies obtained by the diagonalization of the Hamiltonian with the bases are listed in terms of the number of quanta, $E/\hbar\omega - \frac{3}{2}$. See text for the Gaussian parameters.

Exact (2n)	real range	complex range	Exact (2n)	real range	complex range
0.0	0.0000	0.0000	26.0	26.4	26.0001
6.0	6.0000	6.0000	30.0	32.9	30.0003
10.0	10.0000	10.0000	34.0	41.8	34.002
14.0	14.0000	14.0000	38.0	53.8	38.003
18.0	17.998	18.0000	42.0	69.9	42.1
22.0	21.9	22.0000	46.0	91.6	46.3

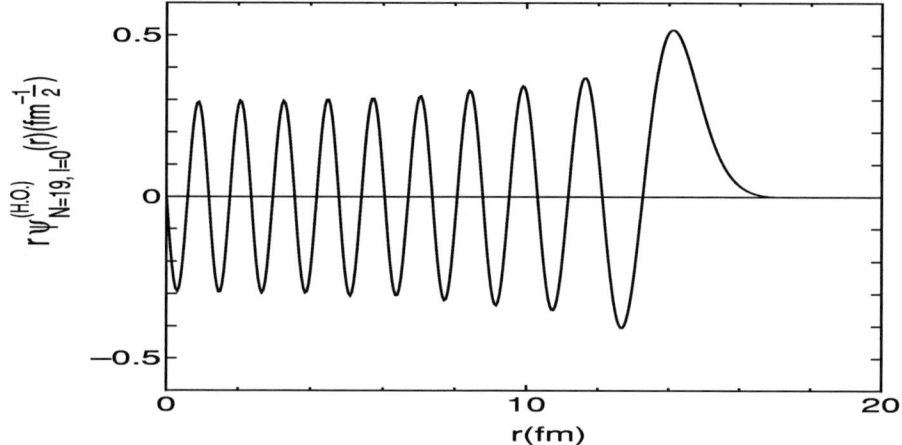

FIGURE 2. Accuracy of the wave function of the $2n+l = 38$, $l = 0$ state obtained by diagonalizing Hamiltonian with a harmonic-oscillator potential for a nucleon using 28 complex-range Gaussian basis functions. It is compared with the exact wave function but the difference is invisible since the error is less than a few % everywhere. See text for the Gaussian parameters. This figure is taken from Ref. [2].

Another advantage of the PS method, in the case of two-body projectiles, is that the discrete breakup S-matrix elements, say $S_{nIL,0I_0L_0}$, for the transition from $\Phi_{0I_0}(\xi)$ to $\Phi_{nI}(\xi)$ can be accurately transformed to smooth S-matrix elements, say $\tilde{S}_{IL,I_0L_0}(k)$, as following [3], since the two-body PS basis functions can form in the good approximation a complete set in the finite region which is important for the breakup processes:

$$\tilde{S}_{IL,0I_0L_0}(k) = \sum_n \langle \tilde{\Phi}_I(k,\xi)|\Phi_{nI}(\xi)\rangle_\xi \, S_{nIL,0I_0L_0} \,, \qquad (7)$$

where $\tilde{\Phi}_I(k,\xi)$ is the exact wave function of the internal motion of the two-body projectile.

Example 1 : ^{6}Li+^{40}C scattering at 156 MeV.

Here, we briefly show results of test calculations done in [3] ^{6}Li+^{40}C scattering at 156 MeV. The $\alpha - d$ continuum of the ^{6}Li projectile is discretized as in Fig. 2 using the real-range Gaussian bases and the complex-range Gaussian bases. The modelspace sufficient for describing breakup processes in this scattering is $k_{\max} = 2.0$ fm^{-1} and $\ell_{\max} = 2$; the modelspace is composed of two k-continua for s-state and d-state. There exists a d-state resonance. The resonance is automatically taken care by the PS method by the lowest-lying several pseudo-states. On the other hand, in the momentum-bin method, the d-state k-continuum is further divided in the momentum-bin method into the resonant part $[0 < k < 0.55\,\text{fm}^{-1}]$ and the non-resonant part $[0.55 < k < 2.0\,\text{fm}^{-1}]$. In the former region the k continuum d-state wave function varies rapidly with k. The momentum-bin method can simulate this rapid change with bins of an extremely small width. In fact clear convergence is found for both the elastic and the breakup S-matrix elements, when

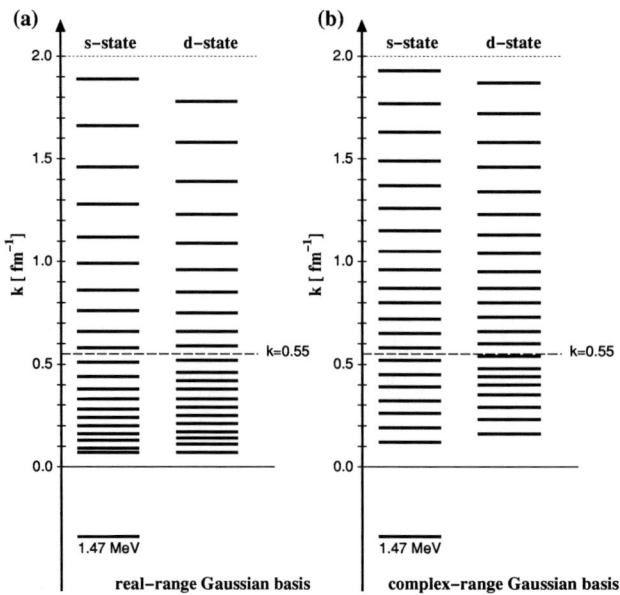

FIGURE 3. Pseudo states (discretized continuum states) for ^{6}Li obtained by using the real-range Gaussian basis functions (left) and the complex-range Gaussian basis functions (right). This figure is taken from Ref.[3].

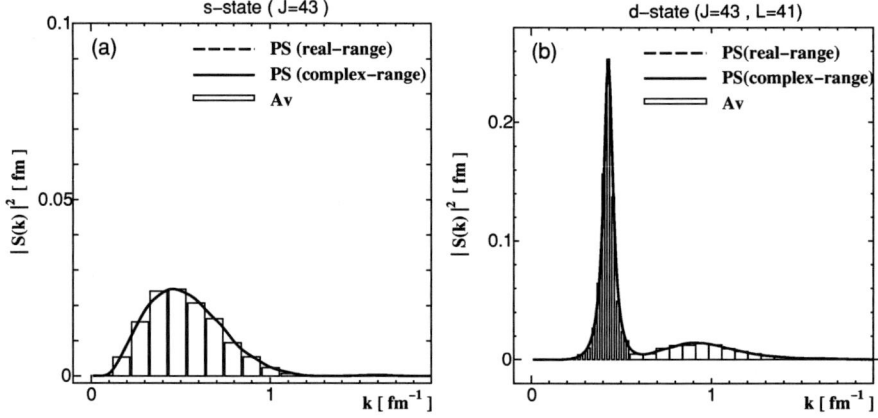

FIGURE 4. The squared moduli of breakup S-matrix elements as a function of k at the grazing total angular momentum $J = 43$ for ^{6}Li + ^{40}Ca scattering at 156 MeV. The step line is the result of the momentum-bin method (Average-methos) with dense bins. s-state breakup (left) and d-state breakup for $L = J - 2$ (right). Note that the difference between the results of the real- and complex-range Gaussian PS methods is not visible since it is less than about 1%. This figure is taken from Ref.[3].

the resonance part is described by 30 bins and the non-resonance part of the d-state and the s-state k-continua by 20 bins.

Figure 3 represents breakup S-matrix elements at grazing total angular momentum $J = 43$; (a) s-state breakup and (b) d-state breakup in the case of $L = J - 2$. The real- and complex-range Gaussian PS discretization well reproduce the "exact" solution calculated by the momentum-bin method with dense bins. The results of the two PS methods turn out to coincide within the thickness of the line. The resonance peak can be expressed by only 8 (12) breakup channels in the complex-range (real-range) Gaussian PS method, while the corresponding number of breakup channels is 30 in the momentum-bin method, as mentioned above. Thus, one can conclude that the real- and complex-range Gaussian PS methods are very useful for describing not only non-resonant states but also resonant ones.

The PS method has at least two advantages over the widely used momentum bin average method. One is that it does not need the exact wave function of the projectile over the entire region of r. This is important from a theoretical point

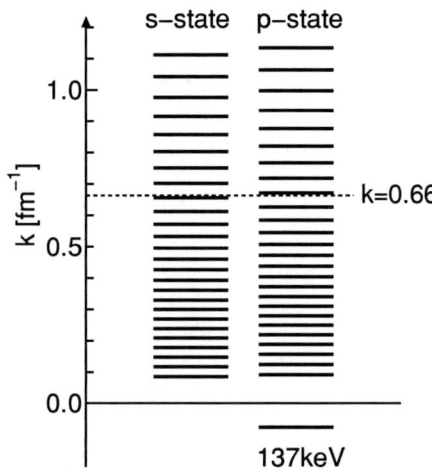

FIGURE 5. Discretized momenta for ^{8}B; the left (right) side corresponds to the s-state (p-state). The horizontal dotted line represents the cutoff momentum $k_{\max}$ taken to be 0.66 fm^{-1} above which is not effective in the reaction. This figure is taken from Ref.[5].

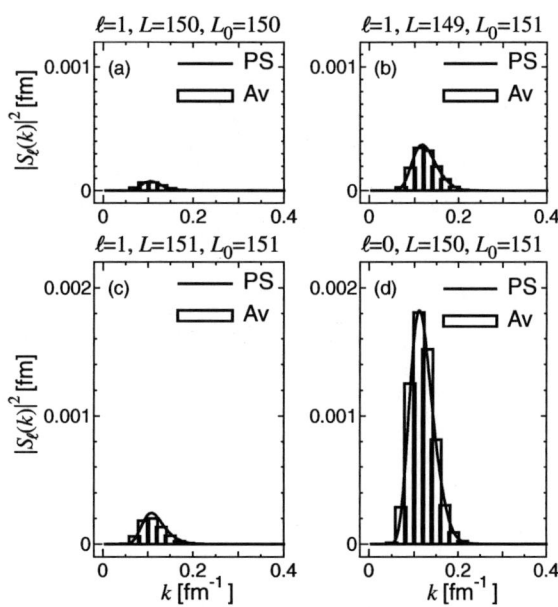

FIGURE 6. The squared moduli of breakup S-matrix elements, as a function of k, at $J = 150$ for ^{8}B+^{58}Ni scattering at 25.8 MeV. The panel (a), (b), (c) and (d) correspond to $(\ell, L, L_0) = (1,150,150)$, $(1,149,151)$, $(1,151,151)$ and $(0,150,151)$, respectively. In each panel, the solid line represents the result of PS-CDCC, while the step line is the result of the momentum-bin (average) method of CDCC that is assumed as the "exact" S-matrix elements. This figure is taken from Ref.[5].

of view. The other is that with the real- and complex-range Gaussian bases one can calculate all the coupling potentials semi-analytically [2], which is very useful in actual calculations; note that the Gaussian bases are very suitable for transforming wave functions and interactions from a Jacobian coordinate system to other ones. Furthermore, if the projectile has resonances in its excitation spectrum, the PS method discretizes the complicated spectrum with a reasonable number of the pseudo-states, without distinguishing the resonance states from non-resonant continuous states. These advantages of the PS method are extremely helpful, sometimes even essential, in applying CDCC to four-body breakup effects of unstable nuclei such as ^{6}He and ^{11}Li.

Example 2 : ^{8}B+^{58}Ni scattering at 25.8 MeV.

Here, we briefly show results of test calculation in [5] for Coulomb breakup process of ^{8}B+^{58}Ni scattering at 25.8 MeV. The ^{7}Be$-p$ continuum in the ^{8}B projectile is discretized as in Fig. 5 by the PS method with the real-range Gaussian bases and the complex-range Gaussian bases. In the PS method, the number of channels included in the CDCC calculation, was 18 for both the s- and p-states at $k < k_{\max} = 0.66$ fm^{-1}, which give a satisfactory convergence of the result. The resulting wave functions with positive eigenenergies turned out to oscillate up to about 100 fm. In the momentum-bin method, the modelspace with $k_{\max} = 0.66$ fm^{-1} and $\Delta k = 0.66/16$ (0.66/32) fm^{-1} for p-state (s-state) gives convergence of the resulting total breakup cross section. The maximum internal coordinate $r_{\max}$ was taken to be 100 fm.

Figure 6 shows the result of the comparison of $|S_\ell(k)|^2$ at $J = 150$, which corresponds to the scattering angle of 10° assuming the classical path. It was found that CDCC calculation with only Coulomb coupling potentials gives a peak at 10° in the total breakup cross section. Thus, it can be assumed that Fig. 6 corresponds to the most-Coulomb-like breakup process; in any case, the feature of the result was found to be almost independent of J. In each panel of Fig. 6, one sees that the result of PS-CDCC (solid line) very well reproduces the "exact" solution (step line by the momentum-bin method) for all k being significant for the ^{8}B Coulomb breakup.

GAUSSIAN EXPANSION METHOD FOR FEW-BODY SYSTEMS

In this section we briefly explain the Gaussian expansion method (GEM) for few-body systems. The method was proposed by Kamimura in 1988 [10] for three-body systems and was much developed by Hiyama using the infinitesimally-shifted Gaussian basis functions even for four-body systems (reviewed in [2]).

A good example to show the accuracy and usefullness of the method is the determination of upper limit of the difference between the masses of proton and antiproton, m_p and $m_{\bar{p}}$, respectively. The first recommended upper limit of $|m_{\bar{p}} - m_p|/m_p$ by the Particle Data Group listed in Particle Listings 2000 [11] was 5×10^{-7}, which could be used for a test of *CPT* invariance. This number was extracted from a high-resolution laser experiment involving metastable states of antiprotonic helium atom (He^{2+} + e^- + $\bar{p}$) [12] by Kino *et al.* [13] through a theoretical analysis of the highly excited states of the Coulomb three-body system using GEM. The ratio was improved to $|m_{\bar{p}} - m_p|/m_p < 1 \times 10^{-8}$, as listed in the Particle Listings 2004, by later, more extensive experiments and additional calculations (cf. Ref.[2])

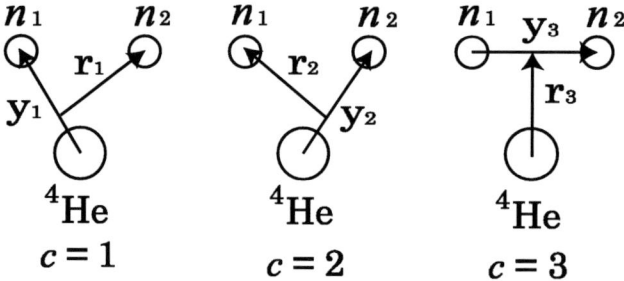

FIGURE 7. Jacobian coordinates of three rearrangement channels ($c = 1$–3) adopted for the $n+n+^4$He model of ^{6}He structure. The two neutrons are to be antisymmetrized.

In the Gaussian expansion method [2], wave functions of the projectile, Φ_{nlm} in (1), is written as a sum of component functions in the Jacobian coordinates for rearrangement channels $c = 1 - 3$ in Fig. 7 as

$$\Phi_{nlm}(\xi) = \sum_{c=1}^{3} \psi_{nlm}^{(c)}(\xi), \qquad (8)$$

Each $\psi_{nlm}^{(c)}$ is expanded in terms of the Gaussian basis functions:

$$\begin{aligned}
\psi_{nlm}^{(c)}(\xi) &= \varphi^{(\alpha)} \sum_{\lambda \ell \Lambda S} \sum_{i=1}^{i_{\max}} \sum_{j=1}^{j_{\max}} A_{i\lambda j\ell \Lambda S}^{(c)nl} y_c^\lambda r_c^\ell e^{-(y_c/\bar{y}_i)^2} e^{-(r_c/\bar{r}_j)^2} \\
&\quad \times \left[[Y_\lambda(\hat{y}_c) \otimes Y_\ell(\hat{r}_c)]_\Lambda \otimes [\eta_{\frac{1}{2}}^{(n_1)} \otimes \eta_{\frac{1}{2}}^{(n_2)}]_S \right]_{Im},
\end{aligned} \qquad (9)$$

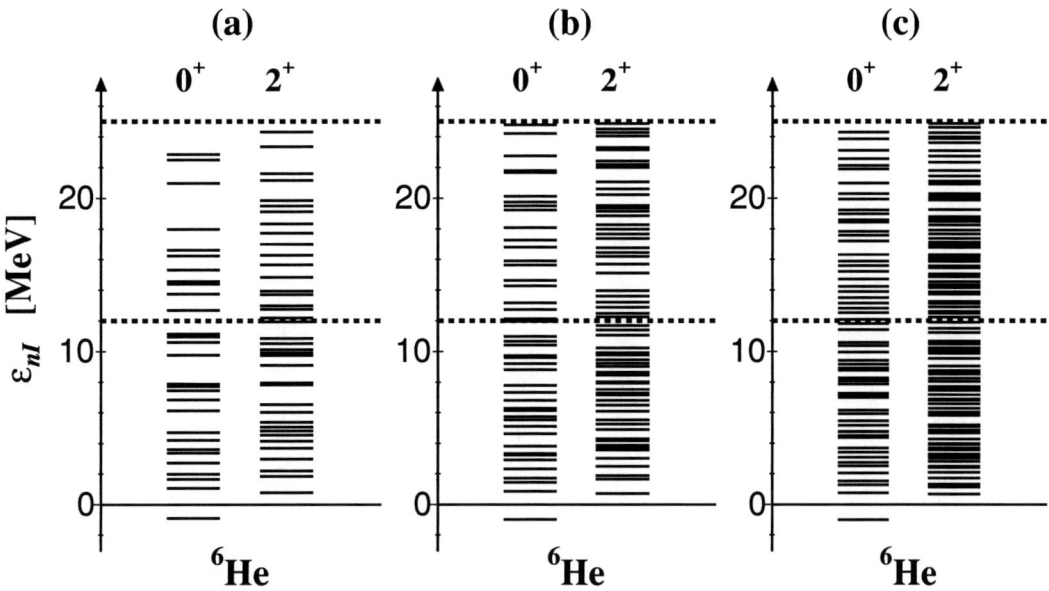

FIGURE 8. Calculated energy levels of the bound and discretized continuum states (pseudo-states) of ^{6}He using three different sets of Gaussian basis functions (see Ref. [4]). The numbers of the pseudo-states used in the ^{6}He+^{12}C scattering at 229.8 MeV to see convergence of the calculated results is (a) 28 (0^+) and 39 (2^+), (b) 44 (0^+) and 64 (2^+) and (c) 60 (0^+) and 85 (2^+) which are located in the region of $\varepsilon_{nI} < 25$ MeV. For 18-MeV scattering, to take pseudo-states of $\varepsilon_{nI} < 12$ MeV is satisfactory. The case (b) was found to be sufficient to obtain a good convergence.

where λ (ℓ) is the angular momentum regarding the Jacobian coordinates $\mathbf{y}_c$ ($\mathbf{r}_c$), and $\eta_{1/2}$ is the spin wave function of each valence neutron (n_1 or n_2). ^{4}He has been treated as an inert core with the $(0s)^4$ internal configuration, $\varphi^{(\alpha)}$. The Gaussian range parameters are taken to lie in geometric progression:

$$\bar{y}_i = \bar{y}_1 (\bar{y}_{\max}/\bar{y}_1)^{(i-1)/(i_{\max}-1)}, \qquad (10)$$

$$\bar{r}_j = \bar{r}_1 (\bar{r}_{\max}/\bar{r}_1)^{(j-1)/(j_{\max}-1)}. \qquad (11)$$

Φ_{nIm} is antisymmetrized for the exchange between n_1 and n_2. Meanwhile, the exchange between each valence neutron and each nucleon in ^{4}He is treated approximately by the orthogonality condition. The eigenenergies ε_{nI} of ^{6}He and the corresponding expansion-coefficients $A^{(c)nI}_{i\lambda j\ell \Lambda S}$ are determined by diagonalizing the Hamiltonian of the interernal motion of ^{6}He [14, 15] using a large number of three-body Gaussian basis functions. Datailed information on the basis is listed in Ref.[4]. The calculated ε_{nI} are -0.98 MeV for the 0^+ ground state and 0.72 MeV for the 2^+ resonance state; here, we took the Bonn A potential between the valence nucleons and increased the depth of the $n-\alpha$ potential by a few percent so that the ground-state energy is reproduced.

In the four-body CDCC calculation of ^{6}He+^{12}C shown in a later section, we take $I^\pi = 0^+$ and 2^+ states for ^{6}He. Here we omit the 1^- state that does not contribute to the nuclear breakup processes (but they are included in the calculation of Coulomb and nuclear breakup in Ref.[7]). In order to demonstrate the convergence of the four-body CDCC solution with increasing the number of the Gaussian basis functions, we prepare three sets of the basis functions, i.e., sets I, II and III listed in Table II of [4]. Resultant energy levels of the ground and pseudo-states are shown in (a), (b) and (c) in Fig. 8, respectively. For ^{6}He+^{12}C scattering at 18 MeV (229.8 MeV) which will be discussed in the next section, high-lying states with $\varepsilon_{nI} > 12$ MeV ($\varepsilon_{nI} > 25$ MeV) are found to give no effect on the elastic and breakup S-matrix elements. Thus, the effective number of the eigenstates of ^{6}He, is reduced much for each of cases (a), (b), (c) as shown in Fig. 8. The case (b) was found to be sufficient to obtain a good convergence. In the GEM, computation time to obtain the wave functions of the bound and pseudo states is very short; for example, all the wave functions of the states in Fig. 8(c) is obatined in 10 minutes on FUJITSU VPP5000, a supercomputer.

It is to be noted that the bound and pseudo-states obtained with the GEM calculations construct an approximate complete sets for each $J(=0,1,2)$ in a finite region which is responsible for the reaction; this was examined by checking that those states (below 100 MeV) satisfies 99.9 % of the energy-weighted cluster sum-rule limit for monopole, dipole and quadrupole transitions.

FOUR-BODY CDCC ANALYSIS OF ^{6}He+^{12}C SCATTERING AT 18 AND 229.8 MEV

In this section, we briefly introduce the results obtained in the work of Ref.[4]. We performed the four-body CDCC calculation for ^{6}He+^{12}C scattering at 18 and 229.8 MeV using the wave functions of the bound state and the pseudo-states of ^{6}He obtained above.

The real part of the CC potentials, say $V^J_{nIL,n'I'L'}(R)$, was constructed by using the double-folding model [18]; the potentials were calculated by folding the DDM3Y NN interaction into the transition densities between the states $\Phi_{nI}(\xi)$ and $\Phi_{n'I'}(\xi)$ (cf. Ref.[4] for details) and the ground-state density of ^{12}C [19]. The imaginary part was assumed, as usually done [1], to be given as (together with the real part)

$$(N_R + iN_I) V^J_{nIL,n'I'L'}(R), \qquad (12)$$

where $N_R = 1.0$ with no renormarization of the real part. The only parameter N_I is searched for to reproduce the observed elastic cross section as well as possible. In the analysis of the ^{6}He+^{12}C scattering, Coulomb breakup effect is ignored since it is negligible for this light target; the Coulomb potential is assumed to work between the center-of-mass of the target and that of the projectile.

Calculated and observed elastic cross sections for ^{6}He+^{12}C scattering at 18 MeV are shown in Fig. 9. The optimum value of N_I is 0.5, which is the same as that for ^{6}Li scattering at various incident energies [1]. The dotted lines represent the elastic cross sections due to the single-channel calculation. Then, the difference between the solid and dotted lines shows the effect of the four-body breakup on the elastic cross section. The effect is sizable and indispensable to explain the behavior of the angular distribution. The case at 229.8 MeV is shown in Fig.10 and the optimum value of N_I is 0.3. The brekup effect in this case is also important in reproducing the data. The origin of the small N_I value for the ^{6}He scattering at 229.8 MeV is not clear at this moment, so more systematic experimental data are highly desirable for ^{6}He scattering.

We calculated the dynamical polarization (DP) potential induced by the four-body breakup processes, in order to understand effects of the processes on the elastic scattering. The DP potential is given by the deviation of the so-called wave-function-equivalent local potential derived using the elastic channel amplitude in the solution of the CDCC equation from the double-folding potential of the elastic channel. From the analysis [4] of the DP potential, one sees that inclusion of the four-body breakup processes makes the real part of the ^{6}He–^{12}C potential shallower and the imaginary one deeper compared with the double-folding potential of the elastic channel. In particular, the latter effect is important and can be assumed to come from the Borromean structure of ^{6}He. This is consistent with the fact that the total reaction cross section is enhanced by the Borromean structure[4].

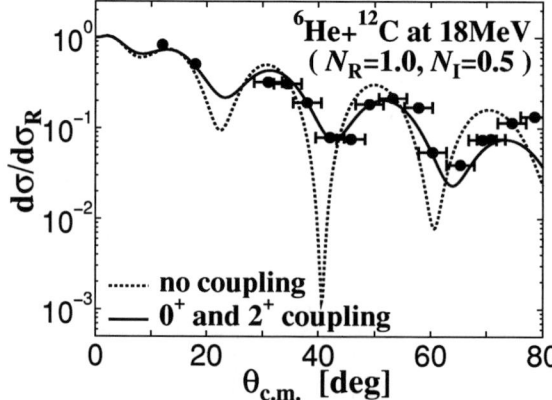

FIGURE 9. Angular distribution of the elastic differential cross section for ^{6}He+^{12}C scattering at 18 MeV. The solid and dotted lines show the results with and without breakup effects, respectively. The experimental data are taken from Ref. [17]. This figure is taken from [4].

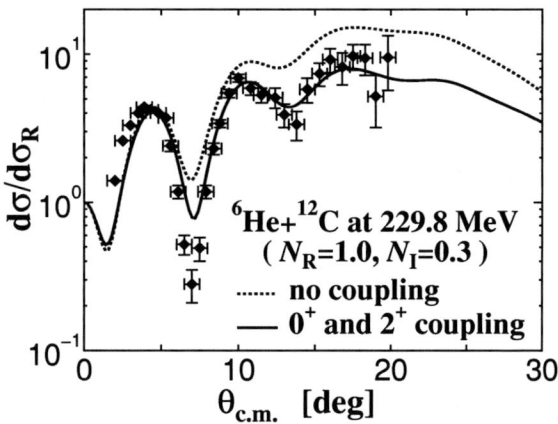

FIGURE 10. The same as in Fig. 9 but for ^{6}He+^{12}C scattering at 229.8 MeV. The experimental data are taken from Ref. [16]. This figure is taken from [4].

CONCLUSION AND NEAR-FUTURE PROBLEMS

In conclusion, a fully quantum-mechanical method of treating four-body breakup is presented by extending CDCC. The validity of the method called four-body CDCC is confirmed by clear convergence of the calculated elastic and energy-integrated breakup cross sections with respect to extending the modelspace. The four-body CDCC is found to explain well the ^{6}He+^{12}C scattering at 18 and 229.8 MeV in which ^{6}He easily breaks up into two neutrons and ^{4}He. For the elastic scattering, the four-body breakup processes make, in particular, the imaginary part of the ^{6}He–^{12}C potential deeper, which is originated in the Borromean structure of ^{6}He.

In the analysis of [4], four-body Coulomb breakup is neglected. However, it is possible to treat it within the four-body CDCC framework (cf. Fig. 1). Actually, after this RIA workshop, we reported in Ref.[7] our four-body CDCC calculation of the ^{6}He+^{209}Bi scattering at 19.0 and 22.5 MeV taking both the Coulomb and nuclear breakup effects into account. The elastic cross sections were well reproduced by the calculation. So, the same framework will be applicable to other cases of three-body projectiles with Coulomb and nuclear breakup.

In order to treat both Coulomb and nuclear breakup processes at *intermediate energies*, Ref. [6] proposed a new method, namely a hyprid calculation with the three-body CDCC method and the eikonal-CDCC (E-CDCC) method. This hybrid calculation is expected to be opening the door to the systematic analysis of Coulomb (plus nuclear) dissociation of projectiles in the wide range of beam energies. For example, the method was recently applied to the analysis of ^{8}B dissociation measurements to determine the astrophysical factor $S_{17}(0)$ accurately [21].

There are some important unstable nuclei that are considered to be composed of four-body constituents. For reactions in which such a four-body nucleus is a projectile, a five-body CDCC calculation is required. The GEM was already severely and successfully tested for the bound states and pseudo-states of four-body systems. A good example is seen in a calculation of four-nucleon system (^{4}He) in Ref. [20]. The four-body GEM calculation with a realistic NN force (AV8') and a phenomenological NNN force (which is adjusted to reproduce the ground-state energy) reproduced the energy of the second 0^+ state and the $^4\text{He}(e,e')^4\text{He}(0_2^+)$ form factor. Furthermore, some 3000 0^+ pseudo-states below 300-MeV excitation satisfied the energy-weighted monopole sum rule by 99.9% (with saturation) and made clear, for the first time, that the major part of the monopole sum rule limit, which had been long unknown, was distributed into low-lying four-body non-resonant continuum states. So, it may be said that it is ready to perform five-body CDCC calculations for reactions induced by four-body projectiles.

REFERENCES

1. M. Kamimura, M. Yahiro, Y. Iseri, Y. Sakuragi, H. Kameyama and M. Kawai, Prog. Theor. Phys. Suppl. **89**, 1 (1986).
2. For a review, E. Hiyama, Y. Kino and M. Kamimura, Prog. Part. Nucl. Phys. **51**, 223 (2003).
3. T. Matsumoto, T. Kamizato, K. Ogata, Y. Iseri, E. Hiyama, M. Kamimura and M. Yahiro, Phys. Rev. C **68**, 064607 (2003).
4. T. Matsumoto, E. Hiyama, K. Ogata, Y. Iseri, M. Kamimura, S. Chiba and M. Yahiro, Phys. Rev. C **70**, 061601 (R)(2004).
5. T. Egami, K. Ogata, T. matsumoto, Y. Iseri, M. Kamimura and M. Yahiro, Phys. Rev. C **70** 047604 (2004).

6. K. Ogata, M. Yahiro, Y. Iseri, T. Matsumoto and M. Kamimura, Phys. Rev. C **68** 064609 (2003).
7. T. Egami, T. Matsumoto, K. Ogata, Y. Iseri, M. Kamimura, E. Hiyama and M. Yahiro, a talk at Annual Meeting of Japan Physical Society, March, 2005, Noda.
8. A. M. Moro, J. M. Arias, J. Gómez-Camacho, I. Martel, F. Pérez-Bernal, R. Crespo and F. Nunes, Phys. Rev. C **65**, 011602 (2002).
9. R. Y. Rasoanaivo and G. H. Rawitscher, Phys. Rev. C **39**, 1709 (1989).
10. M. Kamimura, Phys. Rev. A **38**, 621 (1988).
11. Particle Data Group, D. E. Groom *et al.*, Eur. Phys. J. **C15**, 1 (2000).
12. H.A. Torii *et al.*, Phys. Rev. A59 (1999) 223.
13. Y. Kino, M. Kamimura and H. Kudo, Hyperfine Interact., **119**, 201 (1999).
14. S. Funada *et al.*, Nucl. Phys. **A575**, 93 (1994).
15. E. Hiyama and M. Kamimura, Nucl. Phys. **A588**, 35 (1995).
16. V. Lapoux *et al.*, Phys. Rev. C **66**, 034608 (2002).
17. M. Milin *et al.*, Nucl. Phys. **A730**, 285 (2004).
18. G. R. Satchler and W. G. Love, Phys. Rep. **55**, 183 (1979).
19. M. Kamimura, Nucl. Phys. **A351**, 456 (1981).
20. E. Hiyama, B.F. Gibson and M. Kamimura, Phys. Rev. C **70**, 031001(R) (2004).
21. K. Ogata, S. hashimoto, Y. Iseri, M. Kamimura and M. Yahiro, nucl-th/0505007 (2005).

Special Relativity and Reactions with Unstable Nuclei

C.A. Bertulani

Department of Physics, University of Arizona, Tucson, Arizona 85721

Abstract. Dynamical relativistic effects are often neglected in the description of reactions with unstable nuclear beams at intermediate energies ($E_{Lab} \approx 100$ MeV/nucleon). Evidently, this introduces sizable errors in experimental analysis and theoretical descriptions of these reactions. This is particularly important for the experiments held in GANIL/France, MSU/USA, RIKEN/Japan and GSI/Germany. I review a few examples where relativistic effects have been studied in nucleus-nucleus scattering at intermediate energies.

Keywords: Relativity, direct reactions, exotic nuclei, Coulomb dissociation, halo nuclei, continuum-continuum coupling
PACS: 24.10.-i, 24.50.+g, 25.20.Lj, 25.60.-t

1. INTRODUCTION

The number of radioactive beam facilities are growing fast around the world. Some of these facilities use the fragmentation technique, with secondary beams in the energy range $E_{Lab} \approx 100$ MeV/nucleon. Examples are the facilities in GANIL/France, MSU/USA, RIKEN/Japan and GSI/Germany. Relativity, an obviously important physics concept [1], is often neglected in calculations aiming at relating the reaction mechanisms to the internal structure of the projectiles. For example, popular DWBA codes (FRESCO, ECIS, DWUCK, etc.) useful in the analysis of nuclear reactions, include relativity only in kinematic relations. The effects of relativity in the reaction dynamics (i.e. the interaction) is not accounted for because these codes were intended for lower energies. It is also important to notice that the inclusion of relativistic effects in the nucleus-nucleus dynamics is a very difficult task. A fully covariant treatment of the nuclear many-body scattering (with inclusion of retardation effects between all nucleons) is not possible without approximations.

In this short article I will review a few examples where relativistic effects have been included in nuclear reactions at intermediate energies. It is worthwhile to observe that after 100 years of relativity [1] this still remains a challenge in many aspects.

2. SEMICLASSICAL METHODS AND ELASTIC SCATTERING

Semiclassical methods are a very popular tool for the description of nucleus-nucleus collisions at high energies. As an example, I cite the Coulomb excitation mechanism, in which the inelastic cross section can be factorized as

$$\frac{d\sigma_{i \to f}}{d\Omega} = \left(\frac{d\sigma}{d\Omega}\right)_{el} P_{i \to f}, \qquad (1)$$

where $P_{i \to f}$ is the probability for a nuclear transition between the states i and f when the nuclei scatter through an angle Ω. The collision dynamics enters here in two distinct ways: in the calculation of $P_{i \to f}$ and $(d\sigma/d\Omega)_{el}$. Let us forget about $P_{i \to f}$ for a moment and let us investigate $(d\sigma/d\Omega)_{el}$. The question here is what is the error done by measuring the number of particles scattered to Ω and using the Rutherford formula for $(d\sigma/d\Omega)_{el}$. This has been investigated in ref. [2].

A system of two point charges interacting electromagnetically and moving at low velocities can be described by an approximate Lagrangian which depends only on the degrees of freedom of the particles neglecting those related to the electromagnetic field (the Darwin Lagrangian's). In this approximation it is possible to separate the degrees of freedom associated with the relative position r and relative velocity v of the particles from the center of mass degrees of freedom. For a system of particles with different masses this approximation is only possible up to the c^{-2} order, whereas for a system with equal charge- to-mass ratio (with $Z_1 e/m_1 = Z_2 e/m_2$) the approximation goes up to order c^{-4}. Using Lagrange's equation of motion, it is then straightforward to obtain a numerical result for the deflection

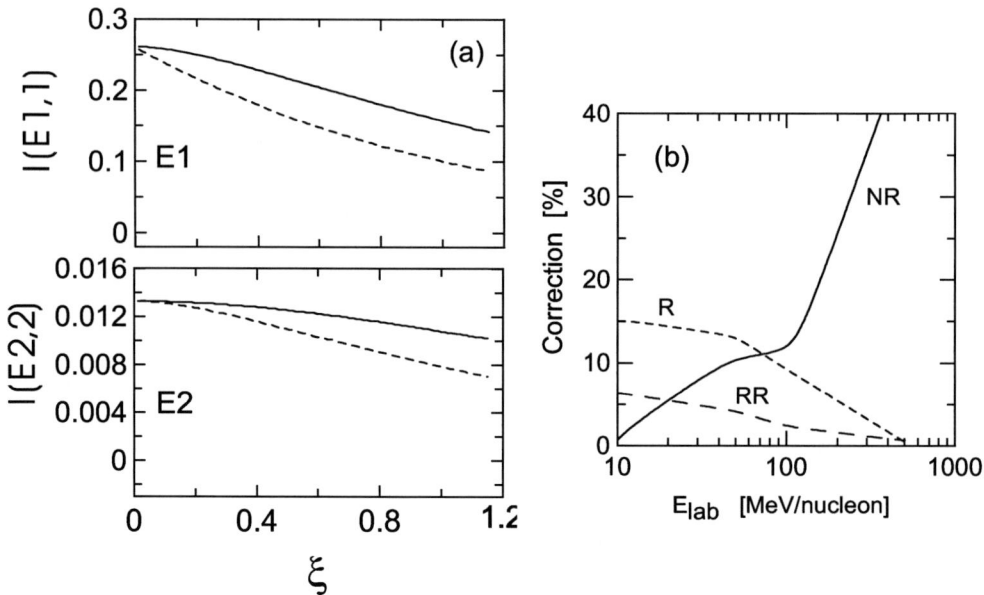

FIGURE 1. (a) Upper panel: Real part (solid line) of the orbital integral $I(E1,1)$ for $\gamma = 1.1$ ($E_{\text{lab}} \simeq 100$ MeV/nucleon), calculated with a theory containing both retardation and relativistic corrections to the Coulomb trajectory. The approximation used in ref. [5], neglecting relativistic corrections to the Coulomb trajectory, is shown by the dashed line. Lower panel: Same plot, but for the orbital integral $I(E2,2)$. (b) Percentage correction in the calculation of the cross sections for Coulomb excitation of the 0.89 MeV state in ^{40}S + ^{197}Au collisions as a function of the bombarding energy. The solid line (NR) corresponds to the use of the non-relativistic orbital integrals [4] compared to the "exact" calculation of ref. [9]. The same is plotted for the other two cases: (R) with retardation effects only [5], and (RR) with the retardation effects plus an approximate recoil correction [5]. For more details, see ref. [5].

angle and the elastic cross section. Explicit expressions for these Lagrangians are given in ref. [2]. The authors show that the scattering angle increases by up to 6% when relativistic corrections are included in ^{208}Pb + ^{208}Pb collisions at 100 MeV/nucleon. The effect on the elastic scattering cross section is even more drastic: $\approx 13\%$ for center-of-mass scattering angles around 0-4 degrees.

Another result obtained in ref. [2] is that the effects of relativity can be taken into account in a much simpler way when the projectile is a light particle scattering on a heavy target (e.g., ^{11}Li + ^{208}Pb). In this case, the main contribution of relativistic corrections (up to 80% of the total effect) is due to the changes in masses, an easily implementable correction. Only about 20% of the relativistic effects are due to magnetic interactions and retardation. Then, a good approximation for the elastic scattering is given by

$$\left(\frac{d\sigma}{d\Omega}\right)_{\text{el}}(\beta, \Theta) = \left[\frac{Z_1 Z_2 e^2}{2mc^2\beta^2 \sin^2(\Theta/2)}\right]^2 \left[1 - h(\Theta)\beta^2 + \mathcal{O}(\beta^2)\right] \quad (2)$$

where

$$h(\Theta) = 1 + \frac{1}{2}\left[1 + (\pi - \Theta)\cot\Theta\right]\tan^2\frac{\Theta}{2}, \qquad \Theta = \pi - 2\frac{\eta}{\sqrt{\eta^2 - \beta^2}}\arctan\sqrt{\eta^2 - \beta^2},$$

with $\eta = \beta L c / Z_1 Z_2 e^2$, L being the angular momentum of the system, and $\beta = v/c$. The function $h(\Theta)$ is always positive for the relevant scattering angles. Therefore one concludes that the relativistic corrections in the elastic Coulomb cross section are always negative if they are parametrized as a function of the initial velocity. On the other hand if one fixes the initial kinetic energy, instead of the velocity, the corrections will be always positive, as shown in ref. [2].

The study mentioned above is useful for an analysis of experimental data on inelastic nucleus-nucleus scattering in intermediate energy collisions. The extension of these calculations to all orders in β is an intriguing problem and an appropriate tool could be the use of the action-at-a-distance electrodynamics of Fokker, Wheeler and Feynman [3]. Certainly, more studies in this direction are needed and the present focus on nuclear reactions at intermediate energies seem to be a good opportunity to investigate such effects.

3. ANOTHER TRACTABLE CASE: COULOMB EXCITATION

The theory of Coulomb excitation in low-energy collisions is very well understood [4]. It has been used and improved for over thirty years to infer electromagnetic properties of nuclei and has also been tested in experiments to a high level of precision. A large number of small corrections are now well known in the theory and are necessary in order to analyze experiments on multiple excitation and reorientation effects.

In the case of relativistic heavy ion collisions pure Coulomb excitation may be distinguished from the nuclear reactions by demanding extreme forward scattering or avoiding the collisions in which violent reactions take place. The Coulomb excitation of relativistic heavy ions is thus characterized by straight-line trajectories with impact parameter b larger than the sum of the radii of the two colliding nuclei. A detailed calculation of relativistic electromagnetic excitation on this basis was performed by Winther and Alder [5]. As in the non-relativistic case, they showed how one can separate the contributions of the several electric ($E\lambda$) and magnetic ($M\lambda$) multipolarities to the excitation. Later, it was shown that a quantum theory for relativistic Coulomb excitation leads to minor modifications of the semiclassical results [6]. In Refs. [7, 8] the inclusion of relativistic effects in semiclassical and quantum formulations of Coulomb excitation was fully clarified.

Recently, the importance of relativistic effects in Coulomb excitation of a projectile by a target with charge Z_2, followed by gamma-decay in nuclear reactions at intermediate energies was studied in details. The Coulomb excitation cross section is given by

$$\frac{d\sigma_{i \to f}}{d\Omega} = \left(\frac{d\sigma}{d\Omega}\right)_{el} \frac{16\pi^2 Z_2^2 e^2}{\hbar^2} \sum_{\pi\lambda\mu} \frac{B(\pi\lambda, I_i \to I_f)}{(2\lambda+1)^3} \mid S(\pi\lambda, \mu) \mid^2, \tag{3}$$

where $B(\pi\lambda, I_i \to I_f)$ is the reduced transition probability of the projectile nucleus, $\pi\lambda = E1, E2, M1, \ldots$ is the multipolarity of the excitation, and $\mu = -\lambda, -\lambda+1, \ldots, \lambda$. The orbital integrals $S(\pi\lambda, \mu)$ contain the information about relativistic corrections on the relative motion between the nuclei, as well as the relativistic effects on the excitation mechanism (e.g. retarded Coulomb interaction). Inclusion of absorption effects in $S(\pi\lambda, \mu)$ due to the imaginary part of an optical nucleus-nucleus potential where worked out in ref. [8]. These orbital integrals depend on the Lorentz factor $\gamma = (1-v^2/c^2)^{-1/2}$, with c being the speed of light, on the multipolarity $\pi\lambda\mu$, and on the adiabacity parameter $\xi(b) = \omega_{fi} b/\gamma v < 1$, where $\omega_{fi} = (E_f - E_i)/\hbar$ is the excitation energy (in units of $\hbar$) and b is the impact parameter. It is often more convenient to write the orbital integrals in another form, e.g. $S(E\lambda, \mu) = \frac{\mathscr{C}_{\lambda\mu}}{va^\lambda} I(E\lambda, \mu)$, where $\mathscr{C}_{\lambda\mu}$ is a geometrical factor, depending only on λ and μ, $a = Z_1 Z_2 e^2/\gamma m v^2$, and $I(E\lambda, \mu)$ is now the orbital integral in an dimensionless form.

Figure 1(a) (from [9]) shows the effect of relativity on the Coulomb interaction and classical trajectory corrections in nuclear collisions at intermediate collisions. The comparison is made in terms of the variable ξ which is the appropriate variable for high energy collisions. Only for $\xi \ll 1$ the expressions in the relativistic limit reproduce the correct behavior of the orbital integrals.

TABLE 1. Coulomb excitation cross sections of the first excited state in ^{38,40,42}S and ^{44,46}Ar projectiles at 10, 50 100 and 500 MeV/nucleon incident on gold targets. The numbers inside parenthesis and brackets were obtained with pure non-relativistic and straight-line relativistic calculations, respectively. The numbers at the center are obtained with the "full" account of relativistic effects, as explained in ref. [9].

Nucleus	E_x [MeV]	B(E2) [e^2fm^4]	10 MeV/A σ_C [mb]	50 MeV/A σ_C [mb]	100 MeV/A σ_C [mb]	500 MeV/A σ_C [mb]
^{38}S	1.29	235	(492) 500 [651]	(80.9) 91.7 [117]	(40.5) 50.1 [57.1]	(9.8) 16.2 [16.3]
^{40}S	0.89	334	(877) 883 [1015]	(145.3) 162 [183]	(76.1) 85.5 [93.4]	(9.5) 20.9 [21.]
^{42}S	0.89	397	(903) 908 [1235]	(142.7) 158 [175]	(65.1) 80.1 [89.4]	(9.9) 23.2 [23.4]
^{44}Ar	1.14	345	(747) 752 [985]	(133) 141 [164]	(63.3) 71.7 [80.5]	(8.6) 17.5 [17.6]
^{46}Ar	1.55	196	(404) 408 [521]	(65.8) 74.4 [88.5]	(30.2) 37.4 [41.7]	(5.72) 10.8 [11]

Table 1 shows the effects of relativistic corrections in the collision of the radioactive nuclei ^{38,40,42}S and ^{44,46}Ar on gold targets. These reactions have been studied at $E_{lab} \sim 40$ MeV/nucleon at the MSU facility [10]. Table 1 shows the Coulomb excitation cross sections of the first excited state in each nucleus as a function of the bombarding energy per nucleon. The cross sections are given in milibarns. The numbers inside parenthesis and brackets were obtained with pure non-relativistic [4] and relativistic calculations [5], respectively. The minimum impact parameter is chosen so that the distance of closest approach corresponds to the sum of the nuclear radii in a collision following a Rutherford trajectory. One observes that at 10 MeV/nucleon the relativistic corrections are important only at the level of 1%. At 500 MeV/nucleon, the correct treatment of the recoil corrections is relevant on the level of 1%. Thus the non-relativistic treatment of Coulomb excitation [4] can be safely used for energies below about 10 MeV/nucleon and the relativistic treatment with a straight-line trajectory [5] is adequate above about 500 MeV/nucleon. However at energies around 50 to 100 MeV/nucleon, accelerator energies common to most radioactive beam facilities (MSU, RIKEN, GSI, GANIL), it is very important to use a correct treatment of recoil and relativistic effects, both kinematically and dynamically. At these energies, the corrections can add up to 50%. These effects were also shown in Ref. [7] for the case of excitation of giant resonances in collisions at intermediate energies.

We conclude that a reliable extraction useful nuclear properties, like the electromagnetic response (B(E2)-values, γ-ray angular distribution, etc.) from Coulomb excitation experiments at intermediate energies requires a proper treatment of special relativity. The effect is highly non-linear, i.e. a 10% increase in the velocity might lead to a 50% increase (or decrease) of certain physical quantities.

4. STRONG INTERACTION: GLAUBER, DWBA AND SEMICLASSICAL METHODS

The treatment of the strong interaction in nucleus-nucleus collisions at intermediate and high energies is evidently much more complicated than the case of Coulomb excitation, described in the previous section. Fortunately, many direct nuclear processes, e.g. nucleon knockout, or stripping, elastic breakup (diffraction dissociation), etc, are possible to study using the optical limit of the Glauber theory, in which the nuclear ground-state densities and the nucleon-nucleon total cross sections are the main input. In fact, this method has become one of the main tools in the study of nuclei far from stability [11]. The reason is that the eikonal (or Glauber) methods only use the dependence of the scattering matrices, $S(b)$, on the transverse direction, b. Transverse directions are always Lorentz invariants. The reason for using $S(b)$, instead of $S(\mathbf{r})$, traces back to the eikonal scattering wavefunction at the asymptotic region ($r \longrightarrow \infty$),

$$\Psi_{scatt} = S(b) \exp(i\mathbf{k} \cdot \mathbf{r}) , \qquad (4)$$

where $\mathbf{k}$ is the particle's momentum, and $\mathbf{r}$ its position. Obviously, the plane wave part of eq. 4 is Lorentz invariant. The S-matrix in eq. 4 is $S = \exp\{i\chi(b)\}$, where $\chi(b)$ is the eikonal phase-shift, given in terms of the interaction potential V by

$$\chi = -\frac{1}{\hbar v} \int_{-\infty}^{\infty} dz\, V(r) . \qquad (5)$$

Under a Lorentz transformation to the target system, the coordinate z transforms as $z \longrightarrow \gamma z$. Thus, strictly speaking, the S-matrix $S(b)$ is Lorentz invariant only if V transforms as the time-component of a four-vector, i.e. $V(r) \longrightarrow \gamma W(b, \gamma z)$.

The relativistic property described above is most easily seen within a folding potential model for a nucleon-nucleus collision:

$$V(\mathbf{r}) = \int d r'^3\, \rho_T(\mathbf{r}')\, v_{NN}(\mathbf{r} - \mathbf{r}') , \qquad (6)$$

where $\rho_T(\mathbf{r}')$ is the nuclear density of the target. In the frame of reference of the projectile, the density of the target looks contracted and particle number conservation leads to the relativistic modification of eq. 6 so that $\rho_T(\mathbf{r}') \to \gamma \rho_T(\mathbf{r}'_\perp, \gamma z')$, where $\mathbf{r}'_\perp$ is the transverse component of $\mathbf{r}'$. But the number of nucleons as seen by the target (or projectile) per unit area remains the same. In other words, a change of variables $z'' = \gamma z'$ in the integral of eq. 6 seems to restore the same eq. 6. However, this change of variables also modifies the nucleon-nucleon interaction v_{NN}. Thus, relativity introduces non-trivial effects in a potential model description of nucleus-nucleus scattering at high energies. Colloquially speaking, nucleus-nucleus scattering at high energies is not simply an incoherent sequence of nucleon-nucleon collisions. Since the nucleons are confined within a box (inside the nucleus), Lorentz contraction induces a collective effect: in the extreme limit $\gamma \to \infty$ all nucleons would interact at once with the projectile. This is often neglected in pure geometrical (Glauber model) description of nucleus-nucleus collisions at high energies, as it is assumed that the nucleons inside "firetubes" scatter independently.

Assuming that the nucleon-nucleon interaction is of very short range so that the approximation $v_{NN}(\mathbf{r} - \mathbf{r}') = J_0\, \delta(\mathbf{r} - \mathbf{r}')$ can be used, one sees from eq. 6 that $V(\mathbf{r})$, the interaction that a nucleon in the projectile has with the target nucleus, also has similar transformation properties as the density: $V(\mathbf{r}) \to \gamma W(\mathbf{r}_\perp, \gamma z)$, i.e. $V(\mathbf{r})$ transforms as the time-component of a four-vector. In this situation, the Lorentz contraction has no effect whatsoever in the diffraction dissociation amplitudes, described in the previous sections within the eikonal approximation. This is because a change of variables $z' = \gamma z$ in the eikonal phases leads to the same result as in the non-relativistic case, as can be easily checked from eq. 5. Of course, the delta-function approximation for the nucleon-nucleon interaction means that nucleons will scatter at once, and Lorentz contraction does not introduce any additional collective effect. This is not the case for realistic interactions with finite range and collisions at intermediate energies.

Using the eikonal approach to account for scattering of particles within the projectile, one obtains the wavefunctions for the initial and final states as

$$\Psi_i = \phi_i(\mathbf{r}) \exp(i\mathbf{k} \cdot \mathbf{R}), \qquad \Psi_f = \phi_f(\mathbf{r}) S(b) \exp(i\mathbf{k} \cdot \mathbf{R}), \tag{7}$$

where $\phi_{i,f}(\mathbf{r})$ are the initial and final probability amplitudes (wavefunctions) that a particle in the projectile is at a distance $\mathbf{r}$ from its center of mass. The particle's S-matrix, $S(b)$, accounts for the distortion due to the interaction. If we now assume that the projectile is a two-body system (e.g. a core+valence particle), we get the *diffraction dissociation* formula as follows. The wavefunction of a two-body projectile in the initial and final states is given by

$$\Psi_i = \phi_i(\mathbf{r}) \exp[i(\mathbf{k}_c \cdot \mathbf{r}_c + \mathbf{k}_v \cdot \mathbf{r}_v)], \qquad \Psi_f = \phi_f(\mathbf{r}) S_c(b_c) S_v(b_v) \exp[i(\mathbf{k}'_c \cdot \mathbf{r}_c + \mathbf{k}'_v \cdot \mathbf{r}_v)], \tag{8}$$

where now $\phi_{i,f}(\mathbf{r})$ are the initial and final intrinsic wavefunctions of the (core+valence particle) as a function of $\mathbf{r} = \mathbf{r}_1 - \mathbf{r}_2$. The relation between the intrinsic, $\mathbf{r}$, and center of mass, $\mathbf{R}$, coordinates is given in terms of the mass ratios $\beta_i = m_i/m_P$. Explicitly, $\mathbf{r}_v = \mathbf{R} + \beta_c \mathbf{r}$ and $\mathbf{r}_c = \mathbf{R} - \beta_v \mathbf{r}$. The core and valence particle S-matrices, $S_c(b_c)$ and $S_v(b_v)$, account for the distortion due to the interaction with the target.

The probability amplitude for diffraction dissociation is the overlap between the two wavefunctions above, i.e.

$$A_{(\text{diff})} = \int d^3 r_c d^3 r_v \, \phi_f^*(\mathbf{r}) \phi_i(\mathbf{r}) \, \delta(z_c + z_v) S_c(b_c) S_v(b_v) \exp[i(\mathbf{q}_c \cdot \mathbf{r}_c + \mathbf{q}_v \cdot \mathbf{r}_v)], \tag{9}$$

where $\mathbf{q}_c = \mathbf{k}'_c - \mathbf{k}_c$ is the momentum transfer to the core particle, and accordingly for the valence particle. The above formula yields the probability amplitude that the projectile starts the collision as a bound state and ends up as two separated pieces, in this case, the core and the valence particle (e.g. a proton or a neutron). All the information for the dissociation mechanism comes from the knowledge of the S-matrices, S_c and S_v. The delta-function $\delta(Z)$ was introduced in the eq. 9 to account for the fact that the S-matrices calculated in the eikonal approximation only depend on the transverse direction.

In the weak interaction limit, or perturbative limit, the phase-shifts are very small so that

$$S_c(b_c) S_v(b_v) = \exp[i(\chi_c + \chi_v)] \simeq 1 + i\chi_c + i\chi_v$$
$$= 1 - \frac{i}{\hbar v} \int V_{cT}(\mathbf{r}_c) \, dz_c - \frac{i}{\hbar v} \int V_{vT}(\mathbf{r}_v) \, dz_v. \tag{10}$$

The factor 1 does not contribute to the breakup. Thus, inserting the result above in eq. 9, we obtain

$$A_{(\text{PWBA})} \simeq \frac{1}{i\hbar v} \int d^3 r_c d^3 r_v \, \phi_f^*(\mathbf{r}) \phi_i(\mathbf{r}) [V_{cT}(\mathbf{r}_c) + V_{vT}(\mathbf{r}_v)] \exp[i(\mathbf{q}_c \cdot \mathbf{r}_c + \mathbf{q}_v \cdot \mathbf{r}_v)], \tag{11}$$

where the integrals over z_c and z_v in eq. 10 were absorbed back to the integrals over $\mathbf{r}_c$ and $\mathbf{r}_v$ after use of the delta-function $\delta(z_c + z_v)$. The above equation is nothing more than the plane-wave Born-approximation (PWBA) amplitude. However, absorption is not treated properly. For small values of $\mathbf{r}_c$ and $\mathbf{r}_v$ the phase-shifts are not small and the approximation used in eq. 10 fails. A better approximation is to assume that for small distances, where absorption is important, $S_c(b_c) S_v(b_v) \simeq S(b)$, where the right-hand side is the S-matrix for the projectile scattering as a whole on the target. Using the coordinates $\mathbf{r}$ and $\mathbf{R}$, and defining $U_{int}(\mathbf{r},\mathbf{R}) = V_{cT}(\mathbf{r}_c) + V_{nT}(\mathbf{r}_n)$, one gets for the T-matrix

$$T_{(\text{DWBA})} = i\hbar v A_{(\text{DWBA})} \simeq \int d^3 r d^3 R \, \phi_f^*(\mathbf{r}) \exp[i\mathbf{q} \cdot \mathbf{r}] \phi_i(\mathbf{r}) U_{int}(\mathbf{r},\mathbf{R}) S(b) \exp[i\mathbf{Q} \cdot \mathbf{R}]. \tag{12}$$

In elastic scattering, or excitation of collective modes (e.g. giant resonances), the momentum transfer to the intrinsic coordinates can be neglected and the equation above can be written as

$$T_{(\text{DWBA})} = \left\langle \chi^{(-)}(\mathbf{R}) \phi_c(\mathbf{r}) \middle| U_{int}(\mathbf{r},\mathbf{R}) \middle| \chi^{(+)}(\mathbf{R}) \phi_i(\mathbf{r}) \right\rangle, \tag{13}$$

which has the known form of the DWBA T-matrix. The scattering phase space now only depends on the center of mass momentum transfer $\mathbf{Q}$. When the center of mass scattering waves are represented by eikonal wavefunctions, one has

$$\chi^{(-)*}(\mathbf{R})\chi^{(+)}(\mathbf{R}) \simeq S(b)\exp[i\mathbf{Q}\cdot\mathbf{R}] . \tag{14}$$

This shows that the PWBA and the DWBA are perturbative expansions of the diffraction dissociation formula 9.

In DWBA (or in the eikonal approximation, eq. 14), b does not have the classical meaning of an impact parameter. To obtain the semiclassical limit one goes one step further. By using eq. 12 and assuming that R depends on time so that $R = (\mathbf{b}, Z = vt)$, the semiclassical scattering amplitude is given by $A^{(i\to f)}_{(\text{semiclass})}(b) = \int d^2 b\, a^{(i\to f)}_{(\text{semiclass})}(b) \exp(i\mathbf{Q}\cdot\mathbf{b})$, where

$$a^{(i\to f)}_{(\text{semiclass})}(b) = \frac{1}{i\hbar} S(b) \int dt d^3 r \, \exp(i\omega_{if} t) \, \phi_f^*(\mathbf{r}) U_{int}(\mathbf{r},t) \phi_i(\mathbf{r}) , \tag{15}$$

where $Q_z Z = \omega_{if} t$ was used.

The semiclassical probability for the transition $(i \to f)$ is obtained from the above equations after integration over $\mathbf{Q}$. One gets $P^{(i\to f)}_{(\text{semiclass})}(b) = \left|a^{(i\to f)}_{(\text{semiclass})}(b)\right|^2$, with b having now the explicit meaning of an impact parameter. Thus, $a^{(i\to f)}_{(\text{semiclass})}(b)$ is the semiclassical excitation amplitude. Equation 15 is well-known (for example in Coulomb excitation at low energies, where $U_{int} = U_C$) except that the factor $S(b)$ is usually set to one. In high energy collisions it is crucial to keep this factor, as it accounts for refraction and absorption at small impact parameters: $|S(b)|^2 = \exp\left[2\chi^{(\text{imag})}\right]$, where $\chi^{(\text{imag})}$ is calculated with the imaginary part of the optical potential. The derivation of the DWBA and semiclassical limits of eikonal methods can be easily extended to higher-orders in the perturbation V. The eikonal method includes all terms of the perturbation series in the sudden-collision limit.

The developments presented in this section show that the DWBA calculations of nuclear excitation, and the higher order terms, are implicitly included in the eikonal models. However, there is a subtle link to the optical potential, which makes the theory Lorentz covariant. Usually, the optical limit of the Glauber-eikonal series is used. In this model, no explicit reference to a nuclear potential is done; only the nucleon-nucleon cross sections and nuclear densities are used as input (see, e.g. ref. [12]). As shown above, this is also not a guarantee of Lorentz covariance. Even worse is the fact that often DWBA (and higher-order, e.g. CDCC [13, 14, 15]) calculations are used without consideration of relativistic effects. In the next section I give an example of a method (relativistic CDCC) which incorporates relativistic corrections in a continuum-discretized basis [16].

5. RELATIVISTIC CONTINUUM DISCRETIZED COUPLED-CHANNELS

Let us consider the Klein-Gordon (KG) equation with a potential V_0 which transforms as the time-like component of a four-vector [17] (here I use the notation $\hbar = c = 1$). For a system with total energy E (including the rest mass M), the KG equation can be cast into the form of a Schrödinger equation (with $\hbar = c = 1$), $(\nabla^2 + k^2 - U)\Psi = 0$, where $k^2 = (E^2 - M^2)$ and $U = V_0(2E - V_0)$. When $V_0 \ll M$, and $E \simeq M$, one gets $U = 2MV_0$, as in the non-relativistic case. The condition $V_0 \ll M$ is met in peripheral collisions between nuclei at all collision energies. Thus, one can always write $U = 2EV_0$. A further simplification is to assume that the center of mass motion of the incoming projectile and outgoing fragments is only weakly modulated by the potential V_0. To get the dynamical equations, one discretizes the wavefunction in terms of the longitudinal center-of-mass momentum k_z, using the ansatz

$$\Psi = \sum_\alpha \mathscr{S}_\alpha(z,\mathbf{b}) \exp(ik_\alpha z) \, \phi_{k_\alpha}(\xi) . \tag{16}$$

In this equation, $(z,\mathbf{b})$ is the projectile's center-of-mass coordinate, with $\mathbf{b}$ equal to the transverse coordinate. $\phi(\xi)$ is the projectile intrinsic wavefunction and $(k,\mathbf{K})$ is the projectile's center-of mass momentum with longitudinal momentum k and transverse momentum $\mathbf{K}$. There are hidden, uncomfortable, assumptions in eq. 16. The separation between the center of mass and intrinsic coordinates is not permissible under strict relativistic treatments. For high energy collisions we can at best justify eq. 16 for the scattering of light projectiles on heavy targets. Eq. 16 is only reasonable if the projectile and target closely maintain their integrity during the collision, as in the case of very peripheral collisions.

Neglecting the internal structure means $\phi_{k_\alpha}(\xi) = 1$ and the sum in eq. 16 reduces to a single term with $\alpha = 0$, the projectile remaining in its ground-state. It is straightforward to show that inserting eq. 16 in the KG equation

$\left(\nabla^2 + k^2 - 2EV_0\right)\Psi = 0$, and neglecting $\nabla^2 \mathscr{S}_0(z,\mathbf{b})$ relative to $ik\partial_z \mathscr{S}_0(z,\mathbf{b})$, one gets $ik\partial_z \mathscr{S}_0(z,\mathbf{b}) = EV_0 \mathscr{S}_0(z,\mathbf{b})$, which leads to the center of mass scattering solution $\mathscr{S}_0(z,\mathbf{b}) = \exp\left[-iv^{-1}\int_{-\infty}^{z} dz' V_0(z',\mathbf{b})\right]$, with $v = k/E$. Using this result in the Lippmann-Schwinger equation, one gets the familiar result for the eikonal elastic scattering amplitude, i.e. $f_0 = -i(k/2\pi)\int d\mathbf{b} \exp(i\mathbf{Q}\cdot\mathbf{b})\{\exp[i\chi(\mathbf{b})] - 1\}$, where the eikonal phase is given by $\exp[i\chi(\mathbf{b})] = \mathscr{S}_0(\infty,\mathbf{b})$, and $\mathbf{Q} = \mathbf{K}' - \mathbf{K}$ is the transverse momentum transfer. Therefore, the elastic scattering amplitude in the eikonal approximation has the same form as that derived from the Schrödinger equation in the non-relativistic case.

For inelastic collisions we insert eq. 16 in the KG equation and use the orthogonality of the intrinsic wavefunctions $\phi_{k_\alpha}(\xi)$. This leads to a set of coupled-channels equations for $\mathscr{S}_\alpha$:

$$\left(\nabla^2 + k^2\right)\mathscr{S}_\alpha e^{ik_\alpha z} = \sum_\alpha \langle\alpha|U|\alpha'\rangle \mathscr{S}_{\alpha'} e^{ik_{\alpha'} z}, \tag{17}$$

with the notation $|\alpha\rangle = |\phi_{k_\alpha}\rangle$. Neglecting terms of the form $\nabla^2 \mathscr{S}_\alpha(z,\mathbf{b})$ relative to $ik\partial_z \mathscr{S}_\alpha(z,\mathbf{b})$, eq. 17 reduces to

$$iv\frac{\partial \mathscr{S}_\alpha(z,\mathbf{b})}{\partial z} = \sum_{\alpha'}\langle\alpha|V_0|\alpha'\rangle \mathscr{S}_{\alpha'}(z,\mathbf{b}) e^{i(k_{\alpha'}-k_\alpha)z}. \tag{18}$$

The scattering amplitude for the transition $0 \to \alpha$ is given by

$$f_\alpha(\mathbf{Q}) = -\frac{ik}{2\pi}\int d\mathbf{b} \exp(i\mathbf{Q}\cdot\mathbf{b})\left[S_\alpha(\mathbf{b}) - \delta_{\alpha,0}\right], \tag{19}$$

with $S_\alpha(\mathbf{b}) = \mathscr{S}_\alpha(z=\infty,\mathbf{b})$. The set of equations 18 and 19 are the relativistic-CDCC equations (RCDCC).

The RCDCC equations have been used [16] to study the dissociation of ^{8}B projectiles at high energies. The energies transferred to the projectile are small, so that the wavefunctions can be treated non-relativistically in the projectile frame of reference. In this frame the wavefunctions are described in spherical coordinates, i.e. $|\alpha\rangle = |jlJM\rangle$, where j, l, J and M denote the angular momentum numbers characterizing the projectile state. Eq. 18 is Lorentz invariant if the potential V_0 transforms as the time-like component of a four-vector. The matrix element $\langle\alpha|V_0|\alpha'\rangle$ is also Lorentz invariant, and one can therefore calculate them in the projectile frame.

The longitudinal wavenumber $k_\alpha \simeq (E^2 - M^2)^{1/2}$ also defines how much energy is gone into projectile excitation, since for small energy and momentum transfers $k'_\alpha - k_\alpha = (E'_\alpha - E_\alpha)/v$. In this limit, eqs. 18 and 19 reduce to semiclassical coupled-channels equations, if one uses $z = vt$ for a projectile moving along a straight-line classical trajectory, and changing to the notation $\mathscr{S}_\alpha(z,b) = a_\alpha(t,b)$, where $a_\alpha(t,b)$ is the time-dependent excitation amplitude for a collision wit impact parameter b (see eqs. 41 and 76 of ref. [18]). The full version of eq. 19 was used in ref. [16], with relativistic corrections in both the Coulomb and nuclear potentials.

If the state $|\alpha\rangle$ is in the continuum (positive proton+^{7}Be energy) the wavefunction is discretized according to $|\alpha;E_\alpha\rangle = \int dE'_\alpha \Gamma(E'_\alpha) |\alpha;E'_\alpha\rangle$, where the functions $\Gamma(E_\alpha)$ are assumed to be strongly peaked around the energy E_α with width ΔE. For convenience the histogram set (eq. 3.6 of ref. [19]) is chosen. The inelastic cross section is obtained by solving the RCDCC equations and using $d\sigma/d\Omega dE_\alpha = |f_\alpha(\mathbf{Q})|^2 \Gamma^2(E_\alpha)$.

Figure 2 shows the relative energy spectrum between the proton and the ^{7}Be after the breakup of ^{8}B on lead targets at 83 MeV/nucleon. The data are from ref. [20]. In this case, the calculation was restricted to $b > 30$ fm. The dotted curve is the first-order perturbation calculation, the solid curve is the RCDCC calculation, and the dashed curve is obtained with the replacement of γ by unity in the nuclear and Coulomb potentials. The difference between the solid and the dashed-curve is of the order of 4-9%.

6. CONCLUSIONS

The consequence of neglecting relativity in nuclear reactions at intermediate energies is not easy to access. The inclusion of relativity introduces non-trivial effects in semiclassical, DWBA, eikonal, and continuum discretized coupled-channels calculations. Nuclear collisions are up to now the most used probe of the internal structure of rare nuclear isotopes. To my knowledge, most experiments have been analyzed using non-relativistic theoretical methods. It might be necessary to review the results of some of these data, using a proper treatment of the relativistic corrections in the theoretical calculations used in the experimental analysis. Other improvements of the formalisms presented here

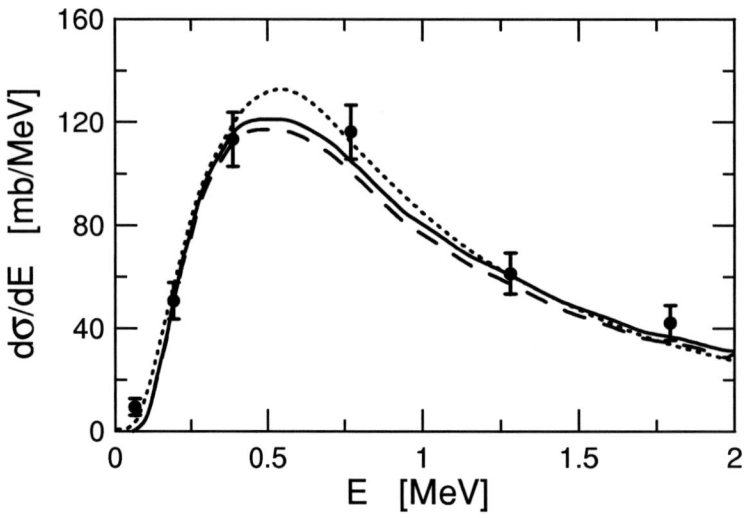

FIGURE 2. Cross sections for the dissociation reaction ^{8}B+Pb $\rightarrow$ p+^{7}Be+Pb at 83 MeV/nucleon and for $\theta_8 < 1.8^0$. Data are from ref. [20]. The dotted curve is the first-order perturbation result. The solid curve is the RCDCC calculation. The dashed curve is obtained with the replacement of γ by unity in the nuclear and Coulomb potentials.

and elsewhere needs to be assessed. The relativistic effects in the nuclear interaction has also to be studied in more depth.

Special relativity [1], one of the most precious theories of Eintein's legacy, still remains a source of intriguing effects, not always easy to tackle. Nuclear physics is full of such examples.

ACKNOWLEDGMENTS

I wish to thank Gerhard Baur, Kai Hencken and Stefan Typel for discussions on various topics in this field.

REFERENCES

1. A. Einstein, Annalen der Physik **17** (1905) 891.
2. C.E.Aguiar, A.N.F.Aleixo and C.A.Bertulani, Phys. Rev. **C 42** (1990) 2180.
3. A.D. Fokker, Z. Phys. **58**, 386 (1929); J.A. Wheeler and R.P. Feynman, Rev. Mod. Phys. **17**, 157 (1945); **21**, 425 (1949).
4. K. Alder and A. Winther, Electromagnetic Excitation, North-Holland, Amsterdam, 1975.
5. A. Winther and K. Alder, Nucl. Phys. **A 319** (1979) 518.
6. C.A. Bertulani and G. Baur, Nucl. Phys. **A 442** (1985) 739; C.A. Bertulani and G. Baur, Phys. Rep. **163** (1988) 299.
7. A.N.F. Aleixo and C.A. Bertulani, Nucl. Phys. **A 505** (1989) 448.
8. C.A. Bertulani and A.M. Nathan, Nucl. Phys. **A 554** (1993) 158.
9. C.A. Bertulani, A. Stuchbery, T. Mertzimekis and A. Davies, Phys. Rev. **C 68**, 044609 (2003).
10. H. Scheit *et al.*, Phys. Rev. Lett. **77** (1996) 3967.
11. P. G. Hansen and J. A. Tostevin, Annu. Rev. Nucl. Part. Sci. **53**, 219 (2003).
12. C.A. Bertulani and P. Danielewicz, "Introduction to Nuclear Reactions", IOP Publishing, London, 2004, p. 419.
13. G.H. Rawitscher, Phys. Rev. **C 9**, 2210 (1974).
14. Y. Sakuragi, M. Yashiro and M. Kamimura, Prog. Theor. Phys. Suppl. **89**, 136 (1986).
15. F.M. Nunes and I.J. Thompson, Phys. Rev. **C 57**, R2818 (1998).
16. C.A. Bertulani, Phys. Rev. Lett. **94**, 072701 (2005).
17. L.G. Arnold and B.C. Clark, Phys. Lett. **B 84**, 46 (1979).
18. C.A. Bertulani, C.M. Campbell, and T. Glasmacher, Comput. Phys. Commun. **152**, 317 (2003).
19. C.A.Bertulani and L.F.Canto, Nucl. Phys. **A 540,** 328 (1992).
20. B. Davids, et al., Phys. Rev. **C 63**, 065806 (2001).

Effective Interactions in Neutron-Rich Matter

F. Sammarruca*, P. Krastev* and W. Barredo*

University of Idaho, Moscow, Idaho, 83844-0903 (USA)

Abstract. We are generally concerned with probing the behavior of the isospin-asymmetric equation of state. In particular, we will discuss the one-body potentials for protons and neutrons obtained from our Dirac-Brueckner-Hartree-Fock calculations of neutron-rich matter properties. We will also present predictions of proton-proton and neutron-neutron cross sections in the isospin-asymmetric nuclear medium.

Keywords: Nuclear Matter, Equation of State, Isospin asymmetry
PACS: 21.30.-x, 21.30.Fe, 21.65.+f

INTRODUCTION

A topic of contemporary interest in nuclear physics is the investigation of the effective nucleon-nucleon (NN) interaction in a dense hadronic environment. Such environment can be produced in the laboratory via energetic heavy ion (HI) collisions and is found in astrophysical systems, particularly the interior of neutron stars. In all cases predictions rely heavily on the nuclear equation of state (EOS), which is one of the main ingredients for transport simulations of HI collisions as well as the calculation of neutron star properties.

Supernova explosions and neutron star formation/stability are phenomena where the nuclear EOS plays a crucial role. The symmetry energy determines the proton fraction in neutron stars in β equilibrium, and, in turn, the cooling rate and neutrino emission. Models of prompt supernova explosion and systematic analyses of neutron star masses provide often conflicting information on the "softness" of the EOS and its incompressibility at equilibrium.

At the same time, collisions of neutron-rich nuclei, which are the purpose of the Rare Isotope Accelerator (RIA), provide a unique opportunity to obtain terrestrial data suitable for constraining the properties of dense and highly asymmetric matter. Such reactions are capable of producing extended regions of space/time where both the total nucleon density and the neutron/proton asymmetry are large. Transport equations, such as the Boltzmann-Uehling-Uhlenbeck (BUU) equation, describe the evolution of a non-equilibrium gas of strongly interacting hadrons. In BUU-type models, particles drift in the presence of the self-consistent field while undergoing two-body collisions, which require the knowledge of in-medium two-body cross sections. In a microscopic approach, both the mean field and the binary collisions are calculated self-consistently starting from the bare two-nucleon force.

The contribution to the mean field from the neutron/proton asymmetry can be measured through isospin-sensitive observables [1] and is one of the focal points of this paper. In the next section, we will discuss the predictions for the single-neutron/proton potentials and the closely related *symmetry potential* as obtained from our Dirac-Brueckner-Hartree-Fock calculations of asymmetric matter [2]. We will compare with empirical information from optical model analyses. The large model dependence of predictions for those observables that depend sensitively on the difference between neutron and proton properties in asymmetric matter calls for additional experimental constraints.

Then, we will shift our attention to microscopic predictions of in-medium isospin-dependent NN cross sections. At this time, we will only be concerned with the case of identical nucleons, and only their strong interaction (Coulomb contributions to the *pp* cross section are not included). As mentioned above, asymmetry considerations are of particular interest at this time due to the opportunity to study collisions of neutron-rich nuclei at RIA energies.

In a simple approach, the assumption is made that the transition matrix in the medium is approximetely the same as the one in vacuum, and the medium effects come in only through the use of the nucleon effective mass in the phase space factors [3, 4]. Concerning microscopic approaches, earlier predictions can be found, for instance, in Ref. [5] and Ref. [6], but asymmetry considerations are not included in those predictions. On the other hand, it is important to investigate to which extent the in-medium cross sections are sensitive to the proton/neutron ratio, one of the purposes of this paper. In-medium cross sections are necessary to study the mean free path of nucleons in nuclear matter and thus nuclear transparency. The latter is obviously related to the total reaction cross section of a nucleus, which, in turn, can be used to extract nuclear r.m.s. radii within Glauber-type models [7]. Therefore, accurate in-medium

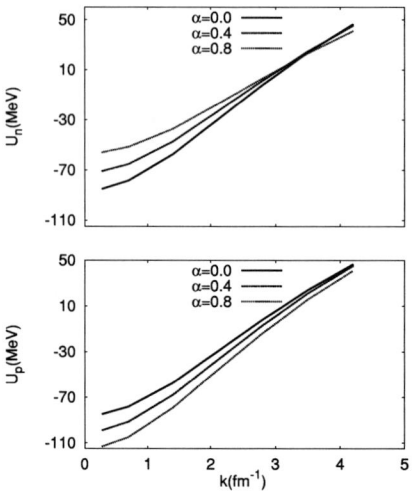

FIGURE 1. The single-neutron (upper panel) and single-proton (lower panel) potential as a function of the nucleon momentum for three different values of the asymmetry parameter. The average Fermi momentum is 1.4 fm^{-1}.

isospin-dependent NN cross sections can ultimately be very valuable to obtain information about the size of exotic, neutron-rich nuclei.

Unless otherwise specified, we use the Bonn-B potential [8] and the relativistic Brueckner-Hartree-Fock (DBHF) model outlined in Ref. [2] and applied to calculations of neutron radii and neutron skins in Ref. [9]. Here, we will concentrate on the aspects of our work that have not yet appeared in the literature.

THE SINGLE-NUCLEON POTENTIALS

Momentum dependence

For the single-particle potential, we use the prescription of Ref. [10, 11]. In the case of unequal Fermi levels for protons and neutrons, that prescription gives, schematically

$$U_i(k) = Re[\sum_{q<k_F^n} <kq|G_{in}|kq-qk> + \sum_{q<k_F^p} <kq|G_{ip}|kq-qk>] \qquad (1)$$

where $i = n/p$ for neutron/proton, and k refers to states below and above the Fermi momentum.

We begin by examining the momentum dependence of $U_{n/p}$, the single neutron/proton potential in neutron-rich matter. In Fig. 1, we show $U_{n/p}$ as a function of the momentum and for different values of the asymmetry parameter, $\alpha = (\rho_n - \rho_p)/(\rho_n + \rho_p)$, with ρ_n and ρ_p the neutron and proton densities. The total nucleon density considered in the figure is equal to 0.185 fm^{-3} and corresponds to a Fermi momentum of 1.4 fm^{-1}, which is very close to our predicted saturation density.

For increasing values of α, the proton potential becomes increasingly attractive while the opposite tendency is observed in U_n. This reflects the fact that the proton-neutron interaction, the one predominantly felt by the single proton as the proton density is depleted, is more attractive than the one between identical nucleons. Also, as it appears reasonable, the dependence on α becomes weaker at larger momenta.

The role of the momentum dependence of the symmetry potential in heavy-ion collisions was recently examined [13] and found to be important. Symmetry potentials with and without momentum dependence and yielding similar predictions for the symmetry energy can lead to significantly different predictions of collision observables [13].

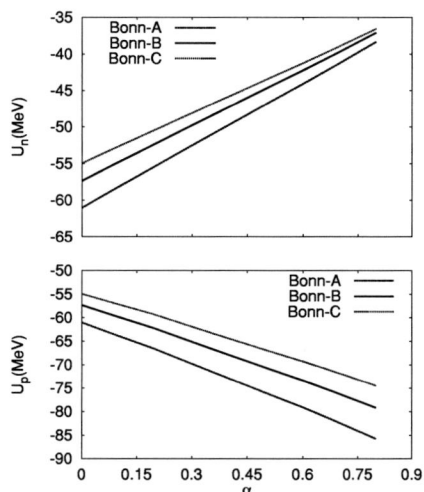

FIGURE 2. The single-neutron (upper panel) and single-proton (lower panel) potential as a function of the asymmetry parameter for fixed average density ($k_F = 1.4$ fm^{-1}) and nucleon momentum ($k = k_F$).

Asymmetry dependence and the symmetry potential

Regarding $U_{n/p}$ as functions of the asymmetry parameter α, one can easily verify that the following approximate relation applies

$$U_{n/p}(k,k_F,\alpha) \approx U_{n/p}(k,k_F,\alpha=0) \pm U_{sym}(k,k_F)\alpha \qquad (2)$$

with the $\pm$ referring to neutron/proton, respectively. Figure 2 displays the left-hand side of Eq. (2) for fixed density and nucleon momentum and clearly reveals the linear behaviour of $U_{n/p}$ as a function of α.

Although the main focus of Fig. 2 is the α dependence, model-dependence is also addressed by displaying predictions for the Bonn A, B, and C potentials [8]. These three models differ mainly in the strength of the tensor force, which is mostly carried by partial waves with isospin equal to 0 and thus should fade away in the single-neutron potential as the neutron fraction increases. Reduced differences among the three models are in fact observed in U_n at the larger values of α.

Already several decades ago, it was pointed out that the real part of the nuclear optical potential depends on the asymmetry parameter as in Eq. (2) [14]. Thus, the quantity

$$\frac{U_n + U_p}{2} = U_0, \qquad (3)$$

which is obviously the single-nucleon potential in absence of asymmetry, should be a reasonable approximation to the isoscalar part of the optical potential. The momentum dependence of U_0 (which is shown in Fig. 1 as the $\alpha=0$ curve), is important for extracting information about the symmetric matter EOS and is reasonably agreed upon [15-23].

On the other hand,

$$\frac{U_n - U_p}{2\alpha} = U_{sym} \qquad (4)$$

should be comparable with the Lane potential [14], or the isovector part of the nuclear optical potential [14]. (Notice that in the two equations above the dependence upon density, momentum, and asymmetry has been suppressed for simplicity.) We have calculated U_{sym} close to nuclear matter density and as a function of the momentum, or rather the corresponding kinetic energy. The predictions obtained with Bonn A, B, and C are shown in Fig. 3. They are compared with the phenomenological expression [14]

$$U_{Lane} = a - bT \qquad (5)$$

where T is the kinetic energy, $a \approx 22 - 34 MeV$, $b \approx 0.1 - 0.2 MeV$.

We include in the figure predictions near saturation density ($k_F=1.3$ fm^{-1}), and at a lower density ($k_F=1.1$ fm^{-1}). (The latter may be more appropriate when comparing with nuclear data.) The differences between the upper and lower

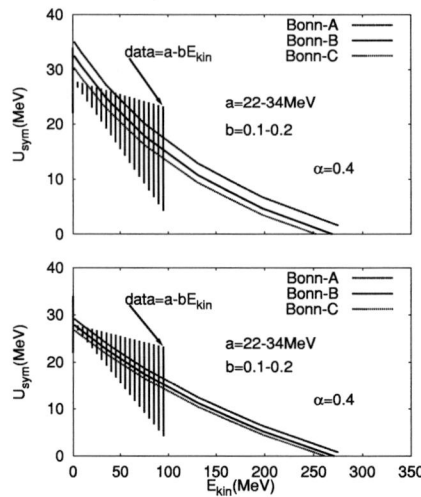

FIGURE 3. The symmetry potential as a function of the nucleon kinetic energy close to saturation density (upper panel) and approximately one-half of saturation density (lower panel). The predictions obtained with Bonn A, B, and C are compared with empirical information from nuclear optical potential data (shaded area). See text for details.

parts of Fig. 3 indicate that the density dependence is strongest at low momentum, as is reasonable. Furthermore, the model dependence is larger at the higher density.

The strength of the predicted symmetry potential decreases with energy, a behavior which is consistent with the empirical information. The same comparison is done in Ref. [24] starting from a phenomenological formalism for the single-nucleon potential [25, 26]. There, it is shown that it is possible to choose two sets of parameters which lead to similar values of the symmetry energy but exactly opposite tendencies in the energy dependence of the symmetry potential as well as opposite sign of the proton-neutron mass splitting. As a consequence of that, these two sets of parameters lead to very different predictions for observables in heavy-ion collisions induced by neutron-rich nuclei [26].

Our effective masses for proton and neutron are shown in Fig. 4 as a function of α and at saturation density. The predicted effective mass of the neutron being larger than the proton's is a trend shared with microscopic non-relativistic calculations [27]. In the non-relativistic case, one can show from very elementary arguments based on the curvature of the single-particle potential that a more attractive potential, as the one of the proton, leads to a smaller effective mass. In our DBHF effective-mass approximation, we assume momentum-independent nucleon self-energies, U_S and U_V, with a vanishing spacial component of the vector part. In such limit, following similar calculations of symmetric matter [28], the one-body potential is written as [2]

$$U_i(p) = \frac{m_i^*}{E_i^*}U_{S,i} + U_{V,i} \qquad (6)$$

where $E_i^* = \sqrt{(m_i^*)^2 + p^2}$, $m_i^* = m_i + U_{S,i}$, and $i = n$ or p for neutrons or protons, respectively. Defining for convenience $U_{0,i} = U_{S,i} + U_{V,i}$, the expression above becomes a two-parameter formula which requires the fitting of two constants, just like in the non-relativistic case. Now, since the single-proton potential is more attractive (see Fig. 1), and both the neutron and proton potentials tend to the same limit at high momenta, it is easy to see from Eq. (6), or rather its derivative, that the proton effective mass obtained in this way must be smaller than the neutron's.

Some comments are in place concerning other DBHF calculations of asymmetric matter where the Dirac neutron effective mass is reported to be larger than the proton's [29, 30]. In Ref. [29], the authors argue that the Dirac mass, $m^* = m + U_S$, should not be compared with the effective mass which can be extracted, for instance, from analyses based on non-relativistic optical models. Instead, an effective mass based on the energy dependence of the Shroedinger equivalent potential should be calculated, in which case one finds that $m_n^* > m_p^*$. The arguments found in Ref. [29] are well known and were already advanced in Ref. [31], where as many as six different definitions for the effective mass are introduced. Again, one must keep in mind the point we already made above. If the nucleon self-energy is written in terms of a scalar potential and *only* the time-like component of a vector potential, and both are taken to be momentum

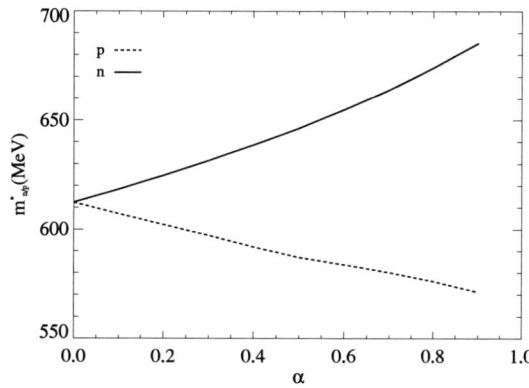

FIGURE 4. The proton and neutron effective mass as a function of the asymmetry parameter and for fixed average density (k_F = 1.4 fm^{-1}).

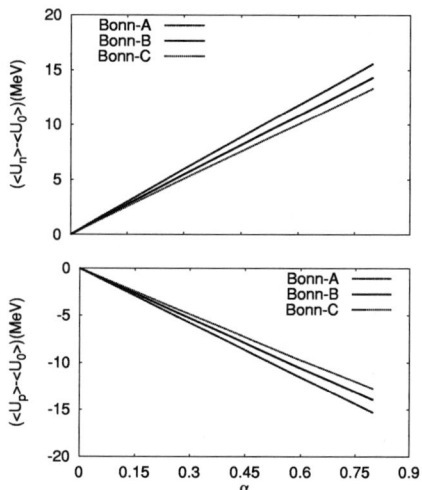

FIGURE 5. Contribution from the asymmetry to the average potential energy per neutron (upper panel) and proton (lower panel). Average density as in the previous figures.

independent, it is then easy to see that the expansion of the single-particle energy is consistent to leading order with the non-relativistic single-particle energy. Thus it is reasonable, and in fact to be expected, that our Dirac masses would be qualitatively consistent with those from non-relativistic predictions, such as BHF calculations. In summary, the effective mass is just part of a convenient parametrization of the single-particle potential. Clearly, how many terms are retained in the nucleon self-energy and how their momentum dependence is handled will impact the parametrization. Ultimately, physical observables depending on the neutron/proton mean field must be correctly described, irrespective of the chosen parametrization.

In closing this section, we also show for completeness the average potential energy per neutron/proton, where the momentum dependence has been integrated out. This is the proton/neutron potential energy contribution to the total energy per particle which then appears in the EOS. Actually, what we show in Fig. 5 are the average potential energies from which the part coming from the symmetric EOS has been subtracted out, that is, just the contribution from the asymmetry to the interaction potential energy,

$$< \Delta U_{n/p} > (\rho, \alpha) = < U_{n/p} > (\rho, \alpha) - < U(\rho, \alpha = 0) > . \qquad (7)$$

Clearly, the contribution from the asymmetry, in both the momentum-dependent and the momentum-averaged potentials, turns out to be large and positive for neutrons, large and negative for protons. This component of the mean

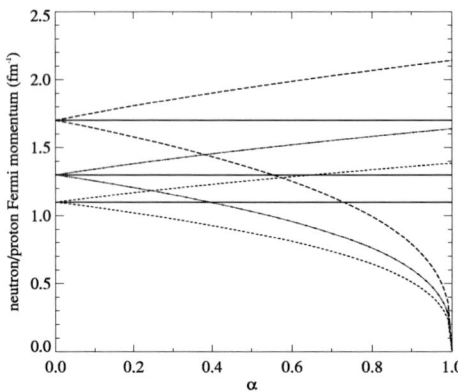

FIGURE 6. Increasing(decreasing) of the neutron(proton) Fermi momentum as a function of the asymmetry parameter. The average Fermi momenta corresponding to the three groups of curves are 1.1, 1.3, and 1.7 fm^{-1}, respectively.

field will then be effective in separating the collision dynamics for neutrons and protons by making more neutrons unbound than protons (or, by making the neutrons more energetic, if already unbound). This effect can be discerned through observables such as the neutron/proton differential flow in heavy-ion collisions [1].

IN-MEDIUM NN CROSS SECTIONS

General aspects

As pointed out in the previous section, the nuclear matter calculation of Ref. [2] provides, together with the EOS, the self-consistent single-proton/neutron potentials as well as their parametrization in terms of effective masses. Those effective masses, and of course the appropriate Pauli operator (depending on the type of nucleons), are then used in a separate calculation of the in-medium reaction matrix (or G-matrix) at positive energies.

Our calculation is again controlled by the total density ρ and the degree of asymmetry, α. For the case of identical nucleons, the G-matrix is calculated using the appropriate effective mass, m_i, and the appropriate Pauli operator, Q_{ii}, depending on k_F^i, where $i = p$ or n. For non-identical nucleons, we would use the "asymmetric" Pauli operator, Q_{ij}, depending on both k_F^n and k_F^p [2]. In Fig. 6, we show the variations of k_F^n and k_F^p with increasing neutron fraction at three fixed densities, according to the relations

$$k_F^n = k_F(1+\alpha)^{1/3} \tag{8}$$

$$k_F^p = k_F(1-\alpha)^{1/3} \tag{9}$$

This, together with Fig. 4, may facilitate the interpretation of results.

Kinematically, the input parameter is the on-shell relative momentum in the c.m. system, q_0. It must be kept in mind, though, that the Pauli operator depends also on the total momentum of the two nucleons in the nuclear matter rest frame. This could be defined in some average (density-dependent) manner, or left as an extra degree of freedom on which the cross section will depend. At the present time, with our goal being to gain some preliminary insight into isospin-dependent in-medium cross section, we make the simple choice of adopting free-space kinematics.

Due to the presence of Pauli blocking, the in-medium scattering matrix does not obey the free-space unitarity relations through which phase parameters are usually defined and from which it is customary to determine the NN scattering observables. As done in Ref. [6] but unlike Ref. [5], we calculate the cross section directly from the scattering matrix elements thus avoiding the use of in-vacuum unitarity relations. That is, we integrate the differential cross section

$$\sigma(q_0,\rho) = \int \frac{d\sigma}{d\Omega}(q_0,\theta,\rho)d\Omega, \tag{10}$$

where $\frac{d\sigma}{d\Omega}$ contains the usual sum of amplitudes squared and phase space factors. An alternative way would be to include Pauli blocking in the definition of phase parameters [32].

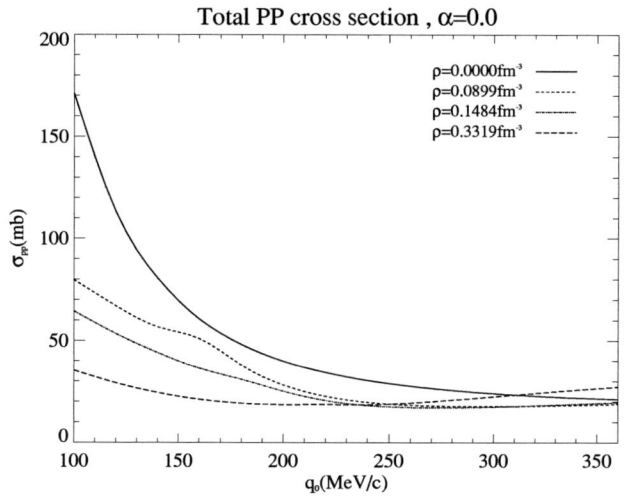

FIGURE 7. Total pp cross section in symmetric matter as a function of the c.m. momentum at the densities indicated in the figure.

Results for pp and nn cross sections

As a baseline, we start with showing the pp cross section as a function of the momentum and at different densities of symmetric matter, see Fig. 7. The given range of momentum corresponds to values of the in-vacuum laboratory energy between approximately 20 and 260 MeV. As compared to the predictions of Ref. [5], and as already pointed out in Ref. [6], the suppression of the cross section with increasing density is less pronounced. This is most likely due to the different handling of the unitarity issue as explained above. Overall, we find reasonable qualitative agreement with previous predictions. We also notice that the high-density cross section tends to rise again at large momenta, a behavior already observed in Refs. [5, 6] particularly for the pp channel.

A simple approach to calculating in-medium cross sections consists of scaling the free-space value according to the relation [4]

$$\sigma(q_0,\rho) = (m^*(\rho)/m)^2 \sigma_{free}(q_0) \tag{11}$$

where m^* is the effective mass. In Fig. 8 we compare this approximation with the predictions from our full calculation. We see that at low momentum, and particularly at the lower densities, the agreement is quite good, but it strongly deteriorates at higher momenta and densities. Medium effects in the interaction are clearly important, particularly the interplay between the (energy-dependent) Pauli blocking and the dressing of the quasiparticle due to the surrounding medium.

We next examine the role of the asymmetry. In Fig. 9, the ratio of the in-medium cross section to the free-space one is shown for both pp and nn scatterings as a function of the asymmetry. The total density is fixed to its value at saturation and the momentum q_0 is equal to the corresponding Fermi momentum. The dependence on α is generally weak, which may be attributed to the competing roles of the effective mass and Pauli blocking. For the pp case, for instance, the smaller effective mass and the lower Fermi momentum would tend to decrease and increase the cross section, respectively. The opposite happens in the nn case. (Compare Fig. 4 and 6.) For the momentum considered in Fig. 9, the role of the effective mass appears to be dominant, lowering the pp cross section and raising the nn one. The relative nn/pp behaviour we observe is in qualitative agreement with predictions based on scaling the cross section as in Eq.(11), with effective masses obtained from the modified Gogny interation recently used in isospin-dependent BUU calculations [34].

Figure 10 displays the cross section ratios for fixed values of the asymmetry and density but changing momentum. There, the ratio is seen to approach unity in the high-momentum limit.

In Fig. 11, the cross section ratios are shown for changing (average) Fermi momentum but fixed asymmetry and momentum. The latter is chosen as in Fig. 9. The cross sections are seen to go down rather quickly with increasing density but then rise again at the higher densities, a tendency that is more pronounced at the higher momenta, As we mentioned earlier, this was already observed, particularly in the pp channel, in previous microscopic calculations of

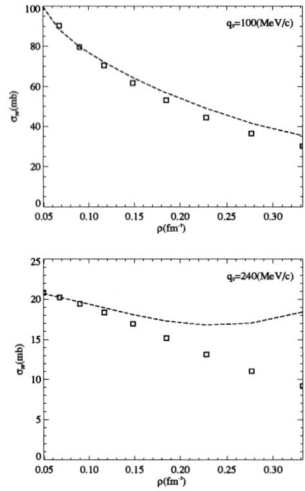

FIGURE 8. Our calculated total pp cross section in symmetric matter compared with the one obtained from Eq. (11) (squares).

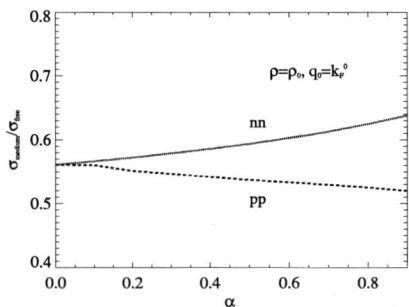

FIGURE 9. Ratio of the total pp and nn cross sections to their free-space values as a function of the asymmetry at saturation density and fixed momentum.

cross sections in symmetric matter [5, 6]. In this region, disagreement with predictions based on Eq. (11) reflects the discrepancy already seen in the high-density part of the lower panel in Fig. 8.

In closing this section, we point out that this is a very limited subselection of results. The phase space to be explored is actually much richer due to the dependence on the total momentum of the two nucleons (through the Pauli operator), see comments at the beginning of this section. We have observed more diverse features and interesting structures show up when that is properly taken into account. Although generally mild, the α-dependence can be non-negligible in some regions of the energy-density-asymmetry phase space. A more comprehensive presentation of results, including np cross sections, will appear elsewhere. Based on the observation of the opposite tendencies of neutrons and protons, we expect the dependence on the asymmetry to be very weak in the np case.

CONCLUSIONS

We have focussed on some of the properties of neutrons and protons in neutron-rich matter. This is a topic of contemporary interest. Its relevance extends from the dynamics of colliding nuclei to nuclear astrophysics.

Different models may be in fair agreement with respect to averaged properties of the EOS, and yet produce very different predictions of properties such as the symmetry potential. ' Clearly, more stringent constraints are needed for the isospin-dependent properties of the EOS.

We have also presented some results from microscopic calculations of cross sections for scattering of identical

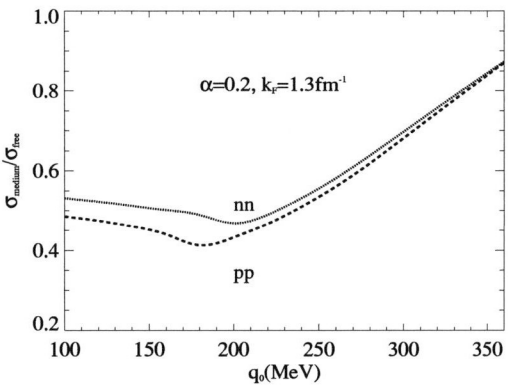

FIGURE 10. Ratio of the total pp and nn cross sections to their free-space values for changing momentum and fixed asymmetry and total density.

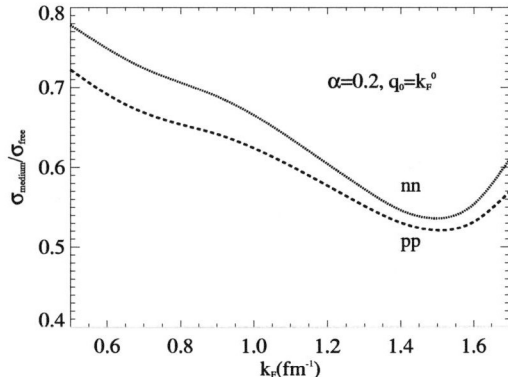

FIGURE 11. Ratio of the total pp and nn cross sections to their free-space values versus the average Fermi momentum for fixed asymmetry and nucleon momentum.

nucleons in neutron-rich matter. Overall, the sensitivity to asymmetry in neutron/proton ratio is mild, which we attribute to the combined effect of Pauli blocking and changing effective mass. As the proton density is depleted, the combined effect for the proton is the result of reduced Pauli blocking and smaller effective mass. The direction of these effects is opposite for neutrons. To test these findings, it would be very useful to identify HI collision observables specifically sensitive to the two-body cross sections.

Very good transport model calculations are available in the literature [33, 1]. However, considerable amount of phenomenology is often involved in the input of these models (for instance, the mean field is based on some phenomenological interaction [12, 13] and/or the elementary cross sections are obtained from empirical data). We calculate all of the above ingredients *microscopically* and internally consistent with respect to the two-body force. We hope this microscopic information (both elementary cross sections and mean field) can be a valuable input for transport model calculations of heavy-ion dynamical observables. This combined effort will complement new data to be taken at RIA and eventually shed light on the less known aspects of the nuclear equation of state.

ACKNOWLEDGMENTS

Financial support from the U.S. Department of Energy under grant No. DE-FG02-03ER41270 is acknowledged.

REFERENCES

1. B.A. Li, C.M. Ko, and Z. Ren, Phys. Rev. Lett. **78**, 1644 (1997); Phys. Rev. Lett. **88**, 192701 (2002).
2. D. Alonso and F. Sammarruca, Phys. Rev. C **67**, 054301 (2003).
3. V.R. Pandharipande and S.C. Pieper, Phys. Rev. C **45**, 791 (1992).
4. D. Persram and C. Gale, Phys. Rev. C **65**, 064611 (2002).
5. G.Q. Li and R. Machleidt, Phys. Rev. C **48**, 1702 (1993); **49**, 566 (1994).
6. C. Fuchs, A. Faessler, and M. El-Shabshiry, Phys. Rev. C **64**, 024003 (2001).
7. See, for instance, R. Crespo and R.C. Johnson, Phys. Rev. C **60**, 034007, and references therein.
8. R. Machleidt, Adv. Nucl. Phys. **19**, 189 (1989).
9. D. Alonso and F. Sammarruca, Phys. Rev. C **68**, 054305 (2003).
10. J.-P. Jeukenne, A. Lejeune, and C. Mahaux, Nucl. Phys. **A245**, 411 (1975).
11. C. Mahaux, Nucl. Phys. **A328**, 24 (1979).
12. C.B. Das, S. Das Gupta, C. Gale, and Bao-An Li, Phys. Rev. C **67**, 034611 (2003).
13. Bao-An Li, Champak B. Das, Subal Das Gupta, and Charles Gale, Phys. Rev. C **69**, 011603 (2004).
14. A.M. Lane, Nucl. Phys. **35**, 676 (1962).
15. L.-W. Chen, C.M. Ko, and B.-A. Li, nucl-th/0407032.
16. C. Gale, G. Bertsch, and S. Das Gupta, Phys. Rev. C **35**, 1666 (1987).
17. G.M. Welke, M. Prakash, T.T.S. Kuo, S. Das Gupta, and C. Gale, Phys. Rev. C **38**, 2101 (1988).
18. C. Gale, G.M. Welke, M. Prakash, S.J. Lee, and S. Das Gupta, Phys. Rev. C **41**, 1545 (1990).
19. Q. Pan and P. Danielewicz, Phys. Rev. Lett. **70**, 2062 (1993).
20. J. Zhang, S. Das Gupta, and C. Gale, Phys. Rev. C **50**, 1617 (1994).
21. V. Greco, A. Guarnera, M. Colonna, and M. Di Toro, Phys. Rev. C **59**, 810 (1999).
22. P. Danielewicz, Nucl. Phys. **A673**, 375 (2000).
23. D. Persram and C. Gale, Phys. Rev. C **65**, 064611 (2002).
24. B.A. Li, Phys. Rev. C **69**, 064602 (2004).
25. I. Bombaci, Chap.2 in *Isospin Physics in Heavy-Ion Collisions at Intermediate Energies*, Eds. B.A. Li and W. Udo Schröder (Nova Science Publishers, Inc., New York, 2001).
26. J. Rizzo *et al.*, Nucl. Phys, **A732**, 202 (2004).
27. I. Bombaci and U. Lombardo, Phys. Rev. C **44**, 1892 (1991).
28. R. Brockmann and R. Machleidt, Phys. Lett. B **149**, 283 (1984); Phys. Rev. C **42**, 1965 (1990).
29. Z-Y. Ma, J. Rong, B.-Q. Chen, Z.-Y. Zhu, H.-Q. Song, Phys. Lett. **B**, in press.
30. E.N.E. van Dalen, C. Fuchs, and A. Faessler, nucl-th/0407070.
31. M. Jaminon and C. Mahaux, Phys. Rev. C **40**, 354 (1989).
32. T. Alm, G. Ropke, and M. Schmidt, Phys. Rev. C **50**, 31 (1994).
33. See, for instance, B.A. Li, Phys. Rev. Lett. **85**, 4221 (2000), and references therein.
34. Bao-An Li, private communications.

PROGRAM of the WORKSHOP

Wednesday March 9. Morning session

9:00 - 9:45. **Daniel Baye** (University Libre de Bruxells). Review of semiclassical models for breakup.
9:45 - 10:30. **Pierre Capel** (TRIUMF).Role of a resonance in the nuclear and Coulomb dissociation of ^{11}Be.
11:00 - 11:45. **Bao-An Li** (Arkansas State University). Transport model for nuclear reactions induced by radioactive beams.

Afternoon session

2:00 - 2:40. **Petr Navratil** (LLNL). No-core shell model and reactions.
2:45 - 3:30. **Jeff Tostevin** (University of Surrey). Exploring the driplines with Eikonal models.
3:50 - 4:30. **Akram Mukhamedzhanov**. (Texas A&M University). Transfer reactions: SF versus ANC.
4:30 - 5:15. Discussion: Absolute spectroscopic factors extracted from reactions (Alex Brown, Betty Tsang and Carlo Barbieri).

Thursday March 10. Morning session

9:00 - 9:45. **Gerhard Baur** (Forschungszentrum Julich). Direct reactions with exotic nuclei.
9:45 - 10:30. **Massimo di Toro** (Laboratori Nazionali del Sud, INFN-Catania). On the splitting of nucleon effective masses at high isospin density: reaction observables.
11:00 - 11:45. **Akira Ono** (Tohoku University). Isoscaling and symmetry energy in dynamical fragment formation.
11: - 11:30. **Jutta Escher** (LLNL). Surragate methods for reactions.

Afternoon session

2:00 - 2:45. **Gentaro Watanabe** (NORDITA). Recent progress on understanding pasta phases in dense stars.
2:45 - 3:30. **Denis Lacroix** (LPC-Caen). Randomness under constraints in cluster formation during nuclear reactions.
3:50 - 4:30. **Maria Colonna** (Laboratori Nazionali del Sud, UNFN-Catania). Fragmentation mechanisms in charge asymmetric systems.
4:30 - 5:15. Discussion: Fragment formation in asymmetric systems (Pawel Danielewicz and Bill Lynch).

Friday March 11. Morning session

9:00 - 9:45. **Ron Johnson** (University of Surrey). Review on adiabatic models.
9:45 - 10:30. **Mahir Hussein** (University of Sao Paulo). Breakup threshold anomaly.
11:00 - 11:45. **Joachim Gomez-Camacho** (Universidad de Sevilla). Is the optical model valid for the scattering of exotic nuclei.

Afternoon session

2:00 - 2:45. **Ian Thomson** (University of Surrey). Reaction models to probe continuum structure.
2:45 - 3:30. **Aksel Jensen** (University of Aarhus). Three-body decay of many-body resonances.
3:45 - 4:30. **Masayasu Kamimura** (University of Kyushu). Continuum-discretized coupled-channels method for four-body breakup reactions.

4:30 - 5:15. Discussion: Fusion with RIB (Mahir Hussein, Jaochim Gomez-Camacho, and Ian Thompson).
6:00. Reception-dinner

Saturday March 12. Morning session

9:00 - 9:45. **Carlos Bertulani** (University of Arizona). Relativistic approach to nuclear reactions with unstable nuclei.
9:45 - 10:30. **Francesca Sammarruca** (University of Idaho). Effective interactions in neutron-rich matter.
11:00 - 12:15. Discussion: What have we learned - what do we need to do? (Ken Nollett and Bob Wiringa).
12:30. Tour of the Laboratory.

Author Index

A

Alvarez, M., 146

B

Baran, V., 119
Barbieri, C., 57
Barredo, W., 193
Baur, G., 61
Baye, D., 1, 12
Bertulani, C. A., 32, 185
Borge, M. J. G., 146

C

Capel, P., 12
Caurier, E., 32
Chamon, L. C., 140
Chen, L.-W., 22
Colonna, M., 70, 119

D

Das, C. B., 22
Das Gupta, S., 22
Dietrich, F. S., 93
Di Toro, M., 70, 119
Durand, D., 112

E

Escher, J., 93
Escrig, D., 146

F

Fedorov, D. V., 164
Fynbo, H. O. U., 164

G

Gale, C., 22
Garrido, E., 164
Goldstein, G., 12
Gomes, P. R. S., 140
Gómez-Camacho, J., 146

H

Hiyama, E., 174
Hussein, M. S., 140

I

Iseri, Y., 174

J

Jensen, A. S., 164
Johnson, R. C., 128

K

Kamimura, M., 174
Ko, C. M., 22
Krastev, P., 193

L

Lacroix, D., 112
Lee, H. C., 49
Li, B.-A., 22
Lionti, R., 119

M

Martel, I., 146
Matsumoto, T., 174
Moro, A., 146
Mukhamedzhanov, A. M., 40

N

Navrátil, P., 32
Nunes, F. M., 40

O

Ogata, K., 174
Ono, A., 83
Ormand, W. E., 32

R

Rizzo, J., 70

S

Sammarruca, F., 193
Sánchez-Benítez, A., 146
Sonoda, H., 101

T

Thompson, I. J., 154
Tsang, M. B., 49
Typel, S., 61

W

Watanabe, G., 101

Y

Yahiro, M., 174
Yong, G.-C., 22

Z

Zuo, W., 22